TAKING-PLACE: NON-REP
THEORIES AND GE

To Edith and Sadie

Taking-Place:
Non-Representational
Theories and Geography

Edited by

BEN ANDERSON AND PAUL HARRISON
Durham University, UK

ASHGATE

Published by
Ashgate Publishing Limited
Wey Court East
Union Road
Farnham
Surrey, GU9 7PT
England

Ashgate Publishing Company
Suite 420
101 Cherry Street
Burlington
VT 05401-4405
USA

www.ashgate.com

British Library Cataloguing in Publication Data
Taking-place : non-representational theories and geography.
 1. Human geography. 2. Culture--Philosophy.
 I. Anderson, Ben. II. Harrison, Paul.
 304.2'01-dc22

Library of Congress Cataloging-in-Publication Data
Anderson, Ben, 1972-
 Taking-place : non-representational theories and geography / by Ben Anderson and Paul Harrison.
   p. cm.
 Includes bibliographical references and index.
 ISBN 978-0-7546-7279-1 (pbk) -- ISBN 978-0-7546-7278-4 (hardback) 1. Geography.
 I. Harrison, Paul. II. Title.

 GN476.4.A63 2010
 910--dc22

2010004879

ISBN 9780754672784 (hbk)
ISBN 9780754672791 (pbk)
ISBN 9781409408963 (ebk)

Reprinted 2012

Printed and bound in Great Britain by
the MPG Books Group, UK

Contents

INTERLUDE

PART III   ETHICS

PART IV   POLITICS

List of Figures and Table

Figures

Table

Notes on Contributors

Ben Anderson is a Lecturer in Human Geography at Durham University, UK. He has published on the geographies of hope and boredom, and utopianism. His current research interests include: the politics of affect, currently with reference to the affects of war and security; theories of matter and materiality; and anticipatory forms of action that govern through the future.

David Bissell is a Lecturer in Sociology at The Australian National University, Canberra. He has published on the corporeal experience of mobilities, non-representational theories of pain, and the relationships between subjectivity and temporality. His current research projects are concerned with phenomenology of stillness; the figure of the passenger and cultures of commuting; and the work of habit.

Jonathan Darling is a Lecturer in Human Geography in the School of Environment and Development at the University of Manchester, UK. He is interested in the spatial politics of asylum, the shifting relationship between geography and ethics, and accounts of cosmopolitanism and responsibility.

J-D Dewsbury is Senior Lecturer in Human Geography at Bristol University, UK. His research centres on bodies, performativity, and the concept of the event in continental philosophy, as well on the performing arts. His recent publications have been on the work of Alain Badiou and post-phenomenological theories as he works towards a book on Performative Spaces: politics, subjectivity, and the event.

Marcus A. Doel is Professor of Human Geography at Swansea University, Wales, UK. His research focuses on spatial theory and continental philosophy. He is the author of *Poststructuralist Geographies* (Edinburgh University Press, 1999) and co-editor of *Jean Baudrillard: Fatal Theories* (Routledge, 2009), and *Moving Pictures/Stopping Places* (Rowman and Littlefield, 2009).

Beth Greenhough is a lecturer in the Department of Geography, Queen Mary University of London, UK. Her research focuses on the ways in which innovations in biomedicine change the relationship between nature and society. Empirically she has studied the how, when and where human subjects become involved in medical research through their participation in genetic and medical research databases and clinical trials. Theoretically, she draws on work in science studies and biophilosophy to critique conventional ethical, legal, social and cultural representations of the experiences, rights and obligations of experimental subjects.

Paul Harrison is a Lecturer in Human Geography at Durham University, UK. His research is concerned with corporeality, ethics and finitude, as well as the relationships between social and cultural theory, space and continental philosophy. He has recently published work on suffering and vulnerability, and is currently working on the politics of passivity, trauma and the immemorial, and the place of the negative in social science.

Steve Hinchliffe is a Professor of Human Geography at Exeter University, UK. He has written and edited a number of books and journal special editions, including, most recently, *Geographies of Nature* (Sage, 2007). He is currently working on biosecurity and on a European Union funded 'Science in Society' project which develops research collaborations with civil society organizations on environmental problems in Europe.

Eric Laurier is a Senior Research Fellow at the University of Edinburgh, UK. His interests are in practical reasoning, ordinary action and aspects of workplaces and everyday life. Quite how ethnomethodology might be connected with Human Geography remains an ongoing interest examined through a number of research projects on variously: cars, cafes, animals, maps, family meals and film editors.

Hayden Lorimer writes about the lives and landscapes shared by human and non-human animals. In the past, his attention has focused on herds, flocks and families, of reindeer, birds and people. He is a Cultural Geographer who teaches at the Department of Geographical and Earth Sciences, University of Glasgow, UK.

Derek P. McCormack is a University Lecturer in Human Geography, Oxford University, UK, and a tutorial fellow at Mansfield College. His research interests lie in a shifting zone between non-representational philosophies, theories of affectivity, and practices that address the moving body as a site of research creation.

Emma Roe is a Lecturer in Human Geography at the University of Southampton, UK. She is interested in materialities, embodied practices and non-human geographies within the agro-food network. She has conducted research in Europe and China on farm animal welfare-related topics – retailing, NGO activity and on-farm practices.

Mitch Rose is a lecturer in the Department of Geography at the University of Hull, UK. His work addresses questions of landscape and material culture using phenomenology and post-structural theory. He has a long standing regional interest in the Middle East and is currently working on a monograph that uses the philosophy of Emmanuel Levinas to re-conceptualise the concept of culture.

Arun Saldanha is Assistant Professor of Geography at the University of Minnesota, USA. He is the author of *Psychedelic White: Goa Trance and the Viscosity of Race* (Minnesota, 2007) and articles in *Environment and Planning A* and *D*, *Social and Cultural Geography*, *Cultural Studies*, *Theory and Event*, and *Annals of the Association of American Geographers*. His research interests include Dutch exploration, poststructuralism, feminism, and evolutionary theory.

Kirsten Simonsen is Professor in Social and Cultural Geography, Roskilde University, Denmark. Research interests include urban, social and cultural studies, practice theory and the philosophy of geography. Her books include *Byteori og hverdagspraksis* (Akademisk Forlag, 1993) [Urban theory and everyday practice], *Byens mange ansigter* (Roskilde Universitetsforlag, 2005) [The multiple faces of the city], and the co-edited volumes *Voices from the North* (Ashgate, 2003) and *Space Odysseys* (Ashgate, 2004).

Nigel Thrift is a Professor of Geography and Vice-Chancellor of Warwick University, UK. Previously he was Pro-Vice-Chancellor for Research at Oxford University. His main interests lie in five areas: non-representational theory, cities, new forms of capitalism, international finance, and the history of time. His most recent books include *Knowing Capitalism* (Sage, 2005) and *Non-Representational Theory* (Routledge, 2008).

Keith Woodward is Assistant Professor of Geography at the University of Wisconsin-Madison, USA. His research explores intersections of affect, politics and ontology as means for developing new understandings of social movements, political change, direct action and autonomous organisation. He is also the co-author of a series of papers devoted to 'site ontology' that, among other things, reconsider the relation between scale, space and politics in Human Geography.

John Wylie is Senior Lecturer in Cultural Geography at the Exeter University, UK. He researches and writes on issues of landscape, performance, subjectivity, haunting and non-representational theory, and is also the author of *Landscape* (Routledge, 2007).

Acknowledgements

In editing a book, especially one like this that has taken a while, many of the conditions that make it possible are amorphous, as much about the intellectual and social milieu in which we work and the atmospheres therein than anything more tangible. Of course, one of the lessons of non-representational theory is that things like atmospheres really matter. With this in mind, we'd like to use this opportunity to acknowledge a bunch of people we work with here at Durham who help make the process of thinking with theory interesting, challenging and sometimes even fun, in no particular order and without assuming they have any interest whatsoever in non-representational theory or theories or bear any responsibility for what follows; Ash Amin, Joe Painter; Gordon Macleod, Adam Holden, Stuart Elden, Colin McFarlane, Angharad Closs-Stephens, Harriett Bulkeley, Mike Crang, Louise Amoore, Steve Graham, David Campbell and Kathrin Hörschelman. Our thinking about non-representational theories has also benefited enormously from encounters with an extraordinary group of visitors to the Geography department, in particular Greg Seigworth, Jane Bennett and Bill Connolly.

Some debts are more specific. As noted this book has been a long time in the making and so we would like to mark our admiration for and thanks to all the contributors not just for their intellectual input but for their on-going commitment, engagement and friendship. In particular, we would like to thank Nigel Thrift for his on-going support and inspiration. Also, we'd like to offer especial thanks to Valerie Rose for the enthusiasm for this book when we first approached her with the idea, her professionalism throughout the editorial process and her patience; Peter Adey and Rachel Colls for offering helpful comments on the introduction at very short notice; and Tim Cresswell, Mike Crang and David Conradson for their thoughtful, critical and supportive engagement with the initial book proposal. Our thanks to Rob Shaw for completing the index and David Shervington for his work on the final text.

Finally thanks to Rach, Edith, Charlotte and Sadie.

Chapter 1

The Promise of Non-Representational Theories

Ben Anderson and Paul Harrison

A Dream

> I can't help but dream of the kind of criticism that would try not to judge but to bring an oeuvre, a book, a sentence, an idea to life; it would light fires, watch grass grow, listen to the wind, and catch the sea foam in the breeze and scatter it. It would multiply not judgements but signs of existence; it would summon them, drag them from their sleep. Perhaps it would invent them sometimes – all the better. All the better. [...] It would not be sovereign or dressed in red. It would bear the lightening of possible storms (Michel Foucault 1997a, 323).

It's the affirmation which gives the quote its force. The affirmation not just of one thing, one subject, one angle, but of many. And beyond this, an affirmation of life, of existence as such, as precarious, as active and as unforeseeable. We will move to a more traditional mode of introduction in a moment however for now let us stay with Foucault's dream. What would 'criticism' have to be to be capable of all these things, of this affirmation and this potential? It seems to us that it would have to be itself multiple, itself composed out of many things. It would have to work out how to move differently, how to step from one topic to the next, one matter to the next, and initiate new ways of relating, walk new routes without tripping, (or at least not often). It would have to take risks, invent new terms, new tones, new objects. It would draw new maps. Perhaps most importantly, it would have to continue changing, not settle in the satisfaction of a judgment but keep experimenting. Further on in the interview from which the quote above comes, Foucault suggests that

> What we are suffering from is not a void but inadequate means for thinking about everything that is happening. There is an overabundance of things to be known: fundamental, terrible, wonderful, insignificant, and crucial at the same time (1997a, 325).

It is our view that non-representational theories[1] are best approached as a response to such a situation. If one single thing can be said to characterise non-representational work in Human Geography over the past 15 years it is the attempt to invent new ways of addressing fundamental social scientific issues and, at the same time, displacing many of these issues into new areas and problems. In doing so we believe that it has multiplied 'signs of existence', helping to introduce all kinds of new actors, forces and entities into geographic accounts and, at the same time, aiding in the invention of new modes of writing and address and new styles of performing Geographic accounts. While the consistency of these attempts may sometimes be hard to see, an issue we will consider below, on a basic level what has linked this diverse body of work is a sense of affirmation and experimentation. In this we believe that they share the ethos of Foucault's dream and, moreover, its invitation to do and think otherwise.

Of course non-representational theories have not done this alone. In the second section of this introduction 'Context' we shall offer a kind of origins myth for non-representational theory in geography, locating its emergence in and from social constructivism in the mid-1990s. However beyond this undoubtedly partial account the main aim of this introduction is to outline three shared commitments or problematics which we believe link together what is a diverse and still diversifying body of work. Our aim here is partly genealogical, taken sequentially one could read these three elements as stages of an evolution and in growing complexity. However the more important (and slightly less artificial) task is that they provide a kind of intellectual 'primer' for the rest of the volume; a chart onto which the reader may map the following chapters and so note their shared concerns and the different routes they plot across common problematics. Thus, following the 'Context', the first of the three substantive sections discusses 'Practices'. Here we describe how and why non-representational theory has a *practical* and *processual* basis for its accounts of the social, the subject and the world, one focused on 'backgrounds', bodies and their performances. In particular this section is concerned with showing how non-representational approaches locate the making of meaning and signification in the 'manifold of actions and interactions' rather than in a supplementary dimension such as that of discourse, ideology or symbolic order. The next section 'Life and the Social' acts as an auto-critique and expansion on the issues just given, charting the movement in non-representational theory from practice based accounts to wider post-humanist accounts of life. Here the influences of Gilles Deleuze and Bruno Latour are most evident, as we attempt to describe the

---

1   Throughout this introduction we will make use of the plural 'non-representational theories' to refer to disparate and potentially loosely connected bodies of thought which do not prioritise the role of representation in their accounts of the social and the subject, and the singular 'non-representational theory' to refer to the specific movement within predominantly British Social and Cultural Human Geography which we are attempting to introduce here. While it may sound a little circular, it should go without saying that non-representational theory is itself diverse, and composed of multiple theories.

consequences of non-representational theory's relational-materialism for thinking about the composition and nature of the social. Following on, 'Event and Futurity' gives the final shared commitment or problematic; here we focus on the 'non' of non-representational theory, and consider exactly how the work gathered by the name is orientated by and to an open-ended future, an orientation through which it attempts to 'bear the lightening of possible storms'. The introduction closes with a brief reflection and a look at the structure of the volume which follows.

Context

Beginnings are always arbitrary, always imagined. One can always extend the genealogy and go back further, or move off sideways seeking the skeleton in the closet, and we will, to some extent. However in this section of the introduction we outline a specific intellectual problematic as the spur behind non-representational theories. In doing so we keep within the recognised genre requirements of an introduction to an edited academic book; 'storying' the emergence of non-representational theories as a successor 'paradigm'. The reasons for this choice are largely pedagogic and heuristic; feeling optimistic, we like to imagine this introduction's primary audience as being composed of people who may not be so familiar with non-representational theories and so the onus is upon us to tell, reductive as it may be, a more or less believable intellectual narrative. However many other beginnings could be plausibly given, not least amongst them; the on-going impact of post-structuralism on the discipline and, in particular, the avenues for thought opened by the translation of the work of Deleuze and Latour; an emergent concern for 'everyday life' and the forms of embodied practice therein; a specific confluence of energies, research interests and institutional setting focused on the School of Geographical Sciences in Bristol in the UK throughout the 1990s; the gathering together and elaboration of non-representational theories by Nigel Thrift; the crystallisation of desires to find new ways of engaging space, landscape, the social, the cultural and the political; the influence of the UK's Research Assessment Exercise through which, in Human Geography at least, value was attached to single author papers and which promoted an academic climate wherein so called 'theoretical' interventions could be valued as highly as more 'empirical' studies; a simple generational shift between the New Cultural Geography and what would follow; an ever more extensive engagement by geographers with other social science and humanities disciplines; a cynical careerist fabulation. As with the account which follows, none of these beginnings are determinate, however all and more probably played a role. We could then classify the emergence of non-representational theories in the discipline as an 'event', (see below), one which, as with all events, arrives somewhat unexpectedly, whose outcome is never guaranteed in advance, and which is composed across but irreducible to a multiplicity of sites, desires, fears, contingencies and tendencies, an event housed within the term 'non-representational theory'.

Still, for now, let's imagine a beginning. It's 1993:

> When it was enthusiastically pointed out within the memory of our Academy that race or gender or nation ... were so many social constructions, inventions, representations, a window was opened, an invitation to begin the critical project of analysis and cultural reconstruction was offered. And one still feels its power even though what was nothing more than an invitation, a preamble to investigation has, by and large, been converted into a conclusion – e.g. 'sex is a social construction', 'race is a social construction', 'the nation is an invention', and so forth, the tradition was an invention. The brilliance of the pronouncement was blinding. Nobody was asking what's the next step? What do we do with this old insight? If life is constructed, how come it appears so immutable? How come culture appears so natural? If things coarse and subtle are constructed, then surely they can be reconstructed as well? (Taussig 1993, xvi).

There can be little doubt that throughout the 1980s and the 1990s social constructivism was the dominant mode of social and cultural analysis, within Human Geography and beyond. 'Social constructivism' is, of course, a convenient shortcut; what is named with this term is less a specific body of work and more a general ontological and epistemological stance, a certain way of delimiting and apprehending 'the social'. In this origins myth, social constructivism plays the somewhat thankless role of context and matrix for the emergence of non-representational theories. So, what traits distinguish social constructivism as an approach and for this dubious honour?

First and foremost social constructivism is distinguished by a preoccupation with representation; specifically, by a focus on the structure of symbolic meaning (or cultural representation). Social constructivism looks to how the symbolic orders of the social (or the cultural) realise themselves in the distribution of meaning and value, and thereby reinforce, legitimate and facilitate unequal distributions of goods, opportunities and power. Thus the primary ontological object for social constructivism is the collective symbolic order understood to be, as the anthropologist Clifford Geertz has it, 'a set of control mechanisms – plans, recipes, rules, instructions (what computer programmers call "programmes") – for governing behaviour' (1973, 44). Or as geographers David Ley and Marwyn Samuels put it five years after Geertz; 'All social constructions, be they cities or geographic knowledge, *reflect* the values of a society and an epoch' (1978, 21 emphasis added). The collective symbolic order is that by which its members make sense of the world, within which they organise their experience and justify their actions. Hence James S. Duncan's characterisation (after Raymond Williams (1981)) of landscape as 'a signifying system through which a social system is communicated, reproduced, experienced, and explored' (1990, 17).[2] An important

---

2   With the selection of this quote and those which follow the reader may well think that by social constructivism we mean the New Cultural Geography; however this both is

point here, one with extensive epistemological (and methodological) implications, is the separation made between the symbolic order and the particular situations within which that order is realised. As Tim Ingold writes; 'Starting from the premise that culture consists of a corpus of inter-generationally transmissible knowledge, as distinct from the ways in which it is put to use in practical contexts of perception and action, the objective is to discover how this knowledge is organised' (2000, 161). Epistemologically, this means that the 'action' is *not* in the bodies, habits, practices of the individual or the collective (and even less in their surroundings), but rather in the ideas and meanings cited by and projected onto those bodies, habits, practices and behaviours (and surroundings). Indeed the decisive analytic gesture of social constructivism is to make the latter an expression of the former. To critically depart, for example, from being 'narrowly focused on *physical artifacts* (log cabins, fences, and field boundaries)' and move towards an understanding of 'the symbolic qualities of landscape, those which produce and sustain social meaning' (Cosgrove and Jackson 1987, 96). A departure through which the objects of investigation – landscape, city space, place – become apprehended as 'texts', where 'the text is seen in terms of the self-realisation or contestation of [ideas, ideologies and] identities, understood as part of the impulse to the self-realisation of the group, class or nation' (Clark 2005, 17).

To sum up, social constructivism's initial impetus and its considerable critical purchase in the 1980s and 1990s lay, in Human Geography at least, in two linked insights. First, in the recognition of the arbitrary nature of symbolic orders, in recognising the fact that they are 'invented' and not 'natural'. Second, in the emphasis placed on the plural and contested (or at least contestable) nature of symbolic orders and the sites at which this occurred. The importance of these insights and the work which followed them is difficult to underestimate; contemporary Human Geographic investigation is unthinkable without them. And so, while we would characterise the emergence of non-representational theory as an 'event', we would also stress that non-representational theory has a debt to, in particular, the New Cultural Geography, one that has to a certain extent gone unacknowledged. There

---

and is not the case. On the one hand, we do clearly implicate the New Cultural Geography within the broad outline of the social constructivism of the 1980s and 1990s; it seems to us that denials to the contrary it was and is wedded to, and indeed gains much of its impetus and insight from, social constructivist assertions about the nature of meaning and its relationship to the world, to matter and to events (see below). However, on the other hand and like non-representational theories, the New Cultural Geography was and is an internally diverse and dynamic movement which, on closer examination, often resists and confounds simplistic reduction. Indeed one may, for example, trace clear continuities between non-representational theory and the ethos and concerns of New Cultural Geography, particular in work on landscape (see Lorimer 2006, Rose 2002 and Wylie 2002), performance (Crang 1994), and mobilities (Merriman 2007; Cresswell 2003). Moreover, we believe that the critical interventions made by those involved in these movements are of ongoing importance and value, not least the founding critique of utilitarianism and functionalism in social and geographic analysis (see for example Cosgrove 1989).

is no doubt that non-representational theory inherits a number of the key insights of New Cultural Geography; that representation matters, that social order is not immutable, and that signification connects to extra-linguistic forces. However, as we shall see, it inherits by rearticulating these insights, framing them otherwise. Why? Because the insight and critical purchase of social constructivism comes at a cost.

Practices

The world and its meanings; this divide is the cost.[3] On one side, over there, the world, the really real, all 'things coarse and subtle', and on the other, in here, the really made-up, the representations and signs which give meaning and value. It's a classic Cartesian divide. Once established there can be no sense of how meanings and values may emerge *from* practices and events in the world, no sense of the ontogenesis of sense, no sense of how real the really made-up can be. Indeed in retrospect it may seem as though, as Ulf Strohmayer (1998, 106) observed, social constructivism's and Human Geography's preoccupation with representation was simply a 'pragmatic' response to the wider, preceding crisis of representation. A response which took critical advantage of the 'constructed' nature of all representation, but which, due to its own anti-realism, was never able to move beyond the crisis and account for the fact that 'if life is constructed, how come it appears so immutable?'. An early, arguably defining trait in the identification and emergence of non-representational theory was a different way of framing and responding to this problem. Indeed this other framing gives us the most literal definition of the term 'non-representational' and the first way of recognising non-representational theories; *they share an approach to meaning and value as 'thought-in-action'*:

> These schools of thought all deny the efficacy of representational models of the world, whose main focus is the 'internal', and whose basic terms or objects are symbolic representations, and are instead committed to non-representational models of the world, in which the focus is on the 'external', and in which basic terms and objects are forged in the manifold of actions and interactions (Thrift 1996, 6).

Before asking of the consequences, it is worth taking a few moments to explore this difference a bit further.

---

3   Non-representational theory is by no means unique in the recognition of this cost; it has been diagnosed in various places, at various times and in various ways across the social sciences and humanities, see for example Bennett (2001); Connolly (2002); Haraway (1991); Ingold (2000); Latour (1993); Law (1993); Massumi (2002b); Seigworth (2003); Stewart (1996); Taussig (1993).

'The manifold of action and interaction'; what does this mean? One way to think about it is as a 'background'. While we do not consciously notice it we are always involved in and caught up with whole arrays of activities and practices. Our conscious reflections, thoughts, and intentions emerge from and move with this background 'hum' of on-going activity. More technically, we could say that 'the background is a set of nonrepresentational mental capacities that enable all representing to take place' and that conscious aims and intentions form, and have the form they do, only against such a 'background of abilities that are not intentional states' (Searle 1983, 143). You are late; you walk quickly into the classroom and sit down. When you walked into the classroom did you think about opening the door, or did you just open it? When you sat down did you have to remember what a seat looked like and how to use one? Of course we can think of examples where people do have to think about these things (a neurological condition may prevent object recognition, one may hesitate and reflect on opening the door due to being nervous, the chair may be an unfamiliar spring-loaded design), however the point is that most of the time in most of our everyday lives there is a huge amount we do, a huge amount that we are involved in, that we *don't* think about and that, when asked about, we may struggle to explain. How did you know to come into the room through the door? How did you know that *that* was a seat? While such reflections may seem somewhat irrelevant to the real business of social and geographic investigation, in many respects nothing could be further from the case. If thinking is not quite what we thought it was, if much of everyday life is unreflexive and not necessarily amenable to introspection, if, as shall be claimed below, the meaning of things comes less from their place in a structuring symbolic order and more from their enactment in contingent practical contexts, then quite what we mean by terms such as 'place', 'the subject', 'the social' and 'the cultural', and quite how 'space', 'power' and 'resistance' actually operate and take-place, are all in question. For now, however, our question becomes how are we to think of this 'background', how are we to characterise it beyond the somewhat limited and limiting definition 'non-representational mental capacities', and so gain some purchase therein?

Insisting on the non-representational basis of thought is to insist that the root of action is to be conceived less in terms of willpower or cognitive deliberation and more via embodied and environmental affordances, dispositions and habits. This means that humans are envisioned in constant relations of modification and reciprocity with their environs, action being understood not as a one way street running from the actor to the acted upon, from the active to the passive or mind to matter, but as a relational phenomenon incessantly looping back and regulating itself through feedback phenomena such as proprioception, resistance, balance, rhythm and tone; put simply, all action is interaction (Ingold 2000, see Gibson 1979; Clark 1997; Thrift 2008). Which is to say that the bodies which populate non-representational theory are, for the most part, *relational* bodies; ecological in form and ethological in apprehension (Lorimer, this volume; Bissell, this volume, Simonsen, this volume). Within such an understanding the world is never an 'out there', a meaningless perceptual mess in need of (symbolic) organisation, nor is it an inert backdrop of brute things projected upon by our hopes, desires and fears, (but see Woodward, this volume, Saldanha, this volume). Rather we are always already 'caught up in the fabric of the world' (Merleau-Ponty 1962, 256); the world is the context from, with and within which what we call subjects and objects emerge, (ibid., see for example Harrison 2000, Hinchliffe, this volume, McCormack, this volume, Wylie, 2002, 2005, and this volume). As Ingold writes:

> For any animal, the environmental conditions of development are liable to be shaped by the activities of predecessors ... The same goes for human beings. Human children, like the young of many other species, grow up in environments furnished by the work of previous generations, and as they do so they come literally to carry the forms of their dwelling in their bodies – in specific skills and dispositions (2000, 186).

Thus we may gain a wider sense of the 'background' described above, one not limited to the (no doubt important) realm of 'non-represententional mental content', but which spills out into and across the body and its milieu. Indeed to speak of practices is to speak precisely of such 'transversal' objects, of arrays of activities which, like musical refrains, give an order to materials and situations, human bodies and brains included, as actions undertaken act-back to shape muscles and hone senses. This is the 'anonymous, pre-personal life of our bodies' which, for the most part, 'remains invisible to us' (Shotter 1995, 2).

What is being described here is a concern with and attention to emergent processes of ontogenesis, how bodies are actualised and individuated through sets of diverse practical relations. A recognisable early and abiding trait of non-representational work in the discipline was a concern for the practical, embodied 'composition' of subjectivities (see for example Rose 2002; Anderson 2004; Harrison 2000; McCormack 2003; Thrift 1996; Wylie 2002; Paterson 2006; 2007). Arguably, what distinguished such accounts was their refusal to search for extrinsic sources of causality or determination, an out-of-field 'power', a symbolic, discursive or ideological order for example. Rather the focus fell on the efficacy and opportunism (or otherwise) of practices and performances. It is from the active, productive, and continual weaving of the multiplicity of bits and pieces that we emerge: out of the 'shapes and contours of our bodies, the recurrent verbal and behavioural patterns' and 'the recurrent diagrams of our emotions, attitudes and posturing' (Lingis 1994, 155).

Equally, it is from such active, productive and continual weaving that 'worlds' emerge. Here, and acknowledging the phenomenological inheritance (see Heidegger 1962, see also Thrift, this volume; Simonsen, this volume), the term 'world' does not refer to an extant thing but rather the context or background against which particular things show up and take on significance: a mobile but more or less stable ensemble of practices, involvements, relations, capacities, tendencies and affordances. A zone of stabilisation within the 'manifold of actions and interactions' which has the form of a holding wave or recursive patterning.

If this sounds abstract and obtuse we do, in fact, use the term world in this sense in everyday life; in, for example, phrases such as 'the world of business', or 'the world of radical politics'. As Alphonso Lingis explains, the term 'world' describes 'not simply an experience of our perceived environment' but, rather, the contexts and fields which are illuminated by our 'movements of concern' and which make 'the multiplicity of beings about us an order, a cosmos' (1996, 13). In this sense 'worlds' are not formed in the mind before they are lived in, rather we come to know and enact a world from inhabiting it, from becoming attuned to its differences, positions and juxtapositions, from a training of our senses, dispositions and expectations and from being able to initiate, imitate and elaborate skilled lines of action. Thus certain embodied gestures and action sequences, certain turns of phrase and idiomatic expressions, certain organisations of objects in space, *do not* 'express' or 'stand-for' certain cultural meanings, values and models; they are not 'vehicles for symbolic elaboration' (Ingold 2000, 283). Rather they are *enactments*; if there is elaboration it is conducted and composed by and in the on-going practical movements and actions, of which the symbolic is a part, but only a part.[4] In this sense non-representational theory may be understood as *radically* constructivist, in that, echoing Latour (1999), it avers that everything is really made-up, but is no less real for this (see Thrift, this volume). Indeed as the distinction between the world and its meaning which sustains social constructivism is collapsed the 'real' and the 'really made-up' are revealed as synonyms, their distinction itself an effect of certain practices. To close this section we want to outline two consequences from the discussion so far.

Firstly, the 'background' itself is hardly inert. If the description of practical bodies and worlds given so far sounds too naturalistic we need only think about the ways in which the human sensorium may be trained, cultivated and entrained. Non-representational theory was not the first to examine this 'pre-personal' dimension of existence. Through its sustained engagement with the phenomenological tradition, Humanistic Geography[5] constantly highlighted the importance of tacit and pre-

---

4 Non-representational theory thus runs along with other turns towards performance and performativity which may be found occurring more or less contemporaneously across geography, the social sciences and humanities. See for example Butler (1990, 1993), Sedgwick (2003), Parker and Sedgwick (1995), Gregson and Rose (2000), Phelan (1993, 1997).

5 What goes by the name 'humanistic' or 'humanistic' is itself a variegated tradition, that still has a force in the present (e.g. Adams, Hoelscher and Till 2001; Mels 2004), particularly given the myriad processes of dehumanisation that damage and destroy humans. We could say the concern of humanistic geographies is something like the composition of environments that can reflect and enhance the variety of human experience (Relph 1976; Seamon 1979) and the means of developing an experientially rich account of lived experience (see Tuan 1977). The critiques are now well known – that a generic and essentialist figure of 'the human' and 'human experience' was centred and celebrated, and that the concept of place ignored process, power relations and remained too bounded (see Massey 1997; Rose 1993). For an account of the cultural politics of place that worked the

cognitive realms in the formation of selves, societies and places, and the myriad ways subjects inhabit the world before they represent that world to themselves and others. However compared to the accounts offered by non-representational theory, humanistic accounts can appear too naturalistic and normative. Perhaps a closer relative is to be found in Pierre Bourdieu's (1977) account of *habitus*, which effectively historicises and politicises phenomenological accounts of the 'background'. However, for all its insight and recognition of contingency and the importance of improvisation, Bourdieu's account of the *habitus* remains curiously inert, constantly supplemented by determinate structural logics at the expense of the 'slight surprise of action' (Butler 1997, de Certeau 1984, Latour 1999). Perhaps closer still are Walter Benjamin's (1992; 1997, see Latham 1999) accounts of our distracted, tactile and habitual means of 'understanding' the city and life in capitalist societies. In his famous city essays Benjamin describes a mobile, embodied, geo-historically specific, sensuous knowing; his object is not an individual but rather modes and moments of subjectification as they emerge across a distracted collective of habits and gestures, buildings and courtyards, speeds and slownesses. It is this account, both more open *and* more specific, which seems to us closest to those given in non-representational theory.

Secondly, if the 'background' is geo-historically specific and generative then it is open to intervention, manipulation and innovation. Thrift (1996, 2008), for example, has traced how many of the spaces of everyday life are increasingly being inhabited, in one way or another, by pervasive intelligent technologies, including biomedical, imaging, storage and recall, track and trace, computation and real-time modelling, as well as mixtures of all of the above:

> Reach and memory are being extended; perceptions which were difficult or impossible to register are becoming routinely available; new kinds of understated intelligence are becoming possible. These developments are probably having most effect in the pre-cognitive domain, leading to the possibility of arguing that what we are seeing is the laying down of a system (or systems) of distributed pre-cognition (Thrift 2008, 164).

We may think, for example, of the increasing role of environmental sensors in the support and care of the elderly, involving new forms of unobtrusive remote monitoring and feedback such as bed and chair occupancy monitors, often coupled intelligent lighting networks, property exit sensors, and fridge content monitors. Through laying down 'awareness' or even 'intelligence' into the environment, each of these technologies makes the delivery of long term care in individual's and family's homes far more feasible, especially for those with dementia or increasing physical frailty. Of course, the development and implementation of such technologies need not be so benign. As the 'background' or pre-cognitive

---

insights of humanistic geographies through a concern with social difference see Cresswell (1996).

realm is rendered visible so it becomes available to be worked on by a whole set of new entities and institutions as, for example, in the increasingly refined attempts to build in kinesthetic and affective experiences into specific commodities, political figures or environing spaces (Thrift 2008; Adey 2008). Here we may think of Jane Bennett's (2001, see also McCormack this volume) analysis of the 'swinging kahkis' GAP advertising campaign, Brian Massumi's (2002b) discussion of the attention to body language and pathic communication in the television appearances of former US President Ronald Reagan, and Thrift's (2008) discussions of the architectures of anticipation at work in urban settings. While such work has been criticised for reintroducing deterministic accounts of social and political action (see Barnett 2008), almost all work within non-representational theory maintains that while 'background', pre-cognitive realms may not always be straightforwardly amenable to conscious reflexivity and representation, this does not make them completely alien and determining. Rather, manipulation, where it is achieved, is always a fragile and contingent achievement, 'prone to failure and always reliant upon being continually reworked in relation to creative responses' (Ash forthcoming). Allowing subjects to become more involved, more complex and less certain of their boundaries and themselves need not lead to functionalism and behaviourism. Indeed, practical existence is clearly available to many forms of self and group 'fashioning'. From the 'techniques of the self' described in Foucault's (1997b) later work, to Ash Amin (2006) on 'everyday cosmopolitanism', and Jonathan Darling's (this volume) examination of practices of hospitality, it is clear the pre-cognitive is not simply a realm for colonisation, domination and control but for cultivation and intervention. Quite simply, however stable they may be at any one time in any one place, background practices are open to change and reconfiguration.

In emphasising practical, lived experience, non-representational theory has been identified as a form of Humanistic Geography, and charged with repeating the same mistakes; the centring of a universal, unmarked, subject shorn of difference (Nash 2000; Saldanha 2005; Tolia-Kelly 2006). However, the comments above should go some way to disabuse this understanding as, insofar as it has a subject, this is a subject that is radically contingent, which is always in and of the mixture of many different elements, but which is also irreducibly specific in its existence (see Harrison, this volume; Wylie, this volume). For us the more pressing question here is what becomes of the subject and the social *as such* once constructivism is radicalised in the manner described above and the human is understood to be part of the on-going becoming of worlds? It is to this question we turn now.

Life and the Social

Thought is placed in action and action is placed in the world. This is the starting point for all non-representational theories. Yet however important these beginnings they are not the sum of non-representational theory. Throughout the 1990s and into this century the initial attention to practices in non-representational theory morphed into a concern with Life, and the vital processes that compose it (see Thrift, this volume). While a concern for practices and 'worlds' provides ways for rethinking the process of ordering, appearance and signification beyond the normative assumptions of humanism and the idealist confines of social constructivism, as well as injecting a degree of action and movement back into the composition of the social, these are still very much practices reckoned in terms of the human; carried out *by* humans in worlds which are *for* humans. And yet, as began to become clear towards the end of the last section, the figure of the human is haunted by all kinds of things, by all that which needs to be excluded for it to maintain its purity and exceptionalism. Humans, their desires and plans, are clearly not the only things active in the world, in fact often we may be very small players in much bigger trans- and non-human systems and complexes. Hence in 1999 we find Thrift writing about places as 'spectral gatherings'; relational-material 'crossroads' where many different things gather, not just deliberative humans but a diverse range of actors and forces, some of which we know about, some not, and some of which may be just on the edge of awareness. The shift to thinking about Life is, therefore, a shift to thinking about how worlds may be arrayed and organised *with* humans, but not only humans. To arbitrarily stop relational understandings of phenomena at the boundary of the human is to re-inscribe precisely the divides between inside and outside, meaning and world, subject and site, which were first in question.[6] If we are to rejoin and rethink these divides, it follows that the 'missing masses' must be allowed back into the social fold and the contingency of the human acknowledged. Hence in this section, the question is what becomes of the 'social' in this process? To start to give an answer this question we will first discuss the general implications of an expanded materialism before turning directly to the question of the 'social'.

In distinction to phenomenologically inclined practice based approaches, we find a wider and wilder sense of a life in Deleuze's joint writings with Guattari (see Dewsbury 2000; 2003). Deleuze's (2001, 29) last piece of published writing – *Immanence: A Life* – is perhaps the touchstone for this work. Likened to a parable, aphorism and testament by John Rajchman, Deleuze writes of a life as

---

6   This is not to suggest there is no debate about and reflection on these issues within non-representational theory, there clearly is. Indeed much recent work under this name has concerned precisely the status of and how to think about the human, but a human defined not by a putative essence or identity, that is to say debate around how to figure the human after or within the broader movement of anti-humanism. Compare for example Harrison (2008, 2009, this volume), McCormack (this volume), Rose (this volume), Thrift (2008), Wylie (2009, this volume). It is also interesting to note that as well as being critiqued for harbouring an implicit normativism humanism, as outlined above, in almost the same instant, non-representational theory has also been criticised for being too anti-humanist, (see for example Bondi 2005; Thien 2005). Without wanting to presage the on-going debates just noted, we would simply note how this situation suggests that, insofar as it has one, non-representational theory may have a new account of the human, one irreducible to either of the terms of critique.

an 'impersonality' that is unattributable to our particular identifications as people or selves:

> A life is everywhere, in all the moments that a given living subject goes through and that are measured by given lived objects: an immanent life carrying with it the events of singularities that are merely actualised in subjects and objects (Deleuze 2001, 29).

Simply put *a* life is not *the* life of an already constituted individual or subject; *a* life is made up of singularities that are both outside and the possibility of the particular identifications that enable us to say 'we' or 'I'. Just as all beginnings are imaginary so are all identifications. As such, the techniques, sensibilities and methods developed in particular through engagements with Deleuze and Guattari, and post-phenomenologists such as Lingis, have taken as their task to attend to a life that occurs *before* and *alongside* the formation of subjectivity, *across* human and non-human materialities and *in-between* distinctions between body and soul, materiality and incorporeality (after Seigworth 2003, 6; see Anderson and Wylie 2009; Latham and McCormack 2004; Greenhough, this volume; Lorimer, this volume; Hinchliffe, this volume; Roe, this volume).

This gives us to the second commitment through which we may recognise non-representational theories; following on from a concern with practices, *non-representational theories work with a relational-material or 'associative' account of 'the social'*. Whilst this definition may not sound very precise this is, in many respects, the point; the social is a weaving of material bodies that can never be cleanly or clearly cleaved into a set of named, known and represented identities. More specifically, non-representational theories are concerned with the distribution of 'the human' across some form of assemblage that includes all manner of materialities.[7] We would suggest that this approach involves three starting points; a commitment to an expanded social including all manner of material bodies, an attention to relations and being-in-relation, and sensitivity to 'almost-not quite' entities such as affects. In order to flesh out non-representational theory's approach to the social and sociality it is worth addressing each of these points in a little more detail.

First, and learning from early explorations in actor-network theory, alongside the various embodied practices and capacities discussed above, the social is

---

7   There are multiple uses of the term 'assemblage' in geography (see McFarlane 2009). For us, assemblage functions as a sensitising device to the ontological diversity of actants, the grouping of those actants, the resulting distribution of agentic capacities, and an outside that exceeds the grouping (after Bennett 2005). This retains the sense of assemblage as *agencement* (in the sense of arrangement) in Deleuze and Guattari (1987), without necessarily repeating the distinction between the actualised and unactualised that is at the heart of DeLanda's (2006) realist development of Deleuze and Guattari's morphogenetic account of life.

repopulated by objects, machines, and animals (see Bingham 1996; Hincliffe 1996, Murdoch 1998; Whatmore 2002).[8] These entities do not exist independently from one another, neatly separated into discrete ontological domains; rather all co-exist on the same 'plane of immanence' (Deleuze and Guattari 1987). Consider, for example, the sheer multiplicity of materialities that are mixed together in non-representational inspired empirical work; beliefs, atmospheres, sensations, ideas, toys, music, ghosts, dance therapies, footpaths, pained bodies, trance music, reindeer, plants, boredom, fat, anxieties, vampires, cars, enchantment, nanotechnologies, water voles, GM Foods, landscapes, drugs, money, racialised bodies, political demonstrations. What gives consistency to this proliferation of whatever matters, what holds together this open ended list, is a simple affirmation; materiality takes many forms (see Anderson and Wylie 2009; see Greenhough, this volume, Roe, this volume; Hinchliffe, this volume). Non-representational theory is unusual, then, in being *thoroughly* materialist. It does not limit *a priori* what kind of beings make up the social. Rather everything takes-part and in taking-part, takes-place: everything happens, everything acts. Everything, including images, words and texts (Doel, this volume; Dewsbury, this volume; Laurier, this volume). Hence a relational-materialist approach departs from understandings of the social as ordered *a priori* (be it symbolically, ontologically, or otherwise) in a manner that would, for example, set the conditions for how objects appear, or as an ostensive structure that stands behind and determines practical action. In the taking-place of practices, things and events there is no room for hidden forces, no room for universal transcendentals or first principles. And so even representations become understood as presentations; as things and events they enact worlds, rather than being simple go-betweens tasked with re-presenting some pre-existing order or force. In their taking-place they have an expressive power as active interventions in the co-fabrication of worlds. Dewsbury, Harrison, Rose and Wylie (2002, 438) put this well in one of the first commentaries on non-representational theory when they stress that

> Non-representational theory takes representation seriously; representation not as a code to be broken or as a illusion to be dispelled rather representations are

---

8   The interest in matter and materiality has occurred as part of a broad concern with the 're-materialisation' of British Social and Cultural Geography. Calls to 'rematerialise' were themselves responses to the perceived overemphasis on signification in the New Cultural Geography (Jackson 2000). It should be noted that there are now significant differences within Social and Cultural Geography around how matter is theorised. Compare, for example, the expansive sense of what counts as a material body in non-representational theories to the concern for a circumscribed realm of objects in material culture studies, or the continued use of 'the material' to refer to an ostensive social structure (for summaries of different theories of matter and materiality see Anderson and Tolia-Kelly 2004; Cook and Tolia-Kelly 2008). The closest connections to non-representational theory can probably be found in the emphasis on the force of materiality in corporeal Feminism (see Slocum 2008; Colls 2007).

apprehended as performative in themselves; as doings. The point here is to redirect attention from the posited meaning towards the material compositions and conduct of representations (Dewsbury, Harrison, Rose and Wylie 2002, 438).

Second, non-representational theory may be characterised by an attention to being-in and being-of relation. An attention which begins from the 'vital discovery' that *relations are exterior and irreducible to their terms* (Deleuze 2006, 41). The key point here is that beginning from relations, 'thinking relationally', opens up 'a world in which the conjunction "and" dethrones the interiority of the verb "is"' (Deleuze 2001, 38).[9] In dialogue with Claire Parnet, Deleuze gives a sense of the strange topologies and topographies that open up if one thinks with AND instead of IS; that is, if one thinks of relations being as real as the different material bodies that populate the social:

> Relations are exterior to their terms. 'Peter is smaller than Paul', 'The glass is on the table': relation is neither internal to one of the terms which would consequently be subject, nor to two together. Moreover, a relation may change without the terms changing ... Relations are in the middle, and exist as such. This exteriority of relations is not a principle, it is a vital protest against principles ... If one takes this exteriority of relations as a conducting wire or as a line, one sees a very strange world unfold, fragment by fragment: a Harlequin's jacket or patchwork, made up of solid parts and voids, blocs and ruptures, attractions and divisions, nuances and bluntnesses, conjunctions and separations, alternations and interweavings, additions which never reach a total and subtractions whose remainder is never fixed (Deleuze 2006, 41).

The emphasis on relations resonates with a broad interest across Human Geography in how everything, from places to identities, is 'relationally constituted' (see 2004 special issue of *Geografiska Annaler B*). The result is an emphasis on the proliferation of diverse relations and a strong sense that the resulting orders are open, provisional, achievements. However, pushing on, any simple definition of 'relation' is immediately undone by the irreducible plurality of relations. Indeed that relations are plural is the main lesson of an 'after' actor-network theory literature, a lesson increasingly being taken up in geography and one that

---

9 There are many emerging questions and unresolved tensions in geography's treatment of 'relations' and 'relationality', including; how to bear witness to the plurality of relations?; how to understand the 'reality' (felt or otherwise) of relations?; are relations internal or external to their terms?; can relations change without the terms also changing?; are actual entities exhausted by their relations?; and how to think what could be termed the 'non-relational'? (see for example Marston, Jones and Woodward 2005; Massey 2005; Harrison 2007; Harman 2009).

has become central to non-representational theory.[10] The consequence is that it is not enough to simply assert that phenomena are 'relationally constituted' or invoke the form of the network, rather it becomes necessary to think through the specificity and performative efficacy of different relations and different relational configurations (see Whatmore 2002; Hincliffe, this volume; Roe, this volume). Somewhat counter-intuitively perhaps, a general affirmation of relations seems to lead to focus on *this* specific relation.

Third, work in non-representational theory has examined how the social is composed of entities that are *both* present and absent; it has drawn attention to the role of 'objects' such as affects, virtual memories, hauntings, and atmospheres in the enactment, composition and durability of the social.[11] There are debates within non-representational work around how to attend to absence (compare Wylie, this volume and Harrison, this volume to McCormack, this volume). Nevertheless, there is a shared concern for 'objects' that are both present and absent, neither one nor the other. Hence the constant attention to questions of affect in non-representational work, or, put differently, the capacities to affect and be affected of human and non-human materialities (Anderson 2006a; 2009; McCormack 2002; 2003; 2008; Thrift 2004; Bissell 2008; 2009; Simpson 2008; see Bissell, this volume and McCormack, this volume).[12] Whilst undoubtedly contested, the term affect has come to name the aleatory dynamics of experience, the 'push' of life which interrupts, unsettles and haunts persons, places or things (Bennett 2010). The social is affective and it is often through affect that relations

---

10   Note, for example, the proper names that are given to just some of the shapes relations can take: 'encounter, arrival, address, contact, touch, belonging, distance, accord, agreement, determination, measuring, translation, and communication are some such forms of relation' (Gasché 1999, 11).

11   The emphasis on the fold between materiality and immateriality chimes with recent work on spectrality, haunting and the peculiar persistence of the past (see Pile 2005; Edensor 2005; Adey and Maddern 2008).

12   Debates around how to theorise 'affect' and 'emotion' have become something of a cipher for engagement with non-representational theory more broadly. We have deliberately downplayed the significance of affect in this introduction (and collection) because non-representational theories do much more than offer an account of worlds of affect. The debate about affect, emotion and their interrelation have turned around three points of concern and critique; the apparent distinction between emphasising an impersonal life and the embodied experience of subjects; the relation between affect and signification; and the crypto-normativism that has arguably been smuggled into work on the politics of affect (see Bondi 2005; Thien 2005; Tolia-Kelly 2006; McCormack 2006; Anderson and Harrison 2006; Barnett 2008). Whilst we have our views on the tone and content of this debate, as well as different positions within it, we will leave it to the reader to navigate their own way through the discussion. What we do want to stress is that there is an 'affective turn' occurring beyond Human Geography where similar issues are being grappled with, in particular by Feminist and queer theory scholars working with a concept of affect (see for example Clough 2007; Puar 2007; Stewart 2007).

are interrupted, changed or solidified. Or so we learn through inventive work that describes how bodies dance together (McCormack 2003), attends to bodies seared by pain (Bissell 2009), or pays attention to the geographies of love (Wylie 2009). The attention to affect as a dynamic process that cuts across previously separated ontological and epistemological domains can be understood as a further repopulation of the social, this time with entities that are both much less and much more than present. We should not, however, be surprised at the intimacy a worldly, materialist thought has with reflections on immateriality. From the void of Epicurean philosophy through to the proletariat in historical materialism, spectres have haunted *all* materialisms (Pile 2005).

To return directly to this sections opening question: if the supposedly unique powers of the human have been problematised by a materialist emphasis on a more-than-human life, what then becomes of the term 'social'? Perhaps we should jettison the term, despite or perhaps because of its current wide currency (Thrift 2008)? However this is, in some senses, to place the cart before the horse. To explain; in offering an associative understanding of the social, and breaking with a focus on collective symbolic orders, non-representational theory has affinities of method and sensibility with a whole series of 'minor' traditions in social geography; most notably, the longstanding attention to practice in time-geography (Hägerstrand 1973, 1982; Pred 1977; Latham 2003), Feminist work on performance and performativity (Gregson and Rose 2000), Erving Goffman's dramaturgical account of social action (Thrift 1983) and Harold Garfinkel's ethnomethodological investigations (Laurier, this volume). As with non-representational theory, all attempt to move away from a distinction between 'individual' and 'society' and all share an emphasis on the ongoing composition of the social from within the 'rough ground' of practices and the concrete richness of life.

Latour (2005) offers perhaps the sharpest account of the refigured notion of the 'social' that non-representational theories share, and which perhaps goes some way to distinguish them from the aforementioned traditions. The social, according to Latour, is a certain sort of circulation, where action is always dislocated, articulated, delegated, and translated; it is not a special domain or specific realm but 'a very peculiar movement of re-association and reassembling' (2005, 7). It is a type of connection between things that are *not* themselves social (ibid., 159):

> At first this definition seems absurd since it risks diluting sociology to mean any type of aggregate from chemical bonds to legal ties, from atomic forces to corporate bodies, from physiological to political assemblies. But this is precisely the point that this alternative branch of social theory wishes to make as all those heterogeneous elements *might be* assembled anew in some given state of affairs. Far from being a mind boggling hypothesis, this is on the contrary the most common experience we have in encountering the puzzling face of the social. A new vaccine is being marketed, a new job description is offered, a new political

movement is being created, a new planetary systems is discovered, a new law is voted, a new catastrophe occurs. In each instance we reshuffle our conceptions of what was associated together because the previous definition has been made somewhat irrelevant. We are no longer sure about what 'we' means; we seem to be bound by 'ties' that don't look like regular social ties (Latour 2005, 5-6, *emphasis original*).

However, it is precisely the 'holding together' of different kinds of bodies that must be explained. 'The social' is, to paraphrase Latour, precisely what must be explained rather than that which can be invoked to explain the durability of this or that practical ordering. Quite simply, there is no order, there is only multiple orderings, and practices are the context for and necessary condition of those orders, each of which must be actively composed or fail (see Laurier, this volume; Hinchliffe, this volume; Bissell, this volume; Simonsen, this volume).

This does not mean, we would stress, that because there is no supplementary dimension to the social that there are no durable orders, or that those orders do not include many forms of damage, loss, suffering and harm. On the contrary, beginning from the social as a practical achievement provides a method for thinking through how systematic processes of harm *become* systematic. Systematic orderings are themselves multiplicities – composed of complex and shifting relations between seemingly discrete elements and types of elements (Connolly 2008). The only way to understand the durability of orderings (or collections of orderings) is to trace the relations between the heterogeneous elements that compose them, to follow how the resultant assemblage functions, and to map the encounters through which the elements within assemblages are brought into contact with forces outside of them. We see this insight being worked through most clearly, although by no means exclusively, in recent work on the formation of race and racisms, where racialised bodies are formed through the agglomeration of diverse elements, including, but never limited to, biological materialities such as phenotypes. Race is here addressed as an assemblage formed from within the heterogeneous materialities of bodies, technologies and places, racial difference being a heterogeneous process of differentiation, as Saldanha (2007) puts it. The task becomes to grasp how race, racial differences and potentially other social differences (Lim 2007), form, become durable and exert a force alongside the many other relations and relational configurations that make up the 'social' (see Saldanha 2006; 2007; Swanton 2008; Lim 2007; Saldanha, this volume; Darling, this volume; Simonsen, this volume).

As noted above, one of the promises of non-representational theory is that it offers a *radically* constructionist rather than *social* constructionist account of the 'social'. As Massumi (2002a) stresses, constructionist accounts of 'the social' wonder about stasis given the primacy of process: how do things fit together and hold together across differences? How to think the irreducible contingency of order? Beginning from the primacy of process opens up the question of change; how are orders disrupted, how do orders fail, and how are new orders coming into being, if only momentarily? It is to a consideration of change that we now turn in order to introduce the third and final way we may recognise non-representational theories; through their concern with events.

Events and Futurity

If non-representational theories begin from practices and advocate a relational-materialist analysis of the social, why the name '*non*-representational theory'? A name that has added to the sense of promise, wariness, and perhaps even irritation that has surrounded non-representational styles of thinking and doing over the past 15 years or so. As we have stressed above, and is hopefully apparent in Part II of this book, non-representational styles of thinking can by no means be characterised as anti-representation *per se*. Rather what pass for representations are apprehended as performative presentations, not reflections of some *a priori* order waiting to be unveiled, decoded, or revealed. But maybe the name was a mistake, maybe it is now time to dispense with it in favour of something more affirmative – 'more-than-representational theory' being one popular suggestion (Lorimer 2005; see Rose, this volume)? Perhaps though, and like actor-network theory, the promise of non-representational theory would have been betrayed by any name that enabled it to be easily summed up and reduced. We think there is something more in the name; a force to the prefix 'non' that hints to something vital to non-representational theories that is worth thinking with and affirming. The 'non' is frustratingly elusive, it cannot be thought as such. It leaves things incomplete. It manages to obscure what it affirms by studiously avoiding positive nomination (see Dewsbury, this volume; Harrison, this volume; Doel, this volume).

In these ways the prefix 'non' opens up the third way that we can recognise non-representational theories; *they are marked by an attention to events and the new potentialities for being, doing and thinking that events may bring forth.* 'The event' has been such an important concept and empirical concern for non-representational theories because it opens up the question of how to think about change. In the previous section we argued that non-representational theories share a reversal of the relation between stasis and process, we can now say more precisely that the task of a materialist analysis of the social is to understand the stability of form amid the dynamism of formation (Massumi 2002b). Within this thinking 'the event' is of importance because it allows the emphasis on the contingency of orders to morph into an explicit concern with the new, and with the chances of invention and creativity. As events have to do with 'lighting fires'; with solicitations or provocations, with promises and threats that create:

> a transforming moment that releases from the grip of the present and opens up the
> future in a way that makes possible a new birth, a new beginning, a new invention
> of ourselves, even as it awakens dangerous memories (Caputo 2007, 6).

Fleshing out these comments requires, however, that we think carefully about what we mean by the term 'event'. There are many occurrences which we might want to understand as events. There are also many ways in which 'the event' is conceptualised, addressed and handled not only within non-representational theories but also by architects, site specific artists, security professionals and other creators of both events and their opposite – recognised occurrences. Given this heterogeneity, let's consider two examples of what we might take to be events in order to unpack what we mean by the term and present a couple of the different ways in which non-representational theories think the relation between orderings, events and change (though see Dewsbury, this volume; Doel, this volume; Woodard, this volume; McCormack, this volume).

First, consider a granite obelisk known as Cleopatra's Needle that sits on the Charing Cross Embankment, London, UK. Placed in its current position on 12 September 1878, it may appear far removed from the dynamism and transitoriness we might want to associate with the concept of event. The process philosopher Alfred Whitehead thought differently. He saw it as a continual event, or better a complex of passing events:

> If we define the Needle in a sufficiently abstract manner we can say that it never changes. But a physicist who looks on part of the life of nature as a dance of electrons, will tell you that daily it has lost some molecules and gained others, and even the plain man can see that it gets dirtier and is occasionally washed (2004 [1920], 167).

Here we find a first sense of the event – the event as a continual differing, if only in modest ways, that takes-place in relation to an ever-changing complex of other events. For, as Whitehead went on to stress, events have always just happened or are about to happen:

> You cannot recognise an event; because when its gone, it is gone. You may observe another event of analogous character, but the actual chunk of the life of nature is inseparable from its unique occurrence. But the character of an event can be recognised. We all know that if we go to the Embankment near Charing Cross we shall observe an event having the character which we recognise as Cleopatra's Needle (2004 [1920], 169).

Here the divergence and discord that events bring is not rare, nor is it some form of caesura, rather 'wherever and whenever something is going on there is an event' (Whitehead 1920, 78). Putting it in the terms of the previous sections (terms which are not necessarily Whitehead's) we could say that events are primary in a world in which the background is open to modification and in which diverse material bodies are constantly being brought into relation. Here the term 'event' describes the escaping edge of any systemisation or economisation; the effects or affects of any 'line of flight' (see Deleuze and Guattari 1987). It is only with effort that any such 'slight surprise' of action can be turned back into a reproduction of an existing order (Latour 1999; Massumi 2002a).

If we are caught within a world of becomings, where events can be found everywhere, then any ordering is always volatile. This is the basic insight at the heart of thinking with the event. However, there are other ways of conceptualising the relation between events, change and order. A slightly different sense of the event as a rare surprise that breaks with how the background is organised, or a specific social-material configuration is assembled, has animated other non-representational theories. Let's consider a second example of what we might want to understand as an event – the event that has come to be housed within a date – September 11th – or a number – 9/11. For Derrida, it is the very brevity of this name and number that indicates that, perhaps, an event in the sense of an absolute surprise may have taken-place:

> 'Something' took place, we have the feeling of not having seen it coming, and certain consequences undeniably follow upon the 'thing'. But this very thing, the place and meaning of this 'event', remains ineffable, like an intuition without a concept, like a unicity with no generality on the horizon or with no horizon at all, out of range for a language that admits its powerlessness and so is reduced to pronouncing mechanically a date, repeating it endlessly, as a kind of ritual incantation, a conjuring poem, a journalistic litany or rhetorical refrain that admits to not knowing what it's talking about (Derrida 2003, 86).

Derrida goes on to stress that the 'impression' that 9/11 was a 'major event' has been reflected on, interpreted and communicated, and that this process is itself an 'event' in the sense of a modification. But is this the same as a 'major event'? Whilst the movement of appropriation is 'irreducible and ineluctable', for there to be an event appropriation must falter at some 'border or frontier' (2003, 90):

> A frontier, however, with neither front nor confrontation, one that incomprehension does not run into head on since it does not take the form of a solid front: it escapes, remains evasive, open, undecided, indeterminable. Whence the unappropriability, the unforeseeability, absolute surprise, incomprehension, the risk of misunderstanding, unanticipatable novelty, pure singularity, the absence of horizon (2003, 90-91).

If we accept this as the minimal definition of the event, then was '9/11' an event? This is less certain, even if we agree with Whitehead that there is something of an event every time something happens, since an event of a 'terrorist attack' was foreseen, there were precedents and the event 9/11 was very quickly captured in geopolitical and biopolitical projects of war and security. Hence here the event is understood a little differently to in our previous example; here the event is an absolute surprise, something that brings 'contingency, unpredictability, and chance into the world' (Dastur 2000, 179). Events, on this understanding, must breach, shatter and overflow horizons of expectation or anticipation, and as such are scarce (Caputo 2007). Faced with this rarity and this alterity, we might, instead, focus on all the ways in which practical orders repeat and reproduce by making the unforeseeable foreseeable and the unrepeatable repeatable, that is all the ways in which events are foreseen, foresaid and foreclosed (see Derrida 2007, see also Harrison this volume; Rose this volume).

In both examples the event does not resemble, conform, or reproduce a set of *a priori* conditions. It does not *represent* those conditions. Rather, and in different ways, events break with their extant conditions, forcing or inviting us to think and act differently (Massumi 2002a, xxiv-xxv). It may be that like the prefix 'non' we can only define the event negatively – the event is the impossible which happens. The event '[a]lways comes to us by surprise, or from that side whence, precisely, it was not expected' (Dastur 2000, 183). The shared sense of 'the event' as that which opens up the chance of something different is expressed well by Rajchman (1991, ix):

> [The event] is not defined by a fixed beginning and end, but is something that occurs in the midst of a history, causing us to redistribute our sense of what has gone before it and what might come after. An event is thus not something one inserts into an emplotted dramatic sequence with its start and finish, for it initiates a new sequence that retrospectively determines its beginnings, and which leaves its ends unknown or undetermined.

The emphasis on the chance of the event means that it is not quite correct to characterise non-representational theory as a type of practice theory, even though, as discussed above, it places thought-in-action, nor as only offering a form of relational materialism, albeit one attuned to affect and other absent-present 'objects'. Although 'the event' is conceptualised in various ways, the concept is so central to non-representational theory because it offers a way of thinking about how change occurs in relation to the on-going formation of 'the social'. Hence the desire that has animated non-representational theory has been to find a means of attending to the difference, divergence and differentiation that events open up, or may open up. We see this across work that has attempted to bear witness to the potential for difference released by the taking-place of a range of events; the fleeting potential that follows the event of a sexually charged glance between two people (Lim 2007), the performative force and sense of mutability found in dance and the performing arts (Dewsbury 2000); the potential for better ways of being touched in moments of hope (Anderson 2006a), and explicitly political events that break with the state of an existing situation (Dewsbury 2007, Woodward, this volume). The lightening of possible storms.

The question of the event opens up a further set of issues about how to *create* and *sustain* events; how to bear and extend the potential that events open up, the sense of promise and futurity that they may hold? How, to put it differently, to relate to the future without capturing it and neutralising it before it happens? Across tangible differences in theory and method, non-representational theories share an affinity of sensibility, what we could call a specific 'existential faith' that crosses various attempts to contribute, if only modestly, and always carefully, to the opening up of different futures (Connolly 2008). This existential faith finds ethical and political import in thinking about methods – understood broadly – as active interventions in the taking-place of events, whether by affirming (generously, hopefully) becoming or waiting (hospitably, anxiously) for the 'to come' (compare McCormack, this volume; Rose, this volume; Woodward, this volume). What this work shares is a commitment to critique as a means of creating turning points in the here and now and a conviction that in any given situation more is needed than critique if (certain) events are to be tended to and cultivated. Critique is necessary but always insufficient. It may be supplemented by a positive attachment to a world of becoming in which 'wherever and whenever something is going on there is an event'. Hence the recent interest in enchantment (Bennett 2001b) or generosity (Diprose 2001) as two such ways of working on the 'background' of thought and life (see McCormack, this volume; Darling, this volume and Roe, this volume). It may also be supplemented by an affirmative, perhaps even utopian, relation with events, everyday or otherwise, that open up traces of radically different futures (Anderson 2006b; Kraftl 2007; Rose, 2007; see Rose, this volume).

Although usually considered to be very different, these ways of relating to the event have a series of affinities with other styles of anticipatory thinking and acting, most notably the attention to disruption that marks queer geographies (e.g. Brown 2008), an emergent Feminist and anti-racist literature attuned to the force of corporeal differences such as gender (Colls 2007), and the explosion of interest in poststructuralist participatory geographies seized by the potential of various micro-economic experiments (Gibson-Graham 2006). All are animated by the question of how better futures may be brought into being. Likewise, the attention to the event in Non-Representation Theory opens up the question of future geographies in a way that returns us to the sense of affirmation and experimentation that we find in Foucault's dream;

> How then can space function differently from the ways in which it has always functioned? What are the possibilities of inhabiting otherwise? Of being extended otherwise? Of living relations of nearness and farness differently? (Grosz 2001, 129).

Openings

To conclude: it seems fair to say that non-representational theories are a set of predominantly, although not exclusively, poststructuralist theories that share a number of questions or problems; how do sense and significance emerge from on-going practical action?; how, given the contingency of orders, is practical action organised in more-than-human configurations?; and how to attend to events – to the 'non' that may lead to the chance of something different or a modification of an existing ordering? In this understanding, non-representational theorists may include ethnomethodologists, (post)phenomenologists, Deleuzians, Corporeal Feminists, and actor-network theorists, amongst others. In other words we take seriously the multiplicity of theorists that Thrift (1996, 1999) identified with non-representational theory, and that when first using the term he uses the plural. This means that the problems and questions that non-representational theories pose are not only being encountered in Human Geography. For example, they are also being taken up in the development of an immanent naturalism in political theory (Connolly 2008), an enchanted materialism in political ecology (Bennett 2001), and a renewed attention to affectively imbued experience in cultural studies (Seigworth 2000).

The four sections we have organised the book around – *Life*, *Representation*, *Ethics* and *Politics* – are designed to draw out a series of problems, questions and imperatives that deepen our introductory remarks, engage in more depth with the debates that have emerged around non-representational theory, and pick up some of the threads we have only been able to touch on or hint at here. When we first invited contributions to the book we asked each author to address a specific concept, problem or question by way of a theory or set of theories that were important to non-representational theories. As you will see, each of the authors interpreted this challenge differently. We have deliberately retained this plurality of tone, style and voice. Differences coexist within non-representational theory, and we wanted to produce a collection that affirmed this in both content and form. Indeed not all of the contributors would agree with how we have characterised non-representational theory in this introduction. These differences mean that each section opens up a set of further questions at a time when non-representational concerns are in the midst of travelling across a range of sub-disciplines within Human Geography, changing as different concepts, sensibilities and methods are taken up in relation to different substantive and theoretical problems.[13] We hope the book makes a modest contribution to this process. By way of a brief summary of each of the four sections, we want to conclude this 'primer' on non-representational theory by

---

13   See, for example, the emerging interest in everyday life, sensory registers and affect in political geography, particularly work on popular geopolitics and the biopolitics of security (Macdonald, Hughes, and Dodds 2009; Adey 2009; Sidaway 2009); nascent work on cultural economy, work and affect (Woodward and Lea 2009; Amin 2007); an attention to the importance of visceral in consumption (Hayes-Conroy 2008); attempts to think the relation between health, therapy and relational bodies (Lea 2008; Conradson 2005); the focus on matters of belief in work on religion (Holloway 2006); the various ways in which the urban is apprehended as an assemblage and architecture as an event (Kraftl 2006; Latham and McCormack 2004); and efforts to enliven children's geographies (Horton and Kraftl 2006; Woodyer 2008). This is in addition to the now huge amount of work concerned with 'everyday practices' of one form or another as reviewed by Lorimer (2005; 2007; 2008).

introducing the set of issues around the encounter between non-representational theories and Human Geography that the chapters address.

Part I – *Life* – is organised around the move from practices to Life. It poses a set of questions that resonate throughout the book and follow from the three shared problematics or commitments that we have argued non-representational theories share. How to attend to the indeterminacy and complexity of the world (Greenhough, this volume)? How to understand the intermingling of different types of lively material bodies (Lorimer, this volume)? How do affects and forms of signification intermix in specific practical orders (Bissell, this volume)? And how to think the relation between life and the formation of subjectivities (Wylie, this volume)? Part II – *Representation* – explores how we might think representation once our attention turns to Life. It offers four partially connected ways of developing the insight that representations enact worlds; through an attention to language-in-use (Laurier, this volume); through an account of representation as transformation and differentiation (Doel, this volume); via the event of language (Dewsbury, this volume); and through a concern with the 'failure' of re-presentation and so the 'failure' of a world (Harrison, this volume). As a whole Part II aims to make the point that non-representational theory does not refuse representation *per se*, only representation as the repetition of the same or representation as a mediation. The hinge between the first half of the book and the second is an interview with Nigel Thrift in which he charts the development of his own interest in practices, reflects on some of the key problematics that open up once one considers Life, and the ethical and political import of non-representational theories in relation to contemporary capitalism and democratic politics.

The second half of the book – *Ethics* and *Politics* – unfolds some of the implications for ethics and politics of non-representational theory's placing of 'thought-in-action', its materialist analysis of the social, and the attention to events. In no case does a politics or ethics simply unfold from a set of theoretical propositions. In each chapter specific problems, concepts, methods or sensibilities are brought into connection with worldly concerns, whether they be UK Asylum Seeker detention policy, the 1999 anti-capitalist protests in Seattle, community gardening groups, or the industrialized mass slaughter of animals. In *Ethics* the concern is with how to respond to social formations as they are *in* formation, where the social includes all manner of material bodies. In each case this involves (but is not limited to) exploring the relations between the affirmative and critical (McCormack, this volume) and experimenting with the corporeal sensibilities that are enfolded into how we learn to affect and be affected by the world, including relations with non-humans (Roe, this volume), and across recognised social differences (Darling, this volume; Simonsen, this volume). The chapters in *Politics* by contrast revolve around a slightly different problematic of difference; how to make a difference if we expand what counts as political and move beyond an exclusively representational politics? As one would expect, the means vary, and this section contains some of the most obvious tensions with chapters in previous sections, but all presume that politics takes-place in a world of differences;

including the force and materiality of social differences (Saldanha, this volume), the multiplicity of partially connected orders (Hinchliffe, this volume); the opening up to the new that events and certain 'abrupt conditions' may herald (Woodward, this volume); and the change the future may bring (Rose, this volume).

Writing an introduction such as this is like trying to 'catch sea foam in the breeze'. As the chapters which follow demonstrate, non-representational theory is on-going, diversifying and disseminating, and so attempting to define such an oeuvre is a largely thankless task. And this is why, in many respects, we have not done so. Rather in this introduction we have attempted to suggest the animating concerns, the conceptual, practical and existential commitments which bring this work to life, but which do not determine or delimit its development. Indeed recognisable across all three elements discussed above is a continual process of de-limiting; of the human, of the social, of the material and of the future. In non-representational theory each becomes multiple and many, contingent and fragile, assembled and scattered. All the better, all the better.

References

Adams, P., Hoelscher, S. and Till, K. (eds) (2001), *Textures of Place: Exploring Humanist Geographies* (Minneapolis: University of Minnesota Press).
Adey, P. (2008), 'Airports, mobility, and the calculative architecture of affective control', *Geoforum* 39, 438-451.
—— (2009), 'Facing airport security: affect, biopolitics and the preemptive securitization of the mobile body', *Environment and Planning D: Society and Space* 27(2), 274-295.
Adey, P. and Maddern, J. (2008), 'Editorial: spectro-geographies', *Cultural Geographies* 15, 291-295.
Amin, A. (2006), 'The good city', *Urban Studies* 43, 1009-1023.
—— (with Thrift, N) (2007), 'Cultural-economy and cities', *Progress in Human Geography* 31(2), 143-161.
Anderson, B. (2004), 'Time-stilled space-slowed: how boredom matters', *Geoforum* 35(6), 739-754.
—— (2006a), 'Becoming and being hopeful: towards a theory of affect', *Environment and Planning D: Society and Space* 35(5), 733-752.
—— (2006b), 'Transcending without transcendence: Utopianism and an ethos of hope', *Antipode* 38(4), 691-710.
—— (2009), 'Affective atmospheres', *Emotion, Society and Space* 2(2), 77-81.
Anderson, B. and Tolia-Kelly, D. (2004), 'Matter(s) in Social and cultural geography', *Geoforum* 35(6), 669-675.
Anderson, B. and Harrison, P. (2006), 'Questioning affect and emotion', *Area* 38(3), 333-335.
Anderson, B. and Wylie, J. (2009), 'On geography and materiality', *Environment and Planning A* 41(2), 318-335.

Ash, J. (forthcoming), 'Architectures of affect: anticipating and manipulating the event in processes of videogame design and testing', *Environment and Planning D: Society and Space*.

Barnett, C. (2008), 'Political affects in public space: normative blind-spots in non-representational ontologies', *Transactions of the Institute of British Geographers* 33(2), 186-200.

Benjamin, W. (1992), *Illuminations*. Trans. Zohn, H. (London: Fontana Press).

—— (1997), *One Way Street and Other Writings*. Trans. Jephcott, E. and Shorter, K. (London: Verso).

Bennett, J. (2001), *The Enchantment of Modern Life: Attachments, Crossings, and Ethics* (Princeton: Princeton University Press).

—— (2010), *Vibrant Matter: A Political Ecology of Things* (Durham, NC and London: Duke University Press).

Bingham, N. (1996), 'Object-ions: from technological determinism towards geographies of relations', *Environment and Planning D: Society and Space* 14(6), 635-657.

—— (2008), 'Slowing things down: Lessons from the GM controversy', *Geoforum* 39(1), 111-122.

Bissell, D. (2008), 'Comfortable bodies: sedentary affects', *Environment and Planning A* 40(7), 1697-1712.

—— (2009), 'Obdurate pains, transient intensities: affect and the chronically-pained body', *Environment and Planning A* 41(4), 911-928.

Bondi, L. (2005), 'Making connections and thinking through emotions: between geography and psychotherapy', *Transactions of the Institute of British Geographers* 30, 433-448.

Bourdieu, P. (1977), *Outline of a Theory of Practice* (Cambridge and New York: Cambridge University Press).

Brown, G. (2008), 'Ceramics, clothing and other bodies: Affective geographies of homoerotic cruising encounters', *Social and Cultural Geography* 9(8), 915-932.

Butler, J. (1990), *Gender Trouble* (London: Routledge).

—— (1993), *Bodies that Matter* (London: Routledge).

—— (1997), *Excitable Speech* (London: Routledge).

Caputo, J. (2007), *The Weakness of God* (Bloomington and London: Indiana University Press).

Clark, A. (1997), *Being There: Putting Brain, Body, and World Together Again* (London: MIT Press).

Clark, T. (2005), *The Poetics of Singularity. The Counter-Culturalist Turn in Heidegger, Derrida, Blanchot and the Later Gadamer* (Edinburgh: Edinburgh University Press).

Clark, N. (2007), 'Living through the tsunami: Vulnerability and generosity on a volatile earth', *Geoforum* 38(6), 1127-1139.

Clough, P. (ed.) (2007), *The Affective Turn* (Durham, NC and London: Duke University Press).

Colls, R. (2007), 'Materialising bodily matter: Intra-action and the embodiment of "Fat"', *Geoforum* 38, 353-365.

Connolly, W. (2002), *Neuropolitics: Thinking, Culture, Speed* (Minneapolis: University of Minnesota Press).

—— (2008), *Capitalism and Christianity, American Style* (Durham, NC and London: Duke University Press).

Conradson, D. (2005), 'Landscape, care and the relational self: therapeutic encounters in southern England', *Health and Place* 11, 337-348.

Cook, I. and Tolia-Kelly, D. (forthcoming), 'Material geographies', in Hicks, D. and Beaudry, M. (eds), *Oxford Handbook of Material Culture Studies* (Oxford: Oxford University Press).

Cosgrove, D. and Jackson, P. (1987), 'New directions in cultural geography', *Area* 19, 95-101.

Cosgrove, D. (1989), 'Geography is everywhere: culture and symbolism in human landscapes', in Gregory, D. and Walford, R. (eds), *Horizons in Human Geography* (Basingstoke: Macmillan), 118-135.

Crang, P. (1994), '"It's showtime!" On the workplace geographies of display in a restaurant in South East England', *Environment and Planning D: Society and Space* 12, 675-704.

Cresswell, T. (1996), *In Place/Out of Place: Geography, Ideology and Transgression* (Minneapolis: University of Minnesota Press).

—— (2003) 'Landscape and the obliteration of practice' in Anderson, K., Domosh, M., Pile, S. and Thrift, N. (eds), *The Handbook of Cultural Geography* (London: Sage), 269-281.

Dastur, F. (2000), 'Phenomenology of the event: Waiting and surprise', *Hypatia* 15(4), 178-189.

de Certeau, M. (1984), *The Practice of Everyday Life.* Trans. Rendall, S. (Berkeley, CA: University of California Press).

DeLanda, M. (2006), *A New Philosophy of Society: Assemblage Theory and Social Complexity* (London and New York: Continuum).

Deleuze, G. and Guattari, F. (1987), *A Thousand Plateaus*: *Capitalism and Schizophrenia.* Trans. Massumi, B. (Minneapolis: University of Minnesota Press).

Deleuze, G. (2001), 'Immanence: A life'. Trans. Boyman, A. In Deleuze, G. *Pure Immanence: Essays on a Life* (Cambridge: MIT Press).

—— (with Parnet, C.) (2006), *Dialogues II.* Trans. Tomlinson, H. and Habberjam, B. (London: Continuum).

Derrida, J. (2003), 'Autoimmunity: real and symbolic suicides – A dialogue with Jacques Derrida', in Borradori, G. (ed.), *Philosophy in a Time of Terror: Dialogues with Jürgen Habermas and Jacques Derrida* (Chicago and London: University of Chicago Press), 85-136.

—— (2007), 'A certain impossible possibility of saying the event'. Trans Walker, G. *Critical Inquiry* 33(Winter), 441-461.

Dewsbury, J-D (2000), 'Performativity and the event: enacting a philosophy of Difference', *Environment and Planning D: Society and Space* 18, 473-496.

—— (2003), 'Witnessing Space: knowledge without contemplation', *Environment and Planning A* 35(11), 1907-1933.

—— (2007), 'Unthinking subjects: Alain Badiou and the event of thought in thinking politics', *Transactions of the Institute of British Geographers* 32(4), 443-459.

Dewsbury, J-D, Harrison, P., Rose, M. and Wylie, J. (2002), 'Enacting Geographies', *Geoforum* 32, 437-441.

Duncan, J.S. (1990), *The City as Text: The Politics of Landscape Interpretation in the Kandyan Kingdom* (Cambridge: Cambridge University Press).

Diprose, R. (2001), *Corporeal Generosity: On Giving with Nietzsche, Merleau-Ponty, and Levinas* (New York: SUNY).

Edensor, T. (2005), *Industrial Ruins: Space, Aesthetics and Materiality* (London: Berg).

Foucault, M. (1997a), 'The masked philosopher'. Trans, Hurley, R. et al. in Rabinow, P. (ed.), *Michel Foucault – Ethics: Subjectivity and Truth* (London: Penguin), 321-328.

—— (1997b), *The History of Sexuality: The Care of the Self*. Trans. Hurley, R. (London: Penguin).

Gasché, R. (1999), *Of Minimal Things: Studies on the Notion of Relation* (Stanford: Stanford University Press).

Gertz, C. (1973), *The Interpretation of Cultures* (New York: Basic Books).

Gibson, J.J. (1979), *The Ecological Approach to Visual Perception* (Boston: Houghton Mifflin).

Gibson-Graham, J-K. (2006), *Post-Capitalist Politics* (Minneapolis: University of Minnesota Press).

Gregson, N. and Rose, G. (2000), 'Taking Butler elsewhere: performativities, spatialities and subjectivities', *Environment and Planning D: Society and Space* 18(4), 433-452.

Grosz, E. (2001), *Architecture from the Outside: Essays on Virtual and Real Space* (New York: MIT Press).

Hägerstrand, T. (1973), 'The domain of human geography' in Chorely, R.J. (ed.), *Directions in Geography* (Andover: Methuen), 67-87.

—— (1982), 'Diorama, path and project', *Tijdschrift voor Economische en Sociale Geografie* 73, 323-39.

Hayes-Conroy, A. (2008), 'Taking back taste: Feminism, food, and visceral politics', *Gender, Place, and Culture* 15(5), 461-473.

Haraway, D. (1991), *Simians, Cyborgs and Women: The Reinvention of Nature* (London: Free Association Books).

Harman, G. (2009), *The Prince of Networks: Bruno Latour and Metaphysics* (London: Re.press).

Harrison, P. (2000), 'Making Sense: embodiment and the sensibilities of the everyday', *Environment and Planning D: Society and Space* 18, 497-517.

—— (2007), '"How shall I say it?" Relating the nonrelational', *Environment and Planning A* 39(3), 590-608.

—— (2008), 'Corporeal remains: vulnerability, proximity, and living on after the end of the world', *Environment and Planning A* 40, 423-445.

—— (2009), 'In the absence of practice', *Environment and Planning D: Society and Space* 27(6), 987-1009.

Heidegger, M. (1962), *Being and Time*. Trans. Macquarrie, J.M. and Robinson, E. (Oxford: Blackwell).

Hinchliffe, S. (1996), 'Technology, power and space: the means and ends of geographies of technology', *Environment and Planning D: Society and Space* 14(6), 659-682.

—— (2007), *Geographies of Nature: Societies, Environments, Ecologies* (London: Sage).

Hinchliffe, S. and Whatmore, S. (2006), 'Living cities: towards a politics of conviviality', *Science as Culture* 15(2), 123-138.

Holloway, J. (2006), 'Enchanted spaces: The séance, affect and geographies of religion', *Annals of the Association of American Geographers* 96(1), 182-187.

Horton, J. and Kraftl, P. (2006), 'What else? Some more ways of thinking about and doing children's geographies', *Children's Geographies*, 4(1), 69-95.

Ingold, T. (2000), *The Perception of the Environment: Essays in Livelihood, Dwelling and Skill* (London: Routledge).

Jackson, P. (2000), 'Rematerialising social and cultural geography', *Social and Cultural Geography* 1(1), 9-14.

Kraftl, P. (2006), 'Building an idea: The material construction of an ideal childhood', *Transactions of the Institute of British Geographers* 31(4), 488-504.

—— (2007), 'Utopia, performativity, and the unhomely', *Environment and Planning D: Society and Space* 25, 120-143.

Latham, A. (1999), 'The power of distraction: Distraction, tactility, and habit in the work of Walter Benjamin', *Environment and Planning D: Society and Space* 17(4), 451-473.

—— (2003), 'Research, performance, and doing human geography: some reflections on the diary-photo diary-interview method', *Environment and Planning A* 35(11), 1993-2017.

Latham, A. and McCormack, D (2004), 'Moving cities: rethinking the materialities of urban geographies', *Progress in Human Geography* 28(6), 701-724.

Latour, B. (1993), *We Have Never Been Modern*. Trans. Porter, C. (London: Harvester Wheatsheaf).

—— (1999), *Pandora's Hope: An Essay on the Reality of Science Studies*. Trans. Sheridan, A. (Cambridge, MA: Harvard University Press).

—— (2005), *Reassembling the Social: An Introduction to Actor-Network Theory* (Oxford: Oxford University Press).

Law, J. (1993), *Organising Modernity* (London: Blackwell).

Lea, J. (2008), 'Retreating to nature: rethinking "therapeutic landscapes"', *Area* 401, 90-98.

Ley, D. and Samuels, M.S. (1978), 'Overview', in Lay, D. and Samuels, M.S. (eds), *Humanistic Geography: Prospects and Problems* (London: Croom Helm), 19-21.

Lim, J. (2007), 'Queer critique and the politics of affect', in Browne, K., Lim, J. and Brown, G. (eds), *Geographies of Sexualities* (Aldershot: Ashgate), 53-67.

Lingis, A. (1994), *Foreign Bodies* (London: Routledge).

—— (1996), *Sensation: Intelligibility in Sensibility* (Atlantic Highlands, NJ: Humanities Press).

Lorimer, H. (2005), 'Cultural geography: the busyness of being 'more-than-representational', *Progress in Human Geography* 29(1), 83-94.

—— (2006), 'Herding memories of humans and animals', *Environment and Planning D: Society and Space* 24, 497-518.

—— (2007), 'Cultural geography: worldly shapes, differently arranged', *Progress in Human Geography* 31(1), 89-100.

—— (2008), 'Cultural geography: nonrepresentational conditions and concerns', *Progress in Human Geography* 32(4), 551-559.

Macdonald, F., Hughes, R. and Dodds, K. (2009), 'Envisioning geopolitics', in MacDonald, F., Dodds, K. and Hughes, R. (eds) *Observant States: Geopolitics and Visual Culture* (London: I.B. Tauris).

Marston, S., Jones III, J-P and Woodward, K. (2005), 'Human geography without scale', *Transactions of the Institute of British Geographers* 30(4), 416-432.

Massey, D. (1997), 'A global sense of place', in Barnes, T. and Gregory, D. (eds), *Reading Human Geography* (London: Arnold), 315-323.

—— (2005), *For Space* (London: Sage).

Massumi, B. (2002a), 'Introduction: Like a thought'. In Massumi, B. (ed.), *A Shock to Thought: Expression after Deleuze and Guattari* (London and New York: Routledge), xiii-xxxix.

—— (2002b), *Parables for the Virtual: Movement, Affect, Sensation* (Durham, NC and London: Duke University).

McCormack, D. (2002), 'A paper with an interest in rhythm', *Geoforum* 33(4), 469-485.

—— (2003), 'An event of geographical ethics in spaces of affect', *Transactions of the Institute of British Geographers* 28(4), 488-507.

—— (2006), 'For the love of pipes and cables: a response to Deborah Thien', *Area* 38(3), 330-332.

—— (2008), 'Engineering affective atmospheres on the moving geographies of the 1897 Andrée expedition', *Cultural Geographies* 15(4), 413-430.

McFarlane, C. (2009), 'Translocal assemblages: space, power and social movements', *Geoforum* 40(4), 561-567.

Merleau-Ponty, M. (1963), *Phenomenology of Perception.* Trans. Smith, C. (London: Routledge).

Mels, T. (ed.) (2004), *Reanimating Places: A Geography of Rhythms* (London: Ashgate).

Merriman, P. (2007), *Driving Spaces: A Cultural-Historical Geography of England's M1 Motorway* (Oxford: Blackwell).

Murdoch, J. (1998), 'The spaces of actor-network theory', *Geoforum* 29, 357-374.

Nash, C. (2000), 'Performativity in practice', *Progress in Human Geography* 24, 653-664.

Parker, A. and Kosofsky Sedgwick, E. (1995) (eds), *Performativity and Performance* (London and New York: Routledge).

Paterson, M. (2006), 'Feel the presence: The technologies of touch', *Environment and Planning D: Society and Space* 24(5), 691-708.

—— (2007), *The Senses of Touch: Haptics, Affects and Technologies* (Oxford: Berg).

Phelan, P. (1993), *Unmarked: The Politics of Performance* (London: Routledge).

—— (1997), *Mourning Sex: Performing Public Memories* (London: Routledge).

Pile, S. (2005), *Real Cities: Modernity, Space and the Phantasmagorias of City Life* (London: Sage).

Pred, A. (1977), 'The choreography of existence: Comments on Hagerstrand's time-geography and its usefulness', *Economic Geography* 53, 207-221.

Puar, J. (2007), *Terrorist Assemblages* (Durham, NC and London: Duke University Press).

Rajchman, J. (1991), *Philosophical Events* (New York: Columbia University Press).

Relph, E. (1976), *Place and Placelessness* (London: Pion).

Rose, G. (1993), *Feminism and Geography: The Limits of Geographical Knowledge* (Cambridge: Polity).

Rose, M. (2002), 'Landscapes and labyrinths', *Geoforum* 33, 455-467.

—— (2006), 'Gathering dreams of presence: a project for the cultural landscape', *Environment and Planning D: Society and Space* 24(4), 537-554.

—— (2007), 'The problem of power and the politics of landscape: Stopping the Greater Cairo ring road', *Transactions of the Institute of British Geographers* 32(4), 460-477.

Saldanha, A. (2005), 'Trance and visibility at dawn: Racial dynamics in Goa's rave scene', *Social and Cultural Geography* 6(5), 707-721.

—— (2006), 'Reontologising race: the machinic geography of phenotype', *Environment and Planning D: Society and Space* 24(1), 9-24.

—— (2007), *Psychedelic White: Goa Trance and the Viscosity of Race* (Minneapolis: University of Minnesota Press).

Seamon, D. (1979), *A Geography of the Lifeworld. Movement, Rest, and Encounter* (New York: St Martin's Press).

Searle, J. (1983), *Intentionality. An Essay in the Philosophy of Mind* (Cambridge: Cambridge University Press).

Sedgwick Kosofsky, E. (2003), *Touching Feeling: Affect, Pedagogy, Performativity* (Durham, NC: Duke University Press).

Seigworth, G. (2005), 'Banality for cultural studies', *Cultural Studies* 14(2), 227-268.

—— (2003), 'Fashioning a stave, or, singing life', in Slack, J. (ed.), *Animations of Deleuze and Guattari* (New York: Peter Lang), 75-105.

Shotter, J. (1995), 'Talk of saying, showing, gesturing, and feeling in Wittgenstein and Vygotsky'. Available at: http://www.massey.ac.nz/~alock/virtual/wittvyg. htm. (Last accessed 21 October 2009).

Sidaway, J. (2009), 'Shadows on the path: Negotiating geopolitics on an urban section of Britain's South West Coast Path', *Environment and Planning D: Society and Space* 27(6), 1091-1116.

Simpson, P. (2008), 'Chronic everyday life: Rhythmanalysing street performance', *Social and Cultural Geography* 9(7), 807-829.

Slocum, R. (2008), 'Thinking race through feminist corporeal theory: divisions and intimacies at the Minneapolis Farmers' Market', *Social and Cultural Geography* 9(8), 849-869.

Stewart, K. (1996), *A Space on the Side of the Road: Cultural Poetics in an 'Other' America* (Princeton: Princeton University Press).

—— (2007), *Ordinary Affects* (London: Duke University Press).

Strohmayer, U. (1998), 'The event of space: geographic allusions in the phenomenological tradition', *Environment and Planning D: Society and Space* 16, 105-121.

Swanton, D. (2008), 'The force of race', *Dark Matter*. Available at: http://www. darkmatter101.org/site/2008/02/23/the-force-of-race/. (Last accessed 20 October 2009).

Taussig, M. (1993), *Mimesis and Alterity: A Particular History of the Senses* (London: Routledge).

Tolia-Kelly, D. (2006), 'Affect – an ethnocentric encounter?: Exploring the "universalist" imperative of emotional/affectual geographies', *Area* 38(2), 213-217.

Thien, D. (2005), 'After or beyond feeling? A consideration of affect and emotion in geography', *Area* 37(4), 450-456.

Thrift, N. (1983), 'On the determination of social action in space and time', *Environment and Planning D: Society and Space* 1, 23-57.

—— (1996), *Spatial Formations* (London: Sage).

—— (1999), 'Steps to an ecology of place', in Massey, D., Allen, J. and Sarre, P. (eds), *Human Geography Today* (Cambridge: Polity Press), 295-321.

—— (2004), 'Intensities of feeling: towards a spatial politics of affect', *Geografiska Annaler* 86B, 57-78.

—— (2008), *Non-Representational Theory: Space, Politics, Affect* (London and New York: Routledge).

Tuan, Y-F. (1977), *Space and Place: The Perspective of Experience* (Minneapolis: University of Minnesota Press).

Whatmore, S. (2002), *Hybrid Geographies: Natures, Cultures, Spaces* (London: Sage).

Whitehead, A. (2004 [1920]), *The Concept of Nature* (New York: Prometheus Books).

Woodward, K. and Lea, L. (2009), 'Geographies of affect', in Smith, S., Pain, P., Marston, S. and Jones III, J-P (eds), *The Handbook of Social Geography* (London: Sage).

Woodyer, T. (2008), 'The body as research tool: embodied practice and children's geographies', *Children's Geographies* 6(4), 349-362.

Williams, R. (1981), *Culture* (London: Fontana Press).

Wylie, J. (2002), 'An essay on ascending Glastonbury Tor', *Geoforum* 32(4), 441-455.

—— (2005), 'A single day's walking: narrating self and landscape on the SouthWest Coast Path', *Transactions of the Institute of British Geographers* 30(2), 234-247.

——(2006), 'Depths and folds: On landscape and the gazing subject', *Environment and Planning D: Society and Space* 24(4), 519-535.

—— (2007), 'The spectral geographies of W.G. Sebald', *Cultural Geographies* 14(2) 171-188.

—— (2009), 'Landscape, absence and the geographies of love', *Transactions of the Institute of British Geographers* 34.

PART I
Life

Chapter 2
Vitalist Geographies: Life and the More-Than-Human

Beth Greenhough

Introduction

> [T]here is a sense in which the life seems to have been sucked out of the worlds
> we variously study under the rubrics of 'society and space' and 'earth surface
> processes and landforms' (Whatmore, 1999, 260).

A series of recent interventions within geography have sought to refocus attention
on matter and materiality (Whatmore 2006; Anderson and Tolia-Kelly 2004). For
Sarah Whatmore these materialist returns have led geographers back to 'the most
enduring of geographical concerns – the vital connections between the *geo* (earth)
and the *bio* (life)' (2006, 601, *emphasis original*). For Whatmore (1999), the current
division of geographical labour into the study of nature (earth surface processes
and landforms) and culture (society and space) has seemingly failed to capture the
liveliness and agency of non-human living beings. Geographical science results in
a deadening of the world and its liveliness, for it separates out humans and agency
(which seems to be an exclusively human capacity) from the more mechanistic
explanations which characterise accounts of geomorphological, hydrological
and atmospheric processes. Yet in the current scientific climate it is becoming
increasing hard to hold human life apart from materiality and to deny agency to
all forms of life except humans. The futility of such an exercise is exemplified by
current debates over the inability of humans to effectively govern and control the
activities of agencies such as Avian Influenza (see Hinchliffe and Bingham 2008),
Bovine Spongiform Encephalopathy (Hinchliffe 2001), Bovine Tuberculosis
(Enticott 2006) and Foot and Mouth disease (Law and Mol 2006). These disease
agents – and the governance practices, regulations and technologies which
humans employ to try and evade or destroy them – blur the boundaries between
human agencies, other forms of life and the material world. Consequently, the life
sciences have been put forward as a key place where geographers might begin to
re-map the relationships between nature, society, life and matter (Castree 1999;
Whatmore 1999; Spencer and Whatmore 2001; Bridge et al. 2003; Greenhough
and Roe 2006). In order to do so, Spencer and Whatmore (2001, 140) suggest
that rather than separating out human and material worlds we should instead be

developing geographical approaches that are better attuned to the ways in which we (as human agents) intervene in and are shaped by the life worlds of others.

This concern with the liveliness of the world is one echoed by the vitalist tradition. Put simply, vitalism argues that there is some exceptional quality, soul, spirit or élan vital (Bergson 2002) that is possessed only by living beings. Underpinning this conviction is a history of resistance to mechanistic and deterministic ways of understanding living beings. Where the sciences sought to define living beings through the application of physical laws and chemical principles, vitalists sought to understand what exactly it was that made life different, thereby laying the foundations for the emergence of the discipline of biology in the nineteenth century. Vitalists drew attention to qualitative differences between living and other phenomena. They noted how instead of staying still and being easily located in time and space, life forms are restless, mobile and constantly changing their relationship to their environment; instead of *being* some *thing*, life forms are constantly evolving, constantly *becoming*, shifting in their composition. Recent years have seen the emergence what has become known as the 'new vitalism' (Fraser, Kember and Lury 2005). Like the materialist return in Geography, this resurgence of interest in vitalist thought is linked to the ways in which developments in technology and the biological sciences are again raising questions about what exactly counts as life. The term bio*technology* reflects the increasingly mechanistic ways in which we understand and intervene in the natural world, conceiving of life as a series of bio-chemical 'building blocks' including genes and DNA. At the same time, advances in information science, cybernetics and the creation of artificial life forms seek to 'introduce information, knowledge or "mind" into social and natural entities, making them less inert, more process-like: bringing them alive' (Fraser, Kember and Lury 2005, 1).

The phrase 'process-like' reflects how for the new vitalists it is process – how things come to be and make their presence felt in the world – which might better mark out the distinction between life and matter. This shift from property-based understandings to process-based ones is one shared by non-representational theory. Indeed non-representational theory's focus on practice and embodiment as modes of engaging with the world might be themselves described as vitalist, as they draw attention to the vitality inherent with diverse material forms that often exceeds and even disrupts the spaces and concerns of human interest. The vitalist imperative to pay attention to the liveliness of material beings finds in non-representational theory's focus on practice a way of making that lively world intelligible (Thrift 2008, 8). This chapter explores this meeting point between non-representational theory and vitalism. I begin by describing how vitalism, and in particular the work of Henri Bergson, offers some interesting opportunities for re-thinking how we practice geography. I then follow by drawing out some of the key facets of a vitalist-inspired geography, exploring how it relates to the move towards non-representational theory (Thrift 2004), before moving to consider three key challenges posed by seeking to adopt a re-vitalised approach to geography.

Vitalist Geographies

Vitalist legacies

> [T]he 'vital principle' may indeed not explain much, but it is at least a sort of label affixed to our ignorance, so as to remind us of this occasionally, while mechanism invites us to ignore that ignorance (Bergson 2002, 190).

While many vitalist scholars were and are sceptical about the more mystical dimensions of vitalism and its ability to explain the meaning and operation of life, for writers such as Bergson (2002) it is a useful way of acknowledging the limitations of our own understanding. Where physics and chemistry seem to offer universally applicable explanations, vitalism acts as 'a sort of label affixed to our ignorance' (Bergson 2002, 190) or perhaps an acknowledgement that '*[t]he world is more excessive than we can theorise*' (Dewsbury, Harrison, Rose, Wylie 2002, 437, *emphasis original*). In other words the important thing about vitalism is not that it assigns an 'exceptional quality' to life, but rather that it makes a particular sort of demand on the researcher to attend to the complexity and indeterminacy of things in the world (Fraser, Kember and Lury 2005). As opposed to both mechanism and positivism, which see things in the world as definable by given material properties, vitalism asserts that

> objects, subjects, concepts are composed of nothing more or less than relations, reciprocal enfoldings gathered together in temporary and contingent unities. Furthermore, since a relation cannot exist in isolation, all entities can be understood in relation to one another (ibid., 3).

Drawing on Fraser, Kember and Lury (2005) we might suggest that the new vitalism makes the following five key contributions. *Firstly*, it insists that we focus on vital processes (becoming) as opposed to essential or given qualities (being) as a way identifying life. *Secondly*, it insists that all entities can be understood only in terms of their relations to one another. *Thirdly* (and of particular interest to geographers), is the conviction that space and time are not external to relations between entities, but rather space and time are co-emergent with entities for the duration of any given event (more on this below). *Fourthly*, an appreciation of the uniqueness and irreversibility of each event, and how this irreversibility leads to the creation of 'stubborn facts', understandings of the world that through their ongoing use take on a semblance of certainty. (Although even these stubborn facts may be radically transformed: they too are constantly in the process of becoming). *Fifthly*, a recognition that if nothing is external to the process of becoming, then the work of geographers and other social and natural scientists is not undertaken in a space outside the phenomena they study. Instead, in seeking to explore and understand events and phenomena in the world, social scientists also intervene in the world. This final point is particularly important when thinking through what it might mean to practice a vitalist-inspired geography.

While the above key points might imply a convergence of vitalist thought, this is to underplay the often sharp distinctions between different vitalist thinkers. Lash (2006), for example, cites two distinct vitalist genealogies: the first tradition (Bergsonian), including Bergson and Deleuze, focuses on understanding vitalism through an attention to perception, sensation and how we experience the world. The second (Nietzschean) tradition, including Simmel and Foucault, focuses on power and how control over others is exercised as a direct intervention in the vitality of others (e.g. through processes which seek to discipline and control bodies like the penal system). The emergence of the new vitalism has seen contemporary scholars engage with a wide range of vitalist thought (see special issues of *Theory, Culture, Society* in 2005 and 2007). Here I choose to focus on what Lash terms the Bergsonian tradition for two key reasons:

Firstly because it is Bergson's concern with not just with what counts as life, but also with the ontological problem of how we encounter life as a process of our living in the world, that I find most resonates with the phenomenological concerns of non-representational theory. Bergson saw a tension between an understanding of the world as consisting of bodies of matter which changed from state to state, as advocated by physics and chemistry, and his sense of the world as lived experience. For Bergson perception, sensation and an understanding of how living beings engage with and make sense of the material world, are central to a vitalist understanding (Lash 2006). It is through the process of perception that the world out there materialises (or is made present). Perception propels us as at once 'into matter' (Deleuze 2002, 25) and the world is a lived world because of our interaction (and that of other living bodies) with it (Bergson 2002, 178). Consequently instead of a knowable world 'out there' waiting to be discovered (the traditional objective of geographical exploration and analysis) we have instead a world which can only ever be partially known as a product of our encounters with it.

My second reason for focusing on the Bergsonian tradition is the influence Bergson had on the work of Gilles Deleuze. Deleuze is perhaps the key figure through which geographers have encountered and engaged with vitalist ideas. Deleuze develops from Bergson's notion of duration new ways of understanding and thinking the world as itself only emerging – or being actualised – in the moment of perceptual encounters between living agents, what he terms an *event*. For Deleuze the world is never completely realised as a final entity, but is better conceived of as an infinite virtual plane of things yet to be (or not be) actualised. What this perspective stresses is the need to see our understandings of life, space and time as emergent properties of the encounters we undertake and study, actualised as a result of a particular encounter. This is distinct from a mechanistic perspective that recognises a limited number of possibilities that are realised according to predefined laws and understandings of space, time and substance. Writing about Deleuze and Guattari's *A Thousand Plateaus*, Bonta and Protevi (2003) suggest it calls for a materialist geography in which 'all "objective" phenomena, no matter how "natural" or "universal" they may seem, should be thought of, and if possible mapped, in terms of their virtual realm'. Importantly for geographers this means space and time are not containers within which events take place whose properties can be defined by physical laws. Rather space and time are brought into being simultaneously with the actualisation of a given phenomenon. As a result 'where something happens, for Deleuze and Guattari, is as important as when it happens; how and why events occur are always entangled with when and where they take place' (Bonta and Protevi 2003, no pagination).[1] This attention to the co-emergence of life and space offers starting-point for developing vitalist-inspired approaches to geography.

Vitalist geographies

If one of the key demands of the materialist return within geography is that we pay attention to the 'livingness' of the world (Whatmore 2006), then vitalism offers some important suggestions as to how this might be achieved. *Firstly*, because it is non-anthropocentric. Vitalism does not restrict agency (the power to sense the world) to humans, but rather extends this capacity to all living beings. This places vitalism in direct conflict with the anthropocentric focus of much phenomenological thought. Indeed, the renewed interest in the vitalism of Bergson has been attributed to Deleuze's 'insistence that Bergsonism is an alternative to the domination of phenomenological thought' (Stanford Encyclopaedia of Philosophy 2008). *Secondly*, because vitalism (unlike mechanism) refuses to separate out material processes from the way in which they are encountered as part of the lived experiences of human and non-human actors. Or as Deleuze summarised, 'for phenomenology consciousness is of a thing, while for vitalism consciousness *is* a thing' (Lash 2006, 327, *emphasis original*). Our understanding or consciousness of the world emerges only though our interactions with the world and other living agents. But Whatmore also suggests Bergson's vitalism does not go far enough in its quest to enliven the world. While a vitalist approach draws attention to the liveliness of *our encounters* (and to an extent those of other life forms) with the material world, the liveliness of that material world seems flattened (see also Massey 2005, 20-24). The world remains dead until other agents or we enliven it. Conversely, Whatmore (2002, 68) argues that the world does not wait passively to be enlivened, but is already lively, active and capable of intruding upon us.

For Whatmore this recognition of a lively material world, which we come to know through active experience rather than passive observation, entails a new way of doing geography. Rather than making, describing or mapping the world it now involves paying attention to, and engaging with, the ways in which dynamic and changing worlds are lived with and performed through the interactions of living and lively beings. For some this insistence that the world and those who inhabit

---

1   For a more extensive exploration of the implications of Bergson's thought for conceptualisations of time and space see Massey 2005, especially Chapter 2.

it need to be understood as contingent and relational is a source of concern. Lash (2006, 328) cites Giorgio Agamben and Gillian Rose[2] as being among those who suggest 'vitalist thinking is thinking without content' and who are concerned by the refusal of vitalism 'to accept human and non-human limitations'. As vitalism describes the world through the possibilities generated in encounters, there are no pre-given understandings or rules that regulate how we interpret those encounters, no social norms or moral codes, only an openness to new possibilities. This leads critics like Agamben and Rose to suggest that vitalism is akin to nihilism, describing a world without meaning, purpose, truth or value. Indeed, we might ask if the world is contingent upon interactions and relations, how do seemingly fixed and external categories – such as definitive territories, spatial boundaries, moral convictions or senses of belonging and identity – endure so as to seem permanent? Whatmore argues that to understand geography as both contingent and dynamic

> is not to ignore the potent affects of territorializations of various kinds, just the reverse. It is a prerequisite for attending more closely to the labours of division that (re-)iterate their performance and the host of socio-material practices – such as property, sovereignty and identity – in which they inhere (2002, 6).

In other words, if we appreciate that the world does not conform to the maps we make of it, then this suggests we should pay close attention to how it is made to appear as though it does. This is geography articulated as one of Fraser, Kember and Lury's (2005) stubborn facts (see above). It is also here we find echoes of Bonta and Protevi's insistence on a materialist geography. Following Deleuze they argue we need to pay attention to exactly how and where things are actualised in the world: to 'trace matter flows across hemispheres and worlds, to map the sedimentation of ideas and the creation of populations and prejudices' (2003, unpaginated). It takes a lot of work to make the world conform to our ideas of what it should be. In illustrating this Whatmore (2002) writes about the multiple ways particular species – including elephants, leopards and alligators – are distributed through lifeworlds past and present, as endangered species, zoo exhibits, skins and exotic souvenirs, breeding databases and natural history documentaries, to name but a few. Here the bodies, body parts and representations of animal bodies are manipulated to fit with our (human) geographies. But these relationships are always also contingent and unpredictable. Other living beings might equally be seeking to remake the worlds they encounter, for example by escaping from zoos or refusing to breed in captivity.

This brief account of Whatmore's work lays the foundations for what she terms a *more-than-human geography* shaped by a focus not on the way the world *is*, but on how the world is *coming to be* through an engagement with our interventions

---

2   The Gillian Rose referred to here is a reader in Sociology in the School of European Studies at the University of Sussex and author of *Dialectic of Nihilism: Post-Structuralism and Law* (1984), not the Gillian Rose who most geographers will be familiar with.

in, and responses to, the world, and those of the other living beings who inhabit it. In the next section I begin outline the key facets of this vitalist-inspired geography developed by Whatmore and others, before turning to ask where might we go from here. How might we move beyond an acknowledgement of the processual and contingent nature of the geographies we study?

Key Facets of Vitalist Geographies

In this section I outline three key elements of a vitalist approach to geography and how these resonate with many of the ideas and methods drawn on by non-representational theory.

Sensing material worlds

Firstly there is a shift recognisable in both non-representational theory and vitalism from epistemology to ontology, in that both move from a concern with explaining and representing the world to focus instead on understanding the world through an engagement with its materiality. Non-representational theory and vitalism share a conviction that *worlds are sensed not just seen* (see also Lash 2006, 324). For Serres (1995) this affective or sensed world could be thought of as though we were surrounded by *noise*.[3] Our every attempt to pull meaning out of specific events faces being drowned out by a sea of *noise* which surround us:

> The background noise never ceases; it is limitless, continuous, unending, unchanging ... Noise cannot be a phenomenon; every phenomenon is separated from it, a silhouette on a backdrop, like a beacon against the fog, as every message, every cry, every signal must be separated from the hubbub that occupies silence, in order to be, to be perceived, to be known, to be exchanged (ibid., 13).

In this sense our first engagement with the world, almost before thought, is that shift in sensory perception which allows us to focus in on one affect, one event, and quieten others: to experience a phenomenon. To understand this process we might borrow heavily from the experimental approach of the physical sciences. Experimentation seems to offer a way of simplifying and clarifying particular moments or engagements with the world in order to focus the senses on one particular moment, aspect or type of engagement. The sociologist of science Bruno Latour exemplifies this when he describes how Pasteur used the techniques of microbe farming in the laboratory to multiply the amount of anthrax bacilli he

---

3   The term noise here is translated from the French, and correctly translated refers not just to sound but in a wider sense to interference and disruption. In this sense it resonates with the notion of excess in non-representational theory, as it seeks to acknowledge that which is excluded through the process of making sense of (encountering) the world.

extracted from farm cattle so as to make the microrganism visible. For Latour (1983, 165) laboratories are spaces through which the scale of phenomena can be reversed 'so as to make things readable'.

From a non-representational perspective what is striking about the process of scientific experimentation is that 'sheer *materiality* of thinking is constantly stressed' (Thrift 2004, 91, *emphasis added*). Latour and other practitioners of science studies show how in order to engage with the material world scientists need to draw on a whole host of other laboratory agents and equipment. Thought becomes 'a kind of *performative material intervention*' (ibid., *emphasis added*) in order to make others kinds of agencies – and worlds – sensible. For Thrift the experimental approaches of the life sciences offer us new ways of posing questions of and with the world. They also reflect how vitalists might differ ontologically as well as epistemologically from their counterparts in physical and chemical sciences. Bergson was concerned not only with knowing and explaining the world (epistemology) but also by how we are already immersed in and dwelling in the world (ontology) and how this shapes experimental practice. Similarly, the biologist's perceptive and sensory capacities to engage with and respond to the phenomena they are studying are key to the success of the experiment,[4] which leads us to a second key facet of a vitalist geography, a focus on practice.

Dwelling perspective: Life as practice

> Rather than just being a response, practice is a living of life itself. While there may be a reactive element to practice, it is also more than that. There is always an excess that cannot be explained. The overabundance of 'life in general' intrudes, taunts and undermines practice, so does it compel, stir and inspire it (Rose 2002, 461).

A second key aspect that emerges from a vitalist focus is the conviction that life is not a defined moment or happening in the world, rather it is always '*going on*' (Ingold 1995, 57, *emphasis original*), evolving (Bergson 2002) or in process (Fraser, Kember and Lury 2005). The result is that 'vital properties are never entirely realised, though always on the way to become so; they are not so much *states* as *tendencies*' (Bergson 2002, 178, *emphasis original*). It is this view of life as a constant process of engaging with the world that finds strong parallels in the notion of practice within non-representational theory. Instead of a world which exists as a series of discrete and observable elements, we have multiple worlds which are constantly being re-made by their living occupants (Thrift 2004; Ingold 1995). The anthropologist Tim Ingold (1995, 58) calls this process 'dwelling', for it is through this process that, animals and humans make themselves at home in the world.

---

4   See for example Evelyn Fox Keller's (1984) description of the scientist Barbara McClintock's 'feeling' for the corn plants she worked with.

Ingold's notion of dwelling stresses the links between a vitalist understanding of life (as being performed through an interaction with the world) and an acknowledgement of the situatedness of life within particular worlds, environments or dwelling spaces. As Pearson argues (drawing on Deleuze's interpretation of Bergson's vitalism):

> [I]t is erroneous to view the organism as an entity entirely separate from, and evolving independent of, its environment ... Organisms cannot be treated as closed models simply subjected to external forces and determinations; rather, they have to be understood in more dynamical terms as open systems that undergo continual flux (1999, 146).

This arguably poses problems for life scientists (like Pasteur) who believe they can get to the truth of an organism by isolating it from its 'natural' environment and re-establishing it in the environment of the laboratory:

> [T]he 'phenomenon' is technically redefined 'in the laboratory' and purified to the extent possible of everything assimilable to noise ... Experimentation in this context, is a *risky* process. It assumes that the phenomenon as isolated and reworked under laboratory conditions is essentially *the same* as the one found in 'nature' (Stengers 1997, 6, *emphasis added*).

Pasteur removes the anthrax bacilli from the site of encounter – the farmyard and those other 'noisy agencies' which drown the bacilli out, 'the smell, the cows, the willows along the pond, or the farmer's pretty daughter' (Latour 1983, 146) – but he also re-situates them in a new environment, the laboratory. The laboratory is an attempt to create an environment which makes certain interactions or relationships between scientists and their subjects endure as repeatable experiments which can, by virtue of their endurance, be put forward as (stubborn) facts-in-the-world (Latour 1983). Ironically however, the work undertaken to try and write-out or purify the specificities of each encounter from the experiment becomes an illustration of the performative material interventions needed to produce these so-called facts-in-the-world. As Thrift (1996, 43) observes (drawing on the work of the sociologist of science, Shapin), the laboratory itself becomes a very highly specified form of encounter 'a structured site of specific forms of social interaction with cultural and physical barriers to entry'. Pasteur (along with the microbes and other laboratory agents) has intervened in the world creating a new kind of microbial environment, distinctive from the one found in nature (Stengers 1997) or on the farm (Latour 1983), but no less complex. As Whatmore argued above, the laboratory is not a flat plain upon which life unfolds, but a dynamic space, shaped by and shaping its interaction with the scientists, microbes, machines and other agents who inhabit it. This brings us to a final key point, the way in which vitalism highlights the role non-human agencies in shaping the worlds we inhabit.

More-than-human agency

Thrift's (1996) account of the laboratory as a formal and structured 'productive local' can be contrasted with a historical account of the 'fly room', the laboratory within which Thomas Hunt Morgan conducted his now famous and Nobel-prize wining experiments with fruit flies on genetic heredity:

> The room itself is pretty small, about five by eight meters – the size of a living room. As we enter we spy shelves crammed with milk bottles. Hanging from a corner is a bunch of bananas. It's a pretty squalid chamber. There are plates with squashed flies on them. There's a distinct smell of yeast and a colony of cockroaches. And there's dust and dirt everywhere (Livingstone 2002, 7).

In Thrift's account of laboratory space in many senses the life (other than a human community of practitioners) seems 'sucked out'. The focus is on human bodies and on how their ability to make sense of the world is channelled in very specific ways through the protocols, procedures and technological devices that guide the process of scientific experimentation. Even Latour, who is keen to emphasise the capacity of machines and bacteria to either co-operate with or confound the experimental process, seems at the same time to enslave those non-human agents to a life-world of human making. In his account of Pasteur, the material needs of microbes are acknowledged, but the central role is given to Pasteur who is the agent 'representing the microbes and displacing everyone else' (1983, 168).

The fly room, by contrast, seems to be inhabited in a different way. The description here evokes smells, sounds, tastes and textures: it hints at a mutual accommodation of flies and geneticists. The fly room is *both* an experimental laboratory *and* a breeding ground for *Drosophila melanogaster* (and yeast, and cockroaches). Ironically this is perhaps what Whatmore means when she suggests that in order to put the life back in, we need to move outside the structured spaces of the laboratory and try to understand how the world becomes liveable, and is made liveable, by species other than our own. This in turn means paying attention to not only the worlds performed by human practices, but also the

> multi-sensual business of becoming animal – a relational process in which animal subjects are configured through particular social bonds, bodily comportments and life habits that are complicated, but neither originated nor erased, by the various ways in which they may be enmeshed in the categorical and practical orderings of people (Whatmore 2002, 37).

In other words the fly room is a space defined not just by human agency and orderings. It is a space defined by the habits and practices that allow flies and scientists to live and work together (see Bingham 2006). In the fly room the flies are not just objects existing in space, instead their agency, their 'constitutive vitality' (Whatmore 2002, 14), has helped define that place, the *fly*room. Stengers

(1997) would go further and point out to us how the constitutive vitality of the flies is also vital to the experiment. Her essays on the practice of science point to how the success of experiment depends on the willingness of the flies or other experimental subjects to co-operate with the experimental process.

Vitalist Challenges

But what implications does this vitalist geography – with its focus on sensed worlds, practice and non-human agency – have for the way in which we 'do' geography? In this final section I want to dwell briefly three key challenges that arise from adopting a vitalist-inspired approach to geographical research and practice.

Co-production

Firstly, attending to the constitutive vitality of non-human agencies as outlined above means we need to take seriously the proposition that research is always a co-production (see Whatmore 2003). It entails and enrols the bodies, capacities, sensibilities and expertise of both researcher and researched. If research is conceived of as being an encounter with the world then research is not achieved in isolation, rather it entails the contributions of numerous other agents operating alongside (assisting, resisting, ignoring) the researcher. This in turn entails a reconfiguration of what we understand by the terms researcher and researched. For example, Stengers argues for a recognition of those who are already engaged in risky and vital engagements with the world that are reconfiguring our understandings of the relationship between life and matter. Stengers (1997, 237) uses the example of the HIV/AIDS community, but importantly she cites them not as communities to be researched, but as researchers (heroes) who,

> explore in their flesh, for pleasure or from passion, what a body is, what it can and cannot tolerate. They tell us and remind us what we are – in this case, producers and consumers of bodily fluids. Living beings, in danger of life.

Likewise, Thrift describes non-representational theory as 'an attempt to change the role of academics by questioning what counts as expertise and who has that expertise' (2004, 81). Non-representational theory demands we pay attention not only to the practices of non-academics, but also the understandings of the world that both inform and are informed by those practices.

This involves re-thinking the practice of research. While recent work in post-structuralism, and in particular feminist geographies, has drawn attention to the ways in which the researcher's identity becomes involved and implicated in the research processes (see for example Rose 1997), here it is not only identities but bodies and behaviours that are a source of intervention. New skills and bodily competencies are needed. As Thrift (2004, 84) suggests co-production requires the re-imagination of practices of 'good' encounter and interaction which we can often only just sense. It requires practices and ethics of listening, talking, metaphorising and contemplation which can produce a feeling of being in a situation together.

This sense of empathy or shared experience, what the microbiologist Barbara McClintock might call 'a feeling for the organism' (Keller 1984, see also Stengers 1997; Whatmore 2003), suggests a moment where research exceeds the process of representation, provoking an emotional, embodied response. In short it acknowledges fieldwork is more than a process of data collection; it is an event through which the researcher and the researched are resituated or repositioned in the world, and thereby are engaged in remaking the world through the process of their encounters. Academic convention often has little tolerance of the kind of self-indulgent autobiography that places the experiences of the researcher at the centre of analysis, but this is not what is argued for here. Rather what is needed is what Haraway calls *modest witnessing* (1997, 269), the kind of research which remains open to and aware of its partiality and subjectivity.

Such an approach is key when research seeks to take into account more-than-human geographies. This is a concern we share with the life sciences in that we both need the co-productive capacities of other than human agents to achieve research. This move is captured in recent endeavours by geographers to ally themselves with watervoles (Hinchliffe, Kernes, Degen, Whatmore 2005) and elephants (Whatmore and Thorne 2000), or to engage with the new relationships with the world being formed by material derivatives in the form of bodily commodities such as bio-information (Greenhough 2006a), DNA (Nash 2004), donated and exchanged organs (Davies 2006) and blood (Morris 2007). Here concerns may be articulated not as neatly framed social commentaries, but as the elusive footprints of a watervole (Hinchliffe, Kernes, Degen, Whatmore 2005), fresh frozen blood plasma (Morris 2007) or as a refusal to engage with a sociologist's research project into genetic disorders (Callon and Rabeharisoa 2004).

Experimental approaches: on being at risk

Secondly we might learn from the life sciences and seek to become more experimental with the ways in which we encounter the world. This in turn requires an 'extended sense of ethnography' (Haraway 1997, 190-191) which entails

> being at risk in the face of practices and discourses into which one enquires. To be at risk is not the same thing as identifying with the subjects of study; quite the contrary ... One is at risk in the face of serious nonidentity that challenges previous stabilities, convictions, or ways of being of many kinds.

Good experimental science entails putting oneself at risk (Haraway 1997; Stengers 1997). This was certainly the case for Nobel prize winning scientist Dr Barry

Marshall when he tested his theory that stomach ulcers were caused by bacteria – and not (as many thought) by stress – by infecting his own body:

> I developed a vomiting illness and had severe inflammation in the stomach for about two weeks', he told The Associated Press. 'I didn't actually develop an ulcer, but I did prove that a healthy person could be infected by these bacteria, and that was an advance because the skeptics were saying that people with ulcers somehow had a weakened immune system and that the bacteria were infecting them after the event (*Isomed News* 2007).

This is not to suggest that all experiments which interrogate the relationship between life and matter involve such bacterial entanglements, rather to stress the point that because research is co-produced it cannot conform to pre-existing theories, convictions or ways of being. There are no guarantees.

The notion of being at risk takes on special aspects when it involves an engagement with living beings. For Stengers, like Bergson, this involves acknowledging that living beings are both '*produced by history and capable of history*' (Stengers 1997, 17, *emphasis original*). This means that they too are involved in making worlds and therefore we need to acknowledge the possibility that the conditions imposed by scientists in the laboratory might lead to them inadvertently silencing the objects of their enquiry. '[I]solation is a dangerous game, and those who believe they can purify their objects in fact intervene actively in the significance of the object they observe' (Stengers 1997, 17). This is a concern highlighted by recent debates over laboratory animal welfare and the concern that the findings of some animal experiments may say far more about the animal's engagement with their laboratory environment than their response to the particular medical intervention on trial. Or to put it another way, we might ask if 'unhappy mice give bad information?' (O'Hanlon 2001).

Importantly, such an insistence on the risky and indeterminate nature of research, and recognition of the limitations of experimentation and representation as ways of recounting our relationship with the world, does not preclude attempts to understand or interpret our research encounters. However it does demand we move away from an understanding of the aim of analysis as being to reduce things to their simplest form. Being at risk means being able to embrace the processes of social science and representation while at the same time holding on to the vitalist reminder that what is there will always exceed our capacities to describe, analyse or otherwise engage with it.

What can such an analysis achieve then? For both Stengers (1997) and Bergson (Deleuze 2002, 15-16), research should seek to achieve problematisation.

> Far from entailing the idea of a more simple world, analysis can lead to the conclusion that we do not know what a being is capable of. One way or another, reductionism always ends up ' ... is only ...'; the analytical method, on the other

hand, may lead to 'this ..., but in other circumstances that ... or yet again that ...' (Stengers 1997, 7).

Research should be about making the world more, not less, complex. If we acknowledge that the world does not consist of a limited and pre-determined set of possibilities, then our task is no longer to describe how the world is, but instead to explore and experiment with the multiple ways in which the world can and is coming to be.

Parasitism and infectious agency

> Knowledge is performative. Science, technology and medicine are ordering and materially productive practices. The sciences (including the social sciences) intervene in nature and politics with their enactments. They do not simply discover, define, interpret or account for these (Moser 2008, 99).

A final methodological challenge comes through an acknowledgement of how geographical research is itself a form of performative material intervention. Recently geographers have argued that geographical research should aim to be more policy relevant by structuring its questions and research activities towards specific policy goals (Martin 2001). But a focus on representing geographical ideas to the political world perhaps causes us (as researchers) to lose touch with the ways in which all geographical research is a form of political and ethical intervention. This is an idea captured in the everyday practices of being an academic geographer, in the environmental footprints left by the coffee we drink or the energy we consume. It also echoes through the practices of doing geographical research, and the need to recognise fieldwork as involving 'the materiality of the field, the contingency of encounters within it, and the embodied practices of the field workers themselves' (Driver 2000, 268).

Elsewhere (Greenhough 2006b) I have suggested that the role of the academic might be compared to that of Serres' (1982) parasite. Whilst the process of political representation is traditionally seen as one of simplification to produce a coherent oppositional standpoint, Brown (2002) suggests the role of Serres' parasite is to act as a catalyst for complexity that prevents and circumvents any kind of political settlement. Indeed, parasitical interventions are distinctly unsettling, for their role is not to provide different political answers to the question of what should be the proper relations between thought and life, but rather to invent '*new relations between thought and life*' (Thrift 2004, 82, *emphasis original*) by impelling those they parasitise to act. They cause their hosts to either include the parasite and the ideas it embodies, or to reject the parasite, redefining their own practice and boundaries in the process. I term this capacity for producing complexity *infectious agency*, for it suggests the role of the academic is not so much to extract information (in the form of a conventional reply) from a given research event, but to incite response in those (human or otherwise) with whom the researcher is engaged. Perhaps the real research outcomes should not be about what we have found (extracted) but what we have done and are doing (infection/excitement).

Conclusions

Drawing on non-representational theory and vitalism we might suggest that in seeking to make sense of the world geographers have somehow also lost many other ways of sensing the world, emphasising our human capacities for visual abstraction and textual communication over other ways on knowing and interacting with the world. As a result we have literally 'written out' the agency of non-human others (including the materials aspects of ourselves), that is articulated through non-representational practices. Drawing on vitalist legacies we might re-focus our some of our geographical endeavours towards taking seriously how worlds are formed through the constant interplay of life and matter. This would allow us to engage critically with those spaces where this relationship is being dramatically reconfigured through the agencies of life science, including biodiversity, biosecurity and conservation; the production and consumption of food and the increasingly molecularised approach to health. Such movements and interventions become increasingly uncomfortable as we begin to position our own bodies alongside those of non-human others as the experimental subjects of biotechnological interventions in the world. As the new kinds of relations between life and matter established in the laboratories of Louis Pasteur, Thomas Morgan and their successors move outside the laboratory, increasingly actions and practices that we conceived of as external (for example, the genetic modification of plants and animals) impact upon our own corporeal registers (see Roe 2006). These concerns are captured in emerging work in geography and anthropology focusing on questions of biosecurity. This work shows how political and scientific agencies seek to control, contain and restrict the movements and associations of a whole host of lively agents who are remaking themselves, each other, humankind and the world in profoundly unsettling ways (see for example Hinchliffe and Bingham 2008; Braun 2007).

However, some authors find in this vulnerability and bioinsecurity the possibility of articulating new kinds of shared political and ethical positions (see for example Clark 2004). Like non-representational theory, an engagement with more-than-human geographies offers 'new means of expression and new modes of agency' (Thrift 2004, 93). Vitalism puts us under obligation to not be contained by existing systems of social and political intervention. It demands we develop experimental approaches to politics which allow others, 'of all shapes, sizes, and trajectories, to object to the stories we tell about them' and 'to intervene in our processes as much as we intervene in theirs' (Hinchliffe et al. 2005, 655-656, see also Hinchliffe 2001). Equally I would argue it puts us under obligation to find new ways of practising geography that acknowledge the collective agency of geographers and those with whom they research in shaping multiple and lively worlds.

Acknowledgements

I am grateful to Jon May, Bronwyn Parry, Dave Parry, Ben Anderson and Paul Harrison for feedback on earlier versions of this chapter.

References

Anderson, B. and Tolia-Kelly, D. (2004), 'Matter(s) in social and cultural geography', *Geoforum* 35, 669-674.

Bergson, H. (2002), *Key Writings* (London: Continuum).

Bingham, N. (2006), 'Bees, butterflies, and bacteria: biotechnology and the politics of nonhuman friendship', *Environment and Planning A* 38, 483-498.

Bonta, M. and Protevi, J. (2003), 'Between geography and geophilosophy: *A Thousand Plateaus* and contemporary earth sciences' (Open University, Offprints for course D834).

Braun, B. (2007), Biopolitics and the molecularisation of life', *Cultural Geographies* 14, 6-28.

Bridge, G., Marsden, T.K. and McManus, P. (2003), 'The next new thing? Biotechnology and its discontents', *Geoforum* 34, 165-175.

Brown, S.D. (2002), 'Michel Serres. Science, translation and the logic of the parasite', *Theory, Culture and Society* 19(3), 1-27.

Callon, M. and Rabaharisoa, V. (2004), 'Gino's lesson on humanity: genetics, mutual entanglements and the sociologist's role', *Economy and Society* 33(1), 1-27.

Castree, N. (1999), 'Synthesis and engagement: critical geography and the biotechnology century', *Environment and Planning A* 31, 763-766.

Clark, N. (2004), 'Infectious generosity: Vulnerable Bodies and Virulent Becomings'. Paper presented at *Social Natures, Natural Relatives III*: *Life Science* conference, London, 2004.

Cloke, P., Crang, P. and Goodwin, M. (2003), *Envisioning Human Geographies* (London: Arnold).

Davies, G. (2006), 'Patterning the geographies of organ transplantation: corporeality, generosity and justice', *Transactions of the Institute of British Geographers* 31, 257-271.

Dewsbury, J-D, Harrison, P., Rose, M. and Wylie, J. (2002), 'Introduction: Enacting Geographies', *Geoforum* 33(4), 437-440.

Deleuze, G. (2002), *Bergsonism*. Trans. H. Tomlinson and B. Habberjam (New York: Zone).

Driver, F. (2000), 'Editorial: Fieldwork in geography', *Transactions of the Institute of British Geographers* 25(3), 267-268.

Enticott, G. (2006), 'Contesting biosecurity: Natural agents and material infrastructures in the Bovine Tuberculosis controversy'. Paper presented at the Annual Meeting of the Association of American Geographers, Chicago, April 2006.

Fraser, M., Kember, S. and Lury, C. (2005), 'Inventive life: Approaches to the new vitalism', *Theory, Culture, Society* 22(1), 1-14.

Greenhough, B. (2006a), 'Decontextualised? Dissociated? Detached? Mapping the networks of bioinformatics exchange', *Environment and Planning A* 38(3), 445-463.

Greenhough, B. (2006b), 'Imagining an island laboratory: Representing the field in Geography and Science Studies', *Transactions of the Institute of British Geographers* 31(2), 224-237.

Greenhough, B. and Roe, E. (2006), 'Towards a Geography of bodily Biotechnologies', *Environment and Planning A* 38, 416-422.

Haraway, D. (1997), *Modest_Witness@Second_Millennium.FemaleMan©_Meets_OncoMouse*™ (London: Routledge).

Hinchliffe, S. (2001), 'Indeterminacy in-decisions – science, policy and politics in the BSE (Bovine Spongiform Encephalopathy) crisis', *Transactions of the Institute of British Geographers* 26(2), 182-204.

Hinchliffe, S., Kernes, B., Degen, M. and Whatmore, S. (2005), 'Urban wild things: a cosmopolitical experiment', *Environment and Planning D: Society and Space* 23, 643-658.

Hinchliffe, S. and Bingham, N. (2008) 'Securing life: the emerging practices of biosecurity', *Environment and Planning A* 40(7), 1534-1551.

Ingold, T. (1995), 'Building, dwelling, living: How animals and people make themselves at home in the world', in Strathern (ed.).

Isomed News 'Two Australians win Nobel for showing bacteria cause ulcers' (published online 2007), http://www.isomed.com/news. Accessed 31 October 2007.

Keller, E.F. (1984), *A Feeling for the Organism: The Life and Work of Barbara McClintock* (New York: W.H. Freeman).

Knorr-Cetina, K.D. and Mulkay, M. (1983), *Science Observed: Perspectives on the Social Study of Science* (London: Sage).

Lash, S. (2006), 'Life (Vitalism)', *Theory, Culture, Society* 23(2-3), 323-349.

Latour, B. (1983), 'Give me a laboratory and I will raise the world', in Knorr-Cetina and Mulkay (ed.).

Law, J. and Mol, A. 'Globalisation in practice: On the politics of boiling pigswill', (published online 2006) http://www.sciencedirect.com/science/journal/00167185, accessed 1 August 2006.

Livingstone, D.N. (2002), *Science, Space and Hermeneutics* (Heidelberg: Department of Geography, University of Heidelberg).

Martin, R. (2001), 'Geography and public policy: the case of the missing agenda', *Progress in Human Geography* 25, 189-210.

Massey, D. (2005), *For Space* (London: Sage).

Morris, B. (2007), 'Bloody Geographies: Series 1'. Paper presented at the Supernatures conference, University of Exeter, September 2007.

Moser, I. (2008), 'Making Alzheimer's disease matter. Enacting, interfering and doing politics of nature', *Geoforum* 39(1), 98-110.

Nash, C. (2004), 'Genetic kinship', *Cultural Studies* 18(1), 1-34.

O'Hanlon, L. (2001), 'Do unhappy mice give bad information?', *Biomednet News*, 9 August 2001.

Pearson, K.A. (1999), *Germinal Life: The Difference and Repetition of the Life Sciences* (London: Routledge).

Pryke, M., Rose, G. and Whatmore, S. (2003), *Using Social Theory* (London: Sage).

Roe, E.J. (2006), 'Material connectivity, the immaterial and the aesthetic of eating practices: an argument for how genetically modified foodstuff becomes inedible', *Environment and Planning A* 38(3), 465-481.

Rose, G. (1997), 'Situating knowledges: Positionality, reflexivities and other tactics', *Progress in Human Geography* 21(3), 305-320.

Rose, M. (2002), 'Landscapes and labyrinths', *Geoforum* 33, 455-467.

Rose, N. (2007), *The Politics of Life Itself* (Princeton: Princeton University Press).

Spencer, T. and Whatmore, S. (2001), 'Biogeographies: putting life back into the discipline', *Transactions of the Institute of British Geographers* 26(2), 139-141.

Serres, M. (1982), *The Parasite* (Baltimore: Johns Hopkins University Press).

Serres, M. (1995), *Genesis*. Trans. Geneviève James and James Neilson (Michigan: University of Michigan Press).

Stanford Encyclopaedia of Philosophy *Henri Bergson*, http://plato.stanford.edu/entries/bergson/#7. Accessed 9 July 2008.

Strathern, M. (1995), *Shifting Contexts: Transformations in Anthropological Knowledge* (London: Routledge).

Stengers, I. (1997), *Power and Invention* (Minneapolis: University of Minnesota Press).

Thrift, N. (1996), *Spatial Formations* (London: Sage).

Thrift, N. (2004), 'Summoning Life', in Cloke, P., Crang, P. and Goodwin, M. (eds).

Thrift, N. (2008), *Non-Representational Theory* (Abingdon: Routledge).

Whatmore, S. (1999), 'Editorial: Geography's place in the life-science era?', *Transactions of the Institute of British Geographers* 24, 259-260.

Whatmore, S. (2002), *Hybrid Geographies* (London: Sage).

Whatmore, S. (2003), 'Generating materials' in Pryke, M., Rose, G. and Whatmore, S. (eds).

Whatmore, S. (2006), 'Materialist returns: practising cultural geography in and for a more-than-human world', *Cultural Geographies* 13(4), 600-609.

Whatmore S. and Thorne, L. (2000), 'Elephants on the move: spatial formations of wildlife exchange', *Environment and Planning D: Society and Space* 18(2), 185-203.

Chapter 3

Forces of Nature, Forms of Life: Calibrating Ethology and Phenomenology

Hayden Lorimer

These past few seasons I've developed a habit. I brood among a strip of sycamore and maple trees located close to the city's edge. Holding fast to the river valley's steep southerly bank, the trees offer a dappled sort of cover, from wind, sun or rain, and the flat-bellied sightlines of my quarry. Thirty feet below, 100 flippers distant: common seals haul up. A cow and her calf; joined very occasionally by a calf-less cow. From this distance, they appear as simple, curved shapes, as if formed by one continuous line. Try to imagine the kind of creature drawn by a child's hand. The seals' softened outlines are in-filled with tones and shades subtle enough, and so changeable, as to give the appearance of colour passing across the skin. Straight from the water, the mother cow's pelt is an inky slick of purplish-black. As it dries, patterns of brown marbling and mottling appear, before finally curdling to clotted cream, stippled thickly with chocolate.

Twice daily, the seals are drawn to the gently sloping shoreline of a small, tear-shaped island, found half a mile upstream from where the river meets the sea. They come ashore once the island's surface area enlarges with the ebbing tide. From my vantage point, backing onto a fencepost that borders a public footpath, the seals look entirely undisturbed. Truth to tell, the bass 'whuur' of tyre meeting tarmac, and sing-song of acceleration and deceleration, is all but incessant and emanates from little more than a stone's throw away. Seldom free from traffic, the bridge over the river channels motorists heading to points north from the city. From triangular turrets jutting out above granite buttresses the seals are clearly in view; though to the less watchful pedestrian, they might register only as beached pieces of driftwood. Lacking their own sure connection to the land, seals are dwellers on the threshold. They roll, loll and, once in a while, lollop to different points along the water's edge. Mostly, they bask quietly in a state of restful alertness. Languorous yawns easily deceive. It would be mistaken to ever regard seals as listless since the intervals are short between lifts or turns of the head. They exhibit spells of restiveness too, monitoring immediate surroundings, starting at sudden noises or stiff gusts of wind. Commonly, an uncomfortable looking, sculptured pose is struck, with head and tail held up proudly, leaving only their blubbery middle to compact the wet sand. When pressed to shift position, they are notoriously lacking in grace. Seals' movements on land are difficult to describe satisfactorily. Let me try. The mother cow's tuberous bulk pushes first downward, then immediately lurches forward. Quickly repeated, several times, the action propels her body, low and scudding over short distances. The muscular effect bears some resemblance to the rippling movement seen when a human neck, strongly proportioned, and of the right girth, swallows.

***

What can we humans discover of ourselves amid the lives of other creatures? How do other creatures inform our sense of what it is to be alive? What cues might our kind take from those many 'kinds' inhabiting the natural world around us? What qualities of animality are shared? And, what forms can togetherness take? Questions such of these – turning on how humans and non-humans relate, or remain autonomous and discrete – raise matters of theoretical and empirical consequence. Recently, geographers have been exercised by the search for possible answers, variously exploring the scope of: animal geographies (Philo and Wilbert 2000), animal landscapes (Matless et al. 2005), animate landscapes (Lorimer 2006), hybrid geographies (Whatmore 2003), cosmopolitan geographies (Bingham 2006; Hinchliffe et al. 2005; Lorimer 2007), post-human geographies (Castree and Nash 2004) and more-than-human geographies (Braun 2005). In spite of this proliferation, the greater part of an encyclopedia of creatures, and possible relations, remain as yet unconsidered: not least those more distant, less immediately familiar and not so easily situated amidst human lives, though nonetheless having affective kinds of association.

In this chapter I explore possible associations between humans and animals, the lore of their likeness, and the consequence of interspecies sociality for the figuring of personhood. Such a task might variously be undertaken. My approach is to consider association and likeness – and, as a corollary, difference and distinctiveness – by casting backwards *and* forwards. Looking both ways allows me to consider, at some length, the emergence of ethology as a branch of mid-20th century science dedicated to the study of animal lives as expressions of dwelling in the physical world, and possible contemporary configurations of ethology as a hybridized life science, fusing biological and geographical knowledge (Thrift 2005; 2007). This task of recovery and revival requires a history of ethological inquiry that is more extensive and critical – if admittedly still selective – than has been attempted hitherto by geographers (though for an early consideration see Tuan 1976; for remarks of recent interest see Hinchliffe and Whatmore 2006; Lorimer 2007). Delving deeper into past practices, anteceding ideas and introducing pioneering personalities from the earliest recognized episodes of this scientific approach, I identify resources to better understand the potentials of an ethological approach (broadly conceived) for geographers currently concerned with giving fuller expression to more-than-human lives, and forging practical, vitalist philosophies of life.

My undertaking then is to chart an intellectual terrain positioned by Frank Fraser Darling, biologist, essayist and conservationist, at 'the borderland of ecology' (Fraser Darling 1939, 103). Initially regarded as a fringe science, ethology offered its few early practitioners certain liberties and brought different dividends. Within the frame of this chapter, these are considered the signature features of ethological praxis, and they are threefold: socio-geographical, textual and experimental. First, ethology found expression in the work of those willing to conduct empirical, observational research in isolation, in the field. Without distraction, embedded ethologists learned to look intently at the workings of the world-in-the-making. If self-imposed exile brought hardship and the most trying conditions for living, the returns were immediate. Being purposefully out of step with one world was to quicken the senses and find attunements and synchronicities with another. Second, ethology was given popular expression in prose bearing witness to the act of being present in the midst of the physical world. Taking on the mantle of story-tellers of 'the social', they mixed description, discovery and revelation to powerful effect. Their prose carefully narrated life science in action and chronicled animal lives with impressive poise. Several of these works can, quite reasonably, be read critically as contributions to the canon of phenomenological thought. Third, ethology was an intellectual space characterized by an open treatment of evidentiary sources normally pushed aside or dismissed by the orthodoxies of scientific reasoning. Fraser Darling and others of his ilk (among them fellow 'New Naturalists' Ronald Lockley and Julian Huxley, ecologist Rachel Carson and ethnologist David Thomson) were willing to put an ear to indigenous voices who spoke otherwise of animality, thereby stretching the possible scope of personhood. Keeping company with reasonable people possessing of sound intellect and humane tongue, but without great formal education or written record, they willingly submitted to 'unreason' and 'uncommon' sorts of observation. The science that resulted was by different measure radical and rational, variously regarded as serious minded and, in its anthropomorphism, quasi-scientific and dangerously maverick.

In isolation, and in aggregate, these signature features – choice of study site, mode of writing and methods of working – require much closer consideration. Read alongside contemporary works of social theory (see for example Latour 2005), early ethologists fostered an acute appreciation of the configuration of a lively commonwealth comprised of interactions between human and animal organisms, and environmental phenomena. Prescriptive suggestions as to exactly *how* their ideas might be of consequence for non-representational theory and method in geography will be held over to the chapter's concluding section. To achieve this task with any measure of success, in the chapter's central sections I consider it helpful to focus attentions on just one sort of creature; substantive and primal, charismatic and reliably present. To that end, I have chosen seals, and concomitantly, a selection of influential seal studies undertaken around Britain and Ireland.

Why seals? Why indeed. The question echoes one posed by George Ewart Evans and David Thomson to introduce their cultural history of the life of the hare. The deceptively simple answer offered, that the animal is 'the focus of so many different points of view' (1972, 13), is just as applicable to other totemic species. Charismatic birds and fish, even certain insects, would feature in any such bestiary (see Lorimer 2007).[1] Literary nature writing, by turns celebrating and interrogating the subject, can be telling enough to confer primary status in the animal kingdom (Baker 1967: White, 1939) as has recently been the case for the crow (Cocker 2007) and the mountain hare (Macfarlane 2007). Likewise, the seal is a creature that has long pre-occupied the human imagination. The seal has weighed us human folk down, and, as often, buoyed us up. It is creature and canvas, working like a trap for our wandering theories and speculations, filtering our creative interpretations of reality. Ancient wisdom and modern science have met to shape the seal (Atkinson 1980; Farre 1957; Lambert 2002; Thomson 1954; Thomson 1976), and, all the while, the seal has stared back at us with a face we think we can read, and sung plaintively to us in songs that we think we know. Where then to begin?

Some initial direction might be taken from Thomas Nagel's (1974) rigorous philosophical inquiry into 'what it is like to be ...' something *else*? In large part Nagel's ontological anxieties can be turned round into practical queries, determined by the limits of what can be known of lives spent, say, on the wing, in the dark, or privileging sound over sight. The existential otherworldliness of animality is most often elemental, in basis and in habit. J.A. Baker found in the peregrine falcon what he felt was the true essence of alterity, essaying in precipitous aerobatic displays 'a beyond world, at work around and beside our own' (Macfarlane 2007, 273). Water is of course the other medium where, if not quite outlawed, humans never can hope to match animals' mastery of movement, depth or range. In search of ways of existing that are so absolutely *other* to human experience as to seem ungraspable, Owain Jones (2001) ponders on lives lived under the sea. Among marine animals there exist worlds of sensory register, sonic and electro-magnetic, so deeply alien as to be fathomless. Frank Fraser Darling – pivotal to those biologist-ethologists scrutinized in this chapter – was inclined to agree: 'The whales and dolphins and porpoises give us only the shortest glimpses of their daily lives and we can never become intimate with them' (Fraser Darling 1939, 11). Seventy years have since passed, and still so much remains unknown of that 'vast country of the oceans' (Fraser Darling 1939, 77). From tourist whale-watching vessels, Cloke and Perkins (2005) describe an aquaculture of anticipation, awe and wonderment found in only the most fleeting glimpses. Meanwhile, Thrift (2006) reports how the latest bio-acoustic research on whale song has revealed how communication reaches across sonic spaces stretched to an oceanic scale; significantly, he identifies comparable capacities for co-presence in spatially distant human-to-human relations. Wherever togetherness happens, likeness and otherness are also to be found in uneasy relation (Tuan 1976). In certain instances, they might be regarded as opposing, even warring, impulses. Tellingly, when framed by these new geographies of lively relations, the grounds for identifying difference and distinctiveness between

---

1   One noteworthy modern-day version of the bestiary, launched by the Reaktion publishing house includes the wolf, tiger, ant, falcon and salmon.

humans and non-human animals are themselves rendered different; no longer fixed as halves, one always reducible to the other's dominion.

By Frank Fraser Darling's measure, seals were of a rather different order than the likes of the whale since 'they are nearer us in the zoological scale and they often spend hours ashore' (Fraser Darling 1939, 11). Following the shifting disposition of sea and land, seals' movements are rhythmic, and thus their proximity to (or distance from) humans is to some degree predictable. Seals were to be found along 'the vivid frontier' (1939, 2), Fraser Darling's memorable description of the littoral margins or tidal zone, a place where the border separating land and sea was ever on the move, subject to daily dowsing and disturbance. Before his and other seal studies are considered in full, the first principles of ethology, and the recent presentation of ethological thought to a geographical audience, merit closer scrutiny.

To whom, or to where, ought geographers' recent spike of interest in ethology be attributed? Gilles Deleuze and Félix Guattari offer one sort of introduction to the curious. An unlikely pair of naturalist-escorts, they are by turns idiosyncratic and edifying, their arguments oftentimes angular and arch, on occasion baffling. The path they cut – through prickly thickets of biological, behavioural and musicological theory – leads on beguilingly, though for stretches remains maddeningly indistinct. So what's to like? As a statement of bio-philosophy, the cosmological-ecological theory they conjure up is kaleidoscopic and pullulating, where the shapes taken by living organisms are successively, relentlessly, expansively and expressively, recombinant. In scope and scale, their theorizing is totalizing and pantheistic, issuing forth a new system of metaphysics. If their primary point of reference is Spinoza's ethics (Deleuze 1988), in spirit the process-based geo-philosophy of Deleuze and Guattari can be aligned with the life project of another eco-visionary, Rachel Carson, who wished to sculpt new descriptions of the world and so expose the 'delicate negotiations of its ingeniously calibrated ecology' (Lear 2007, xiv; see also Lear 1997). And tellingly, the version of ecology coined by Deleuze and Guattari rests on a direct engagement with other seminal works in the biological and life sciences.[2]

The Deleuzo-Guattarian eco-system holds certain appeal (to say nothing of a certain notoriety) for possessing its very own turbulent topology, arranged according to technical-spatial concepts and systematized by a singular terminology.[3] From derivative terms, new conditions are created and standards set. Deleuze and Guattari unsettle the discrete, compartmental structuring of organisms and the classificatory taxonomic labels around which biological-zoological knowledge convenes and operates. To fully disassemble this atomistic tradition of thought, animal behaviour has to be understood as kinds of expressiveness within greater

---

2   The pioneering ideas of comparative ethologists Konrad Lorenz, Niko Tinbergen, William Thorpe and Jakob von Uexkull are subject to fairly close, critical consideration.

3   For the uninitiated, Bonta and Protevi (2004) offer a primer, field guide and glossary all-in-one.

fields of relations. Thus, sentience is understood relationally and dynamically, where forms of awareness are always in formation. Thus, intelligence is a competence reconfigured by new kinds of sociality and co-existence (Thrift 2005; Whatmore 2002). The rippling dynamics of extended or distributed organisms and adaptive environments demand different sorts of practical detection so that the sheer vibrancy of the world taking place becomes more readily apparent: through organic rhythms, where actions have a pulse, creating multiple tempos and differential paces or durations of life; through territories, shaped by habits, customs and range of movement; through proximities, which flex and contract according to boundaries and spaces and their changing properties and functions; through the forces, urges and passions fired by particular sorts of relation and episodes of association; and, through the sensual textures and material markers of place-memory. On these terms, for sure it is possible to find smaller, non-individuated unities bound together within the greater whole, but these are continually shifting in relative status and emphasis. Only ever partially enclosed, the Deleuzo-Guattarian eco-system spills over with complex, indeterminate expressions of life-*becoming*. The 'refrain' – or alternatively the 'assemblage' – is its transcendent (and regulative) motif, and immanence its primary motor (and unifying constant). '1837: Of the Refrain', appearing in *A Thousand Plateaus* (1988) is especially notable – at least for the purposes of this chapter – in taking as its pivot the ethological domain. A thumbnail version is supplied here; not so much for the purposes of direct instruction (and far less so, doctrinal adherence) but rather to open up possible lines of connection, and to show up shades of contrast, in the chapter's later sections.

The cosmological system is arranged into what appears first as a series of tiers, and later as aligned terrains, and later still developmental stages; none of which, we should note carefully, are conventionally hierarchical or scalar: 'it is less a question of evolution than of passage, bridges and tunnels' (1988, 322). The initial descent from wholeness begins with *forces of chaos*. Chaos gives way to *forces of the earth*, comprised of organic and inorganic substances and energies; enrolling the elements (air, water), animal life, features (mountains, forests, vegetation). Imparting in correspondence, forces of the earth enable the emergence of *milieus*, identifiable by diverse types of organization and arrangement, and shaped from the exterior-in and the interior-out. From hereon in, our intrepid voyagers-cum-field scientists are best equipped to take up the task of explanation:

> ... all kinds of milieus, each defined by a component, slide in relation to one another, over one anther. Every milieu is vibratory, in other words, a block of space-time constituted by the periodic repetition of the component. Thus the living thing has an exterior milieu of materials, an interior milieu of composing elements and composed substances, an intermediary milieu of membranes and limits, and an annexed milieu of energy sources and actions-perceptions. Every milieu is coded, a code being defined by periodic repetition; but each code is in a perpetual state of transcoding or transduction. Transcoding or transduction is

the manner in which one milieu serves as the basis for another, or conversely is established atop another milieu, dissipates in it or is constituted in it. The notion of the milieu is not unitary: not only does the living thing continually pass from one milieu to another, but the milieus pass into one another; they are essentially communicating. The milieus are open to chaos, which threatens them with exhaustion or intrusion. Rhythm is milieus' answer to chaos … There is rhythm whenever there is a transcoded passage from one milieu to another, a communication of milieus, coordination between heterogenous space-times (Deleuze and Guattari 1988, 313).

By this stage, the less experienced or less determined reader might be tiring of the mixture of opacity and technicality exhibited; perhaps even given to wonder 'whatever can they mean?' Wholly different in tone, the following passage penned by Paul Evans, country diarist for *The Guardian*, can be read in parallel and offers such transcendent thinking an empirical dimension:

The hawk lands in a small tree. It settles on a branch only four feet off the ground and arranges its wings, shrugging shoulders under a dark overcoat. Its chest is pale, drizzled with fawns and browns, and yellow legs end in talons which nail into the bark. The sparrowhawk keeps very still – but for its head, which switches from side to side so its eyes can watch the traffic of low autumn sunlight through the bluster of leaves. A breeze shoves stiffly, twisting the leaves as their stalks strain against branches to follow the autumnal migration into the earth. Ash leaves – last to come, first to go – are turning lime-green and falling. Linden trees and hazel are showing yellow ochre. Elder burns red from the bottom up. The sparrowhawk remains still, watching the details of a small world get smaller: speckled wood and small tortoiseshell butterflies, moths over bending grass stems, shadows which belong to nothing. There are other raptors in the sky: bigger, blunter, more powerful. Buzzards are sliding along the breeze, turning slowly with one wing pressed against an invisible column. They are dark and heavy with the light behind them, but when it spills under their wings they are pale, bronze and tawny. It is the autumn equinox, a kind of balance of day and night in a year whose seasons have slewed a bit. But this feels right: the buzzards turning silently through the wind, leaves spiralling to the ground – kinds of balance within kinds of light. Small birds avoid the place where the sparrowhawk sits in the tree. It has ducked out of the wind to watch the world move at its own pace, without its own blurring speed. But that is about to change. The hawk turns on the branch, and in one movement, as its wings and tail feathers open, it has spun away through the branches (Evans 2007).

To reiterate then, *rhythms* create multivariate, possible relations amongst living things, expressed within, and operating across, different *milieus*. For Deleuze and Guattari, the emergence of the *territory* is a powerful variation in the spacing of life. A territory is an expressive occurrence, an entity to be explained according to certain functions (in birdlife, say for example, courtship, nesting, rearing young) and their own associated rhythms:

> In a general sense *we call a refrain any aggregate of matters of expression that draws a territory and develops into territorial motifs and landscapes* (these are optical, gestural, motor, etc., refrains). In the narrow sense, we speak of a refrain when an assemblage is sonorous or 'dominated' by sound – but why do we assign this apparent privilege to sound? (1988, 323, *emphasis in original*)

Crucially, territorialization – those active processes and performances of shaping territory – is regarded the outcome (and not the cause) of shifts in relative position and expression; where particular socialities create grounds for clustering *and* for separateness. The expressive life of the territory is rich and complex; most often – though never exclusively – it is exercised among members of a single species. For Deleuze and Guattari, humanness is enveloped into, and altered by, processes of re-territorialization such that the corporeal prospect of becoming-animal is possible.

***

Read today, Frank Fraser-Darling (1939, 1940, 1943, 1948) and Ronald Lockley (1930, 1932, 1934, 1935, 1938, 1947a, 1947b) remain absorbing advocates of island life, as passionate about the exacting demands it placed on the body as they were appreciative of its rich intellectual rewards. Along the Atlantic's archipelagic edge (Fraser Darling re-located to North Rona, the Treshnish Isles and the Summer Isles, Lockley to Skomer and Skokholm) space and time were freed up for enquiring minds, since here 'the web of experience is largely of your own weaving' (Fraser Darling 1939, v). Creature comforts foregone, a defiant tone is sounded in their self-conscious statements of separation, sufficiency and containment. Maverick status, it seemed, was most fully formed and properly expressed with a drift away from the centre ground. Fraser Darling's striking title for the polished version of his field diary – *Essays of a Biologist in Isolation* – cleared rhetorical space for his creative impulses and analytical insights. The smallest islands have of course for long time been regarded as a perfectly formed and pristinely kept proving ground suited to scientific study and self discovery (Lowenthal 2007). Such was the case for Fraser Darling and Lockley. Geographical isolation from fellow man warranted the development of independent and unorthodox approaches to social inquiry. However, it was not only by virtue of its island setting that ethology emerged as a most supple and agile form of science.

Experiments with the structure and scope of ethological study were just as formative. Far from being solitary in nature, their inquiries were intensely social exercises, requiring that the closest of company be kept with non-human collectives (families, herds, flocks, rookeries and colonies). Fraser Darling and Lockley were notable figures in the advancement of the single-species model of study. Isolationists but not individualists, their ethological model was founded on a principled commitment to understand animal types as undisturbed but never autonomous, always immersed in a natural habitat and surrounding socio-environmental relations. Through the systematic rigours of field science they built on the relational principles and woodland (pre)ambles for ethology which inspired Jakob von Uexküll (1957) to conceive of core behaviour in terms of *umwelt*, and the intersecting activities of living beings as world-making, or *unwelten*. Anxious to know their chosen organism 'in the round', and prepared to travel the extra mile, they also sought out the seasoned wisdom of countrymen steeped in a 'social context in myth, story and superstition' (Evans and Thomson 1972, 238). On North Rona and Skokholm it was the great annual assemblies of Atlantic grey seals that fell under the spotlight. Composing portraits of seal behaviour meant devising hybrid sorts of method: 'The task imposes development of more cunning techniques, a fusion of the cleverness of the laboratory with the elemental craft and awareness of the primitive hunter. There is here a rich field of research which will satisfy aesthetic, academic and practical ends' (Fraser Darling 1939, 76). The organization of seal colonies was observed on land in exact and first-hand detail, and once the animals were at sea, by keeping watch from overhanging cliffs. Study was concentrated and sustained, taking place according to the seals' existence, arranged by *their* haunts, forms of sociability, customary habits, sequences of behaviour and hours kept. And yet, much about the seals' world did not bend to human will or faculties.

Animals create their own terrestrial, atmospheric and aquatic '-scapes' where the sensation of mediums, flows and currents is particular, and where attributes, speeds, ranges and distances are differently apprehended. Even so, for the observant human substantial signs of life abound. Massings and assemblies are still etched into place. Migratory passage makes routes stand out proud. Patterns of living repeat: for over-wintering, summering and breeding. Fraser Darling sensed these spectra and deep, customary structures on the seal island of North Rona: 'How ancient must be the civilization of the seals in this place! The island was called Rona long before St. Ronan adopted the name for himself when he went to live there in the eighth century. Here, without doubt is a capital city of well-marked ways older than the cultures of Sumer and the valley of the Indus' (Fraser Darling 1939, 80). The gulf separating humankind from the seal tribe could be partially overcome by experiments in becoming more animal-like, *or*, by becoming less predictably human. Lessons learned were sometimes the practical outcome of failed practice. Fraser Darling noted that seals became visibly agitated when he raised his field glasses to eye-level, eventually concluding that they associated the action with a hunter sighting his prey down the barrel of a gun. Closer union was possible through experiments in forms of inter-species communication: 'I have found the way to speak to them so that they are not afraid, but pleased … and in their confidence they have come out of the water to my feet as I have sat there on a rock, using my voice the way I have learned' (Fraser-Darling 1939, 79). If conditioned wariness was a fate that humans and seals had come to share, only with the slow brokering of trust could it be partially shed. By trial and error, sometimes for precious moments only, uncertain situations were arrived at where status and type were scrambled. Such knowledge practices, based on vocal and physical dialogue, would find favour with Alphonso Lingis for whom the possibilities of open, proximate encounter remain underappreciated among scientific communities: 'The noble impulses are nowise contrived to serve human needs and wants, human whinings. The impulses and the external appearance of the noble animals lend themselves to the utilitarian explanations of biologists. But when a human animal comes to inhabit other animals' territory with them, or even inhabit their bodies as they his, the movements released by the excess energies in his body are composed with the differentials, directions, rhythms, and speeds of their bodies' (Lingis 2000, 56). By gradually yielding to animal landscapes and to states of wildness, ethologists found the means to physically embody their gift for empathy.

Alongside experiment in body language and mimicry, systematic aspects of the trained biologist's technical repertoire retained a place in the structure and design of field studies. Through local area survey, mapping and closest observation, note was taken of seals as gregarious, relating subjects: through patterns of association, movement, competitiveness, territoriality, and different episodes of the lifecycle. As a result, informed speculation and tentative theory-building were possible: concerning demographics and kinship (e.g. seasonality, coupling, courtship, breeding, parenthood, rearing young, care and mourning) and the binding, non-familial forces at work in extended forms of social organization (e.g. local geography, playfulness, performance, curiosity, spontaneity, rest, hunting, care and sexuality). Not content to narrowly delimit functions of animal behaviour to instinctual drives and neural responses, and citing instances of individuality, self-awareness, co-operative arrangement and collective action, ethologists sought to map out a complex matrix of perception and action. The claims were divisively controversial. Lockley offered interpretations of attitude, intent and purposeful action. Functional utilitarianism seemed ill-suited to explain animal cultures, specifically events such as the seals' 'surf dance', a sensual ritual taking place outside the annual mating cycle: 'I have known it go on, waltz and figure-of-eight and minuet and shallow-dive in close embrace, face to face or pick-a-back, for over an hour; with little breaks for kissing' (Lockley 1954, 127). Descriptions of other coastal creatures consciously coming-into-being, even expressing a joy in living, were stylistically bold and lyrical, daring readers to wonder: 'The shag comes up and shakes the jewel-like drops from the burnished green of his back. I am not justified in saying he is pleased, but he looks it all the same' (Fraser Darling 1939, 14). The knotty question of anthropomorphism was seldom far removed from non-conformist descriptions of animal intelligence or emotion, not least when these were exposed to opinion in mainstream sociobiology.

Source data stemming from the outward world of social and physical relations was generally regarded insufficient for authoritative explanations of life history and ecology, and, if considered in isolation, very likely to cloud judgment. H.R. Brewer (1974), Professor of Zoology at University of London and career-long seal scientist, argued that field observations feed into laboratory-based studies of anatomy and histology, themselves disclosing internal physiological processes. The need for an appreciation of environmental intelligence *and* physiological process was carefully calibrated by Rachel Carson, here describing the memorable properties of a marine habitat for migrating fish: 'By the younger shad the river was only dimly remembered, if by the word "memory" we may call the heightened response of the senses as the delicate gills and sensitive lateral lines perceived the lessening saltiness of the water and the changing rhythms and vibrations of the inshore waters' (Carsen 1941, 16). Konrad Lorenz, comparative ethologist and commercially successful popular scientist, was wary. Cautioning fellow authors, he figured a price to pay for heavily stylized expression, reminding that a gift for description carried with it responsibility:

> The creative writer, in depicting an animal's behaviour, is under no greater obligation to keep within the bounds of exact truth than is the painter or the sculptor in shaping an animal's likeness. But all three artists must regard it as their most sacred duty to be properly instructed regarding those particulars in which they deviate from the actual facts ... There is no greater sin against the spirit of true art, no more contemptible dilettantism than to use artistic license as a specious cover for ignorance of fact (Lorenz 1952, 20).

The version of ethology shaped by Lorenz was one where human sacrifices made whilst sharing in animal lives could be a source of joy, affection and good humour, but warmth or depth of description ought not to divert attention from primary explanations of behaviour rooted in a plain, hard language of mechanisms, inhibitors, releases, triggers, switches, instincts and drives (Lorenz 1952, 1954, 1971, 1972). Thus, for Lorenz shows of aggression were regarded an absolute and unflinching expression of animals' territoriality.[4]

At the advent of ethological practice, the extent to which human lives were incorporated into the study of animal life differed. Consequently, configurations of animality varied, and so too explanations for feelings of trans-species sameness. Lorenz explained away human empathy with large mammals by our recognition of facial form and comparable features, into which are read a range of changing moods, or an inner state of mind. Seals were especially remarkable in this regard for their expressive mien and capacity to shed tears. For other observers, the deep, dark pools seen in animals' eyes suggested complex emotions, and thus greater intensities of relation. The narratives of affiliation and mutuality that Fraser Darling and Lockley fashioned out of their seal colony encounters were based on levels of familiarity with charismatic individuals and their on-going life histories. Togetherness extended to emotional attachment, fondness and lasting friendship:

---

4   Deleuze and Guattari questioned Lorenz's theory for its reductionism, and too easy transference into the human realm of political autocracy.

My heart is with them always, and when I come up against that barrier of indifference and readiness to fear, I suffer for the generations of men who have harried and slaughtered. If I could speak and the animals turn their heads in pleasure; if I could offer my hand and they touch it with their delicate muzzles, I would be happier and less lonely. It seems our lot to watch the lives of animals only by stealth and artifice, and our right has gone to approach them in amity as lowlier brethren of the same earth (Fraser Darling 1939, 78).

For Fraser Darling the great promise of learning-to-relate through affective communication was that new civilities and ethics might emerge. Finding a mode of being (or becoming) that picked its pace and purpose from animal lives, required that conditioned aspects of human-ness fall away. Investigating aspects of 'seal-ness' was thus to experiment in becoming an*other* kind of human: valuing life differently, training more finely-tuned senses, in a social world patterned by seasons, orientated towards basic elements. It was according to these ethical and ecological principles, that Fraser Darling arranged field studies as a family affair. Raising children, knowing territory, making a home, finding food, each aspect of the domestic round existed amidst the conducting of research. This version of dwelling was not so much science-at-home and rather science-*as*-home, lived out day-to-day. Organically led, close to the land and sea, stripped of ornament (coarsened to an extent), yet thickened in texture, sanctuary was found in an ideal of moral improvement traced along the interspecies boundary. With sovereignty between the personal and natural scrambled, private lives intruded on each other, co-existing families learning to care and conduct themselves within eyesight and earshot. So began a narrative tradition of *dramatis personae* in popular natural history – now well established – where the life histories of animals speak volubly of their biographers' own transformations.[5] On finally departing the island, betterment was a more-than-human achievement and – to adopt two felicitous phrases coined by Hinchliffe and Whatmore (2006, 125, 135) – its ethological principles were born both of 'corporeal generosity' and 'the accommodation of difference'.

***

'It's no wonder they were thought to be like us', he said. 'For the seals and ourselves were aye thrown together in our way o' getting a living, and everything we feel, they feel, ye may be sure o' that' (Thomson 1954, 152)

Since pre-history, on the northernmost and westernmost edges of the British Isles and Ireland, the lives of seals and humans were closely interwoven: an aquaculture

---

5   It is worth noting how domestic hospitality was reconfigured by Konrad Lorenz who chose to turn his family home into an adoptive and adaptive place for animals to dwell, and co-opted his extended family as hosts. Here, experiments in learning took place under one roof.

populating coastal reaches and waters, sharing a dependency on the sea for livelihoods. Along the North Atlantic archipelago, patterns of settlement, harvest, movement and association have been forged by 'the people of the sea'. This most suggestive phrase was used by David Thomson, writer and broadcaster, as the title for a remarkable social inquiry exposing a particular and regional geography of human-animal sociality (Thomson 1954). Evoking a powerful sense of language, time and place, he succeeded in chronicling a half-world, styling himself the writer-traveler caught up in the very moment of its passing. His methods were those of the itinerant oral historian, the folklorist and ethnologist. His poetics of field study are wondering and unhurried, paced out on foot, stilled by patient waits for the returning ferry man. His lyrical narrative is comprised of brief encounters, sea crossings, return visits, steady observations and gentle opportunism. The keenest of ears enabled his stylistic use of reported speech acts and lengthy dialogue:

> 'My grandfather says they do weep and he says they do caress one another with kisses. They throw stones too'.
> 'They throw stones, do they?'
> 'He says it is dangerous to be below them on a rocky slope'.
> Mairi watched me unsteadily.
> 'It is only what he says', she said stifled. Covering her face with her hands, she sat down at the table. She was sobbing, inwardly, with very little sound.
> When it was possible to speak again, I tried to change the subject, but she stopped me.
> 'It is all lies', she said. 'You know well it is lies'.
> 'What do you mean, Mairi?'
> 'It is well for you to come and ask about the seals. And away home with you, then, to the mainland'.
> 'But I don't think of the stories that way – as lies or truth. I like to hear them; that's all'.
> She stared.
> 'Like reading a Western?'
> 'Perhaps'
> 'But the old people believe them'.
> 'Well, I don't see any harm in that, do you?'
> 'On the mainland they wouldn't believe them'.
> 'No'.
> 'Not even the old people?'
> 'Very few of them would. But they believe lots of other things, just as strange'.
> (Thomson 1954, 172)

Navigating passage by the testimonies of witnesses and storytellers, skirting the doubts arising between generations, and ever sensitive to the effects of a greater geography on local currents of belief, much of Thomson's writing has a translucent, dreamlike quality. Elsewhere, his encounters are more direct, unflinching and matter-of-fact, seemingly of harsher, visceral meaning. Either way, the lives of seals materialize out of observable realities and humans' imaginative engagement with events and interactions. Throughout, Thomson is a receptive and sympathetic listener. Observing what is customary in lifestyle and labour, in legend and lore, he exposes a vernacular culture that is at once the province and the product of co-habitation. His immersion into indigenous tradition was deep enough to be entrusted with knowledge of the seal-person, or *selkie*, as an uncanny presence, or harbinger for the 'unco strange' (Thomson 1954, 142). Hereabouts, seals were storied as sentient, having powers of speech and recall. Transactions with humans were complex, oftentimes coercive in nature. Hard bargains could be driven and the heaviest of tolls taken. Intimacies shared might lead to traumatic, desperate, even fateful kinds of outcome. Thomson retold these tales without judgment: of wives, unsatisfied with husbands, who consort with seals; of offspring born with webbed fingers which must be slit apart; of forced marriages and unions; of changelings and shape-shifting; of skins borrowed, inhabited or stolen; of episodes of lives spent in exile on land or under the sea; of rightful returns (never) made; and, of human families said to be descendents of the seal tribe. Magical realism immediately springs to mind.

Seamus Heaney is not so easily persuaded. Figuring Thomson's achievement as 'luminously its own thing', the poet (1996, ix) concedes that genres and idioms are touched on in the prose but asserts that these were never intended to be read prescriptively or systematically. Rather, he divines in the work a revelation: that in spite of scholarly training or temptation, the exercise of interpretive mastery might actually be headed off:

> In the presence of such alluring inventions and such substantial voice, analysis and appreciation feel superfluous. Talk of the willing suspension of disbelief, of the salubrious effect of imaginative narrative, characterizations of the mental habits of pre-industrial societies, conjectures about how the sociological facts got displaced in earlier days into the parallel universe of the mythological – all such commentary seems to lead in the wrong direction (Heaney 1996, x).

Though emergent from the cultural world of the Atlantic edge, the cosmology of the grey seal is not limited to fixed cartographic co-ordinates. It ripples outwards from a regional mythic tradition towards an open-ended, animistic realm where the promise of wonderment can still be felt, and where a lasting, soulful kind of enchantment is possible. Gavin Maxwell, British author of notable works in travel and nature writing, was as favourably disposed to *The People of the Sea*. Discovering a kindred spirit in Thomson, he paid tribute to piercing insights into 'the lost world of childhood, of the individual or the race – vision undimmed, sense of wonder unconfined; yet this is the very antithesis of a childish book'. (Maxwell 1965, ix). Seeming innocence of outlook, and fieldcraft at first-hand, produced farsighted findings.

Fellow travellers found a different language in their quest for areas of congruence in the lives of humans and seals. For Ludwig Koch (1952), pioneering wildlife sound recordist, seals' familiarity and foreignness were held together, and given voice most powerfully in music and song. On the island of Skomer he lowered microphones into a seal-cave to obtain a sound portrait. His field recordings captured the creatures' voices at play, in anger, upset and love. Among the peoples of the sea, vocal mimicry has long been a lure, creating an irresistible tug either towards those on land or sea. To the human ear, the sound of a seal pup crying is perceived as all but indistinguishable from that of a human baby in distress; thus, mimicry figures commonly as a narrative hook in stories and legends. The modulations and melodies of adult seals' calls were thought to bear close resemblance to human singing. Consequently, seal calls were learned by hunters, and used to draw the most curious creatures in towards waiting guns on the shore. The same method was put to more peaceable ends by Fraser-Darling; though fearing less scrupulous impersonators he chose not to disclose his techniques for selkie-speak. Other treatments of vernacular secrets were less protective. Koch's recorded song of 'a bachelor seal, lonely and disconsolate on a rock' was transcribed from disc by the folklorist Francis Collinson, and published in the *Journal of the English Folk Song and Dance Society*.[6]

In *Seal Woman* (1974) Ronald Lockley interrupted writings on natural history, employing fictional fantasy as a literary device to animate personal reminiscence. For source material he drew on his personal diaries kept during wartime service for the Navel Intelligence Division of the Admiralty containing a combination of surveillance information and ethnological-ethological observations. On a posting to Ireland's far south-west – ostensibly patrolling for submarines and enemy landings along the coastal margins – the novel's narrator encounters a different sort of outsider. Shian is a young woman, free-spirit and goatherd. She, it is said, was born to a seal mother in a seal cave, and as a human foundling was adopted into a local family of ancient lineage. On having sustained injuries during the Normandy landings, the narrator revisits Ireland to convalesce. Shian is re-discovered, and a story of supernatural love unfolds, she having recognized the naval officer as her sea-prince whose arrival has long been anticipated. They journey to a faraway kingdom populated by the people of the sea. Following their coronation as mermaid and merman, they raise a young girl (*Mor-lo*), before experiencing the heartbreak of an enforced parting. Although in form and tone *Seal Woman* seldom departs from the most standard of literary conventions, it was the most creative work in Lockley's corpus on seals.[7] Though his prose nowhere matches the emotional

---

6   The notation was included in an appendix on seal song, added to later editions of Thomson's *People of the Sea*.

7   That Lockley strove to experiment, dramatising memories originating in field experiences from decades earlier, was no accident. The best-known of his single species studies, *The Private Life of the Rabbit* (1965), was the inspiration for Richard Adams fantasy world of *Watership Down* (1972). In turn, Lockley followed Adams example for

intensity of Thomson's, their ambition can be regarded as shared: 'What could have been a matter of field work written up into a casebook becomes a matter of memory and its contents being liberated into a new and transfigured pattern. The book recovers and revives the old trope of human beings as creatures dwelling in the middle state, caught between the world of the angels and the animals' (Heaney 1996, xii).

The spacing and crossing of different subjectivities along a continuum of life, is compelling for the harmonies it presents with the version of a relational ontology advanced by anthropologist Tim Ingold (2000, 2005, 2006). In his culturally diverse studies of the social and cosmological interactions that bind the lives of humans and animals, Ingold unpicks the orthodox belief in existence as sealed-in and sovereign. Instead he recognizes lives as an occurrence; on-going, many-sided and inter-dependent, happening *outwards* and *through* the world. This sort of emergence is possible when life is lived by skills and knowledge practices which are themselves incorporations of environmental interaction. Ranging across the life-worlds of many indigenous peoples, Ingold finds commonality, and conceptual purchase, in the figure of personhood. Crucially, personhood is a condition – better still, it is a convivium – affording a welcome to beings who need not necessarily be human. To recognise personhood is to inhabit a relational space where people understand the place of animals in their lives, and their place in the lives of animals. Generative and associative, humane and reciprocal, personhood is a natural *and* cultural phenomenon. Ingold's summary of the practical, linguistic processes where recognition and creation of an animal occurs amid an extended field of life is instructive: 'The name of an animal as it is uttered, the animal's story as it is told, and the creature itself in its life activity, are all forms of this occurrence. Animals happen, they carry on, they *are* their stories, and their names … are not nouns but verbs' (Ingold 2005, 172). There will be those unwilling to endorse such thinking, perceiving in it romanticism or primitivism, or a troubling requirement to suspend disbelief of phenomena otherwise destined to rank only as 'the unexplained'. Nadasdy's (2007) dissatisfaction with Ingold is wholly different. He argues for a still *more* radical mode of encounter placing absolute trust in aboriginal people's ontological explanations of sociality between humans and animals.

Where scientists once learned to intuit forms of life through observational studies of animal sociality, and where anthropologists now appeal for greater trust to be placed in knowledge as practiced through a nearness of interspecies relations, it is equally possible to find areas of overlap with the Spinozan ethics which inspired Deleuze's version of ethology:

Seal Woman where he was at liberty to tell a story that 'arises naturally' and where some, but not all characters, are acknowledged as fictional. Enigmatically, no greater level of detail is supplied.

It is not that one may be a Spinozist without knowing it ... He is a philosopher who commands an extraordinary conceptual apparatus, one that is highly developed, systematic and scholarly; and yet he is the quintessential object of an immediate, unprepared encounter, such that a non-philosopher, or even someone without any formal education, can receive a sudden illumination from him, a 'flash'. Then it is as if one discovers that one is a Spinozist; one arrives in the middle of Spinoza, one is sucked up, drawn into the system or the composition (Deleuze 1994, 631).

It is by these ethical terms of reference that Deleuze proposed 'transcendental empiricism' as the most far-reaching mode of praxis. It is the same terms of reference that suggest a way to recall and rework the seminal seal studies of naturalists, biologists and ethnologists.

***

It's my habit to approach the seals on foot; whilst running. Depending on the route I've taken along the river valley, the stop-off-look-out point to sit and watch occurs reasonably soon after setting out, or alternatively, towards the end of exercise. Generally, I prefer the former. After 20 minutes, I've built up enough warmth to afford a halt in progress. With movement stilled, alertness is a guaranteed side-effect of the prickle felt as sweat slowly dries on skin. The length of my stay depends on circumstance, departure sometimes being judged by when discomfort overcomes curiosity. From my high station, I presume that the seals perceive me as a human presence that is safely distant, and of little threat. When we exchange looks (which at times can seem like steelier stares) ordinarily they exhibit only disinterest, seeming to confirm our attachment to be all of my making. Then, from time to time, it appears I have outstayed my welcome: the cow's departure announced with a look flashed up in my direction before she ploughs headlong into the river's current, her pup in hurried pursuit. Proximity also has its limits. Tempted to experiment with the distance between us, one ill-judged slither down to the water's edge met with a gruff snort of alarm. Good-natured tolerance was brought to an abrupt end, any shred of animal curiosity jettisoned. Mostly events are less dramatic in their unfolding. Between riverbank and home, the greater part of the run remains for me to ponder what I've seen, been most affected by and to find adequate words for their description: the sharp, sibilant 'spat' that sounds when the seal surfaces to snatch air; the elongated swoop of its form in motion sub-surface, barely perceptible, slipstreaming one way, then streamlining the other; the expressions and dispositions that appear to shape mothercraft between cow and pup; the lived-in look of the sand flat territory, soon to be rinsed clean of shallow impressions and spoor marks by the encroaching waters; the interplay of stolen sidelong looks between the seals and a nearby heron, neighbours and rivals for fish stocks during the long run of the tide.

My 'seal diary' is, of course, a most partial and experimental chronicle of life-*beside*, never properly achieving the status of knowing life-*with*. But systematic study was never the intention. Rather what I have compiled are tentative examples of learning-by-witnessing, born of as much vigilance as can daily be mustered, each giving voice to momentary intensities of relation. And the summons out-of-doors, I should be frank, is not solely of the seals' making. A quickening in my pulse – I've learned from years of running as a feature of the daily round – allows me to gather a different sort of perspective on events, on work, (on this piece of writing) and to slow the sometimes too frantic, jittery pace of thoughts. It's a means to get outside my self, and then, to better get back inside my skin too. When the setting is right, and for precious spells, senses seem to reach beyond their normal range, such that inner and outer worlds become more closely aligned. Science recognizes this as the 'runner's high', a short-lived neurological phenomenon brought on by the body's production of endorphins during exercise. Though no sports scientist, I'm most familiar with the immediate heady rush, and lingering sense of well-being. And so, my encounters with the seals, and the timing of any afterthoughts, are purposefully choreographed and engineered so as to be bio-chemically affected and materially realized. The simple pleasure of 'going out running' can be said to operate simultaneously as an individuated, molecular 'release mechanism' *and* as a catalyst for a less autonomous, trans-personal sensing of life.

In still other respects, the physical promise of liberating encounters demands advance planning. The seals' appearance on the island's foreshore is always coincident with the period of low water in the sea's tidal range. Their regular patterns of feeding, foraging and hauling-up for rest depend on the water's continual passage, sliding landwards and washing out seawards. Consequently, closest observance of the tides has become my habit too. Change occurs twice daily. Measured according to clock-time and wall calendar, the high-water and low-water marks inch ever forward: by approximately 70 minutes at each new cycle. The local tide tables specify two low-water options for any seal visit; or one only when hours of darkness make observation impractical. My own body clock (to say nothing of my humours, or working routine, or family arrangements) must re-set repeatedly to the shifting pattern of the tides. Their movement is in fact the differential gravitational effect caused by the orbital cycles of the sun and moon in relation to the earth. I've discovered how these physics of motion generate spring tides (higher and heralded by the new and full moon) and neap tides (lower and at the first and third quarter of the moon). In this day and age it is peculiar – at least it seems so at first – to regularly observe a ritual ordained by the passage of the planets. Sometimes, these astronomical forces can seem truly Copernican and unearthly, and then, just as likely, the play of natural laws is felt deep inside, as gut instinct. Heading out to see the seals light-headed before breaking fast, or too soon following an evening meal, creates a mode of encounter and the tempo of experience. Besides which, well-springs of energy and levels of patience can fluctuate according to different kinds of register. Sometimes they seem to track changes in the field of visibility. Outings made in the dewy, grey murk of dawn feel markedly different in tone from the flat brightness of a mid-afternoon haze, or those visits held back until the evening's shadowy world of contrasts, low angled rays and light drawing in.

The shifting bio-rhythms and the bodily mediation of animality are complexly layered, variously textured and circumscribed, by greater and lesser forces of nature. And much the same can be said for the seals.[8] If the phenomenological experiment of encounter is pushed far enough, a portrait of shared existence emerges encompassing more-than-human lives and habits, repeatedly emerging into the world. Pushed this far, the ontological status traditionally afforded to the sovereign subject begins to unravel. I'm put in mind of rhythm analysis: a theoretical approach to the study of everyday life once proposed by Henri Lefebvre (2004) but never quite fleshed out to include methodological principles, and seldom turned into a tool for routine use (though see Spinney 2006), or towards the physical world. Perhaps these descriptive passages amount to a revival in miniature, a generative and participative ventilation of the original idea.

What would such an experiment disclose? Certainly that living space-times cannot be bracketed off to the specific habitat of the tidal island at the river mouth, since co-existent ecological relationships emerge from a trans-local milieu and have followed patterns and rhythms ceaselessly, in timeworn fashion.[9] If this ranks as revelation, the literary-ecologist Rachel Carson (1941, 23) was on to the perennial qualities of life some while ago:

> To stand at the edge of the sea, to watch the flight of shore birds that have swept up and down the surf lines for untold millions of years, is to have knowledge of things that are as nearly eternal as an earthly life can be. These things were before man stood on the shore of the ocean and looked out upon it with wonder; they continue year in, year out, through the centuries and ages, while kingdoms rise and fall.

Figured thus, a different order of persons and powers in the world does become palpable, taking place through fields of variations, relations, sensations and affects: life felt on the pulse, in the turning of seasons, in mass movements of water and air, in depths, and surfaces, inhalations and exhalations, in the quickening and slackening of energies, in the pacing and duration of encounters, in the texture of moods and casts of light, in washes that are bio-chemical and tidal, and currents that twine the personal and impersonal, the substantial and immaterial, the perpetual and occasional, the territorial and transitory. With so much of life drawn into the

---

8   Today, zoologists studying the temporalities of animal behaviour attempt to discriminate between the impacts of 'masking rhythms' (a category into which tidal forces and prevailing weather patterns would fall) and primary 'biological rhythms', which are endogenous and innate.

9   My very first encounter with this 'seal-place' is, give or take a year, now half a lifetime away; their island haul-up originally caught my notice during daily trips over the bridge between a student flat and university classes.

orbit of so much else, it might already be too late to counsel for a little caution. I still ought to. It is warranted. Appeals made to generous sympathies and receptive sensibilities run the risk of seeking, and then always finding, that complexes of connectivity everywhere abound. To give full voice to the vital and irrepressible promise of on-going relations between human and non-human organisms, and to speak of co-existence between actions, forces and elements, *can* sound exultant, rhapsodic, and just a touch too blissed-out; a song of nature worship seemingly in need of a lower key. Seekers should be under no illusion, the recalibration of relations necessary for greater levels of intimacy, still requires an on-going acceptance that disjuncture occurs in the midst of connectivities, and that there remain untold forces of nature always occurring, and non-human lives always existing; unannounced, unadorned, wholly on their own account.

***

No endgame, this chapter might be regarded as, at best, an opening set of remarks, working the shifting tideline between ethology and phenomenology (see also Lestel, Brunois, Gaunet 2006). It presses together matters of methodology and manner, and, of praxis and principle. What of substance has accreted from its several layers of description? For answers, it is appropriate that comments return to the kind of process philosophy currently preferred in non-representational theory, and associated appeals for geographers' affective apprehension of life's on-going occurrence. Spinozan-Deleuzian thought offers one starting point, and a means for conversion. So too, does the relational ontology which Ingold prefigures in the condition of personhood. However, non-representational theories can be further augmented by histories of experimental science. Considered from the present moment, early episodes of ethological enquiry into social interaction throw up different dimensions of the non-representational: learning how to be affected, the limits of verbal and non-verbal communication, and the primacy afforded to the event of encounter.

The benefits of holding together modern experimentalism with older versions of ecological wisdom are multiple, each concerned with the apprehension of phenomena. First, harmonies and tensions exist between seemingly disparate fields of concern, notably theory-making and life outdoors. Second, human-animal relations can be refigured between, around (and sometimes outside) discrete, individuated organisms. This is also to consider future research on the 'beyond worlds' of other living things and elemental phenomena of life: varieties of movement, suddenness, simultaneity, purposive action, blind motion, and, the affects of fluids and currents. And third, novelty is not so much a quality fixed tight in the ethological idea, and rather is revealed only once the idea is understood through the ordinary circumstances and trans-personal capacities that bring things into being.

To date, non-representational argument has placed greatest emphasis on the immediacy and direct impact of practice, rather than through filters of discourse or cultural signification. One implication of such a stance on embodied intelligence is that co-presence can become a requisite of learning (Merriman, Revill, Cresswell, Lorimer, Matless, Rose and Wylie 2006). However, for the learner or understudy, finding texts of reference for technical training in the rudiments of biological field science is still necessary to achieve basic levels of competence. 'It has never been easy to learn life from books' (1939, 13) is one of many epigrammatic lessons to be sourced in T.H. White's account of his efforts to learn falconry by taming and training a goshawk. Attention to such primary studies helps us to make sense of the past in the present, and the present in the past. Yi Fu Tuan (1976, 274) considers watchfulness the most immediate benefit of such conduct: 'From ethology [... we] learn techniques of observation'. Bearing witness to life's momentary acts and their multivariate expression need not imply any speeding up of method. Rather, by slowing the world down we find means to quicken the senses. To bear the weight of comparison with past practice, contemporary experiments in the rich description of life must be fostered through quiet and patient scheme.

Acknowledgements

For critical comments, thanks to Jamie Lorimer, Owain Jones, Leah Gibbs, Jennifer Lea, Chris Philo, Lou Cadman, Jo Norcup, Merle Patchett, Jonas Fox, and audiences in Glasgow, Bristol and London.

References

Atkinson, R. (1980), *Shillay and the Seals* (London: Collins and Harvill).
Baker, J.A. (2005), [1969] *The Peregrine* (New York: New York Review of Books).
Braun, B. (2005), 'Environmental issues: writing a more-than-human urban geography', *Progress in Human Geography* 29(5), 635-650.
Carson, R. (2007), [1941] *Under the Sea Wind* (London: Penguin).
Castree, N. and Nash, C. (2004), 'Posthumanism in question', *Environment and Planning A* 36(8), 1341-1343.
Cloke, P. and Perkins, H.C. (2005), 'Cetacean performance and tourism in Kaikoura, New Zealand', *Environment and Planning D: Society and Space* 23(6), 905-924.
Cocker, M. (2007), *Crow Country: A Meditation on Birds, Landscape and Nature* (London: Jonathan Cape).
Deleuze, G. (1988), *Spinoza: Practical Philosophy.* Trans. R. Hurley (San Francisco: City Lights Books).
Deleuze, G. (1994), 'Ethology: Spinoza and us', in Crary, J. and Kwinter, S. (eds), *Incorporations* (New York: Zone Books).
Deleuze, G. and Guattari, F. (1988), *A Thousand Plateaus: Capitalism and Schizophrenia.* Trans. B. Massumi (London: Athlone).

Evans, P. (2007), 'Country diary', *The Guardian*, 26 September 2007.

Ewart Evans, G. and Thomson, D. (1972), *The Leaping Hare* (London: Faber and Faber).

Farre, R. (1957), *Seal Morning* (London: Hutchinson).

Fraser Darling, F. (1939), *A Naturalist on Rona: Essays of a Biologist in Isolation* (London: Clarendon Press).

Fraser Darling, F. (1940), *Island Years* (London: Bell).

Fraser Darling, F. (1943), *Island Farm* (London: Bell).

Fraser Darling, F. (1948), 'Science or skins?', *The New Naturalist: A Journal of British Natural History*.

Heaney, S. (2001), 'Introduction' in Thomson, D. *The People of the Sea: Celtic Tales of the Seal-folk* (Edinburgh: Canongate).

Hewer, H.R. (1974), *British Seals* (London: Collins).

Hincliffe, S., Whatmore, S., Degan, M. and Kearns, M. (2005), 'Urban wild things: a cosmopolitical experiment', *Environment and Planning D: Society and Space* 23(5), 643-658.

Hinchliffe, S. and Whatmore, S. (2006), 'Living Cities: Towards a Politics of Conviviality', *Science as Culture* 15(2), 123-138.

Ingold, T. (2000), *The Perception of the Environment: Essays in Livelihood, Dwelling and Skill* (London: Routledge).

Ingold, T. (2005), 'Animal names', *Instituti Veneto di Scienze Lettere ed Arti*, 159-72.

Ingold, T. (2006), 'Rethinking the animate, re-animating thought', *Ethnos* 71(1), 9-20.

Jones, O. (2000), '(Un)ethical geographies of human-non-human relations: encounters, collectives and spaces', in Philo, C. and Wilbert, C. (eds), *Animal Spaces, Beastly Places: New Geographies of Human-Animal Relations* (London: Routledge).

Koch, L. (1952), *Memoirs of a Birdman* (London: Science Book Club).

Lambert, R. (2002), 'The grey seal in Britain: a twentieth-century history of a nature conservation success', *Environment and History* 8, 449-474.

Latour, B. (2005), 'How to talk about the body? The normative dimension of science studies', *Body and Society* 10(2-3), 205-229.

Lear, L. (1997), *Rachel Carson: Witness for Nature* (London: Allen Lane).

Lear, L. (2007), 'Introduction' in Carson, R. (2007) [1941] *Under the Sea Wind* (London: Penguin).

Lefebvre, H. (2004), *Rhythmanalysis: Space, Time and Everyday Life*. Trans. S. Elden and G. Moore (London: Continuum).

Lestel, D., Brunois, F. and Gaunet, F. (2006), 'Etho-ethnology and ethno-ethology', *Social Science Information* 45(2), 155-177.

Lingis, A. (2000), *Dangerous Emotions* (Berkeley: University of California Press).

Lockley, R.M. (1930), *Dream Island Days: A Record of the Simple Life* (London: Witherby).

Lockley, R.M. (1932), *The Island Dwellers* (New York: Putnam).

Lockley, R.M. (1938), *I Know an Island* (London: Harrap).

Lockley, R. M. (1943), [1941] *The Way to an Island* (London: Dent).

Lockley, R.M. (1947a), *Letters from Skokholm* (London: Dent).

Lockley, R.M. (1954), *The Seals and the Curragh: Introducing the Natural History of the Grey Seal of the North Atlantic* (London: Dent).

Lockley, R.M. (1964), *The Private Life of the Rabbit: An Account of the Life History and Social Behaviour of the Wild Rabbit* (London: Deutsch).

Lockley, R.M. (1966), *Grey Seal, Common Seal* (London: Deutsch).

Lockley, R.M. (1974), *Seal Woman* (London: Rex Collings).

Lorenz, K. (1952), *King Solomon's Ring: New Light on Animal Ways.* Trans. M.K. Wilson (London: Methuen).

Lorenz, K. (1954), *Man Meets Dog.* Trans. M.K. Wilson (London: Methuen).

Lorenz, K. (1970), *Studies in Animal and Human Behaviour, Volume I* (London: Methuen).

Lorenz, K. (1971), *Studies in Animal and Human Behaviour, Volume II* (London: Methuen).

Lorimer, H. (2006), 'Herding memories of humans and animals', *Environment and Planning D: Society and Space* 24(4), 497-518.

Lorimer, J. (2007), 'Nonhuman charisma', *Environment and Planning D: Society and Space* 25(5), 911-932.

Lowenthal, D. (2007), 'Islands, lovers and others', *Geographical Review* 97(2), 202-228.

Macfarlane, R. (2007), *The Wild Places* (London: Granta).

Matless, D., Merchant, P. and Watkins, C. (2005), 'Animal landscapes: otter and wildfowl in England, 1945-70', *Transactions of the Institute of British Geographers* 30(2), 191-205.

Maxwell, G. (1965), 'Foreword' in Thomson, D. *The People of the Sea: A Journey in Search of the Seal Legend* (London: Barrie and Rockliff).

Merriman, P., Revill, G., Cresswell, T., Lorimer, H., Matless, D., Rose, G. and Wylie, J. (2008), 'Landscape, mobility and practice', *Social and Cultural Geography* 9(2), 191-212.

Nadasdy, P. (2007), 'The gift in the animal: the ontology of hunting and human-animal sociality', *American Ethnologist* 34(1), 25-43.

Nagel, T. (1974), 'What is it like to be a bat?', *Philosophical Review*, LXXXIII(4), 435-450.

Philo, C. and Wilbert, C. (2000), (eds) *Animal Spaces, Beastly Places: New Geographies of Human Animal Relations* (London: Routledge).

Spinney, J. (2006), 'A place of sense: a kinaesthetic ethnography of cyclists on Mont Ventoux', *Environment and Planning D: Society and Space* 24(5), 709-732.

Thomson, D. (1954), *The People of the Sea* (London: Turnstile Press).

Thomson, D. (2001), *The People of the Sea: Celtic Tales of the Seal-Folk* (Edinburgh: Canongate).

Thomson, F. (1976), *The Supernatural Highlands* (London: Robert Hale).

Thrift, N. (2005), 'From born to made: technology, biology and space', *Transactions of the Institute of British Geographers* 30(4), 463-476.

Thrift, N. (2006), 'Space', *Theory, Culture and Society* 23(2-3), 139-146.

Thrift, N. (2007), *Non-Representational Theory* (London: Sage).

Tuan, Y-F. (1974), *Topophilia: A Study of Environmental Perception, Attitudes and Values* (New Jersey: Prentice Hall).

Tuan, Y-F. (1976), 'Humanistic geography', *Annals of the Association of American Geographers* 66(2), 266-276.

Von Uexküll, J. (1957), 'A stroll through the worlds of animals and men', in Schiller, C. (ed.) *Instinctive Behaviour* (New York: International Universities Press).

Whatmore, S. (2003), *Hybrid Geographies: Natures, Cultures, Spaces* (London: Sage).

Chapter 4

Placing Affective Relations: Uncertain Geographies of Pain

David Bissell

Deferred decision; delayed judgement. This was the way that the writer responded to being in pain. Submitting to the medical gaze would be sure to confirm what he suspected all along. Narrating this sensation and revealing it to others would surely lead to collapse. Endless hours twisted into years spent wrestling with the decision. Better to live with doubt than to surrender to the verdict. For the writer, the hospital constituted a place of absolute certainty. The certainty that the pain in his head was sinister. Crossing the threshold here would be the beginning of the end. It would put in motion the all-too familiar sequence of decomposition from diagnosis to failed treatment: from disease to decease. The force-field of the hospital was intense. Thoroughly embedded within the fabric of the city yet somehow removed, it was an otherworldly realm of entrance and exit. The tall, mottled orange brick building loomed on the horizon of the city. Twelve floors of strip-lit judgement. Next to it, pale yellow smoke silently drifting from the chimneystack of the hospital mortuary, calmly dispersing people who had responded to being in pain. This is where bodies melt away. This is where he would melt away if he came here. This was not a beacon of optimism or hope, but a place of sentence. Yet this place was unavoidable. He had no choice but to pass it on the number 11 bus every day on his way to college. He would choose a seat on the opposite side so he didn't have to look. Eyes averted to prevent the hospital from grabbing him and taking possession. Fists clenched and heart beating fast. The wailing of an ambulance siren would set nerves reeling. The pain intensified and blood rushed to his face. Looking would bring him closer to this certainty. If he accidentally turned his head, he would see the signs which bore witness to painful procedures: oncology, radiology, accident and emergency. Each of them a powerful synecdoche, possessing a gravity that would engulf their hapless victims, dispatching them along a predictable but unstoppable trajectory. This place promised escape.

***

Deciding whether to see a doctor or a hospital consultant can be a surprisingly difficult dilemma. Whilst conventional logic would suggest that to not seek medical attention at the onset of particularly distressing or unexpected bodily symptoms might be somewhat counterintuitive or even foolish, fear of learning something unsettling about one's body often prevents people from disclosing their symptoms straight away (for example see Burgess, Hunter and Ramirez 2001; Chapple, Ziebland and McPherson 2004). Indeed such a dilemma might be familiar to many people who have experienced persistent, unwilled pain of some kind. Yet a fear of finding out what one suspects is the matter, coupled with a relatively pessimistic or fatalist disposition, serves to generate a host of what might be really rather crippling anxieties that have the capacity to smoulder. Deciding whether to seek advice and who to seek it from might therefore be a highly charged experience and one that generates a host of embodied responses.

In considering these responses, we could think about the various emotions that emerge in this event of deferral. Interpretations of fear or anxiety. Following in the footsteps of other geographers' research on emotions, we could consider how these emotions discernable in the opening paragraph wax and wane at different times and in different places, most notably the experience of anxiety whilst on the bus which intensifies in proximity to the hospital (see Anderson and Smith 2001; Davidson, Bondi and Smith 2005, for example). Indeed there are some striking similarities between the fearful emotions suggested here, and other geographical research which has explored some of the everyday place-based practices and critical geopolitical topographies that emerge from particularly negative emotional experiences – and fear might be just one of these – for different groups of people (Kwan 2008; Pain and Smith 2008). Yet a problem with thinking about emotion in the situation is that it does not describe the complexities of this situation adequately enough. This is not to say that thinking through emotions is uncomplicated. Indeed considering the tensions that emerge when talking about 'mixed feelings' (Bondi 2005), emotions are far from clear cut. Neither are emotions inert or sedimented. Massumi acknowledges these complexities when he talks about how emotions are disorientating, since they are 'described as being outside of oneself at the very point at which one is most intimately and unshareably in contact with oneself' (2002, 35). Yet to think through the 'emotional' dimensions of this experience presumes the existence of a body that is able to reflexively interpret and *make sense* of his or her world. Emotions in this sense constitute the 'proof' of the subject and the qualification of their existence. To draw on Massumi once more, we might consider that emotion is a 'subjective content, the socio-linguistic fixing of a quality of experience which is from that point onward defined as personal' (2002, 28). But bodily pain pushes at the limits of sense and the ability for the body in pain to understand and account for itself. As Harrison writes, 'the suffering of suffering unworks the ordering of such structures of intention and meaning' (2007, 594).

So how to approach this experience that does not take recourse to making sense of the body? To think through the problematic of how to approach the experience of pain through 'affect' rather than 'emotion' might help to illuminate rather different aspects of this experience. Attending to the affects that are going on performs a radical decentring of the body by taking a more *relational* ontology where neither the figure of the singular body nor individual, reflective subjectivity take centre-stage. Though affect has been approached through and translated into number of different competing perspectives (see Thrift 2004), central to its comprehension is its *im*personal autonomy prior to capture and expression through regimes of signification (which could be conceptualised as 'emotion'). If we consider affect as impersonal 'force' that precedes processes of cognition, this provides us with the opportunity to explore the movements and consequences of these forces and how they impact on and shape the body in pain.

To return to the initial illustration, rather than thinking about what emotions might best describe and address this experience, we can understand how this situation emerges through a series of complex and overlapping movements between different affects. First, there is the affective force of the physical painfulness of the pain itself impressing through the body: the sheer bite of pain through the body itself. Then there is the affective power of the materiality of the hospital that is significantly enhanced by the presence of the pain, demanding a bodily response through choosing to sit on a particular side of the bus. Indeed this might be supplemented by the communication of the 'bite' of pain through facial expressions and bodily comportments. But then there is the affect of the hospital that *intensifies* the embodied sensation of physical pain, the clenched fists, the thumping of the heart, through its proximity. Furthermore, there is also the affective power of the condition of uncertainty itself. This pertains to how uncertainty over the nature of the physical pain serves to generate a tension between a (limited) sense of freedom afforded by this self-imposed ignorance and a parallel sense of constraint generated by the anxiety. Thinking through this event through the lens of affect therefore allows us to think about the affective relations that comprise both human and non-human worlds and to consider the complex distributed agency that emerges from the blurring of subject-object distinctions. The affects emergent in this situation, in contrast to the emotional response, are characterised by an autonomy that does not reside in some 'internal' world of the body, but flow between the bus, seats, hospital and bodies. And it is these affects that make space. Rather than being transmitted between bodies, as Thrift notes, 'transmission is a property of particular spaces soaked with one or a combination of affects to the point where space and affect are often coincident' (2008, 222). Whilst the popular currency of emotions get their durability from the process whereby sensations are made conscious, reflectively classified and assigned according to a limited vocabulary of semiotic identifiers, affect on the other hand is uncaptured, unqualified intensity that pushes at the limits of signification. Affect always exceeds understanding and conceptualisation. There is so much going on here that cannot be squeezed into knowable or representational form.

Thinking through and alongside these different movements of affect invites us to attend to these experiences, not as some peripheral grievance or occasionally-gratifying add-on that haunts everyday corporeal existence, but as the very means by which we navigate our way through life and negotiate our encounters with others. Indeed the so-called 'affective turn' within the social sciences more broadly (Clough 2007; Sedgwick 2003; Greco and Stenner 2008) has been instrumental in according a renewed importance to pre-discursive dimensions of experience that are not necessarily bound up with discourse and meaning. Affect is first and foremost a pre-personal phenomena. It *precedes* signification and the formation of meaning. Yet if we consider meaning to be the gluey viscosity that stabilizes and sediments identities, biographies and other symbolic systems that, from time to time, give impressions of coherence, this is not to say that meaning is absent from experiences of pain. Indeed it is often through this movement whereby meaning is coupled with and ascribed to such sensations that affects are felt most tangibly and made most present. Pain and how we talk about pain to others is bound up with particular meanings, although the processes by which meaning *becomes* attached to particular sensations and the affects that emerge from these becomings is by no means straightforward. If we take physical pain to *mean* danger, damage and destruction; the affective charge might be intensified. Conversely, we change the meaning of pain, perhaps to stand for empowerment, warmth and love; negative affects such as anxiety or fear *might* just be quiesced.

Whilst these complex reciprocal relations between the affective and discursive have been a central problematic within the 'affective turn' at large, geographers working through these relationships have been key in drawing attention to the timespaces of these affects. Indeed in emphasising the temporally and spatially differentiated characteristics of affect, geographers are well placed to explore the 'affective topologies' of everyday life (see Rose and Wylie 2006). Or put simply, how different affects are intensified or quiesced at different times and in different spaces. What is particularly significant about the account at the start of this chapter is how these affects emerge in spatially and temporally specific ways, emerging most forcefully in proximity to the hospital. But this is not to say that spatiality and temporality determines the movements of affect. Rather, and considered more processually, different timespaces may provide the conditions and the *possibility* for different affects to emerge. It should be pointed out that certain aspects of this are not new. For example, humanist accounts of place stressed how sensate attachments can make particular places become significant for individuals (see Tuan 1977; Relph 1976). But thinking about the geographies of affect moves away from such a determinate, sedentary metaphysics where the primacy of place is assumed.

Instead, geographies of affect allow us to consider the how the complex interplay between sensations, percepts and affects plays out over timespace. It is these contingent relations which constantly transform the dimensions of our possible field of action and change the realm of possibility for the body in pain. This is a probabilistic, not a determinist geography. Thinking with and through affect forces us to consider the effects and capacities of pain that go beyond the fleshiness of individual bodies in pain. This might at first sound counterintuitive. Pain is surely something that is individual; something that the body in pain *possesses*. Indeed this individualism when approaching pain is, in part, why narrating one's own pain to others might be such a difficult undertaking; sparking, at best, brief glimpses of recognition and, at worst, mutual incomprehensibility. 'Suddenly or over time', Harrison argues, 'pain, loss, and affliction tend toward the erosion and depletion of the capacity for speech and communication, toward the unravelling of our words and sentences into stutters and inchoate cries, into moans, and into the catatonia of silence' (2007, 593). But affects are generated and quiesced through entanglements with people, physical locations, material objects, at different times and in different spaces. In the opening paragraph, the hospital had the capacity to generate such intense affectual movements, in part, because of its configuration with other objects and bodies: the route of the bus, the layout of the interior of the bus, the relative location of the college and home, the avoidance of doctors at the time, and so on.

Affects are transmitted between bodies and objects in ways that are often unpredictable and unforeseen. As such, a key benefit of thinking through affect over other ways of attending to the body in pain is how different affects are intensified and quiesced over time and space. Importantly, attending to pain in different places, to different people, surrounded by different objects, embedded in different systems of governance and value can dramatically alter these affective configurations. Or to put it a different way, a reciprocal relationship of affecting and the ability to be affected depends on the nature of these configurations. Whilst the obduracy of the pain might be inert, we might feel more *comfortable* with the pain in some places over others. Our affective orientation towards pain shifts contingently and contextually. Key here is not however to distinguish *between* the individual or human and the non-human dimensions of pain. Rather, it is about recognising how different sets of things, their configuration, their assemblage and spacing; their energy, have different capacities to do different things. Put simply, this is not about what the body in pain *is*, but about what the body in pain has the *capacity to do*, and how it can go on, depending on the configuration of individuals, objects and places. With this, it is important to consider the *potential* of an event, a set of objects – or rather the multiplicity of potentials that might emerge at any time for the body in pain. Different configurations of things in different places create a different 'field of potential' in the sense that the *effects* that emerge through these affects take spatially and temporally specific forms. As Massumi describes, 'potential is the immanence of a thing to its still indeterminate variation, under way' (2002, 9). To take the event at the start of the chapter, the relations between the body, the hospital, the time, the bus and the seat set up a field of potential, embedded within larger regions since 'the field of potential is the *effect* of the contingent intermixing of elements' (2002, 76). Yet this field of potential is mobile. Whilst a particular shape or coherence begins to form, as it does in the opening paragraph, it 'no sooner dissolves as its region shifts in relation to the others with which it is in tension' (2002, 34). Life has moved on. Thinking through affect therefore not only decentres the body from analysis but also liberates it from the notion of a singular, predictable and fixed trajectory. This provides the possibility for transition, where different affects might be circulated. Thinking through affect therefore generates the space for something different to emerge; for something to change (see Harrison 2000; Thrift and Dewsbury 2000).

***

Something changed. The writer gave in. The charge was too intense and he could stand it no more. Biting and ripping into every moment, the pain had finally transported him into the tunnel of diagnosis. On crossing the threshold, the waiting was excruciating. The possible was now waiting for him. A set of futures had been accumulated and condensed through two pieces of paper and a black and white image. This is all that was required. Somewhere, sometime these three powerful objects; a neurological report, a medical record and a brain scan; would be fused together and would eventually find him. These objects promised escape. He arrived home, taking the number 11 bus which passed the place that already knew. Standing in the hallway, one word changed everything. A word uttered from his mother's lips. Negative. The guns had stopped. His body froze. Universes were colliding in silence and stillness. Incomprehensibility and disbelief. Overwhelmed. Tears. But gently rising through the sunlight spiralled a thin plume of pale yellow smoke. Smouldering yet unextinguished, the doubt which had given rise to this event in the first place had mutated into something more sinister.

***

The event of receiving medical results can be charged with a heightened affective intensity. The anxieties and insecurities that are threaded through and become the familiar refrain of everyday life during what can be months of waiting are brutally and often unexpectedly interrupted. The escape of affect. These mundane materialities; the faint ink on a piece of paper, the fragility of the messenger's utterance; contrast dramatically with the weight of their significance. The message that they convey has the capacity to orientate and align bodies along new trajectories, sending them spinning. This event reconfigures the field of potential for the body in pain. The spacing and timing of potential movements and events are changed. Events are reprioritised: some quietly fall off the horizon; others draw closer or flicker into being in a world where hue and saturation is fuller; deeper. The view is stunning. Yet the elation and euphoria of potential can be at once juxtaposed with a sinking feeling. The intelligibility of the judgement jars with the indeterminate 'what if?'. A series of reflective, cognitive processes might be partly responsible for this. A refusal to invest trust in the judgement perhaps stemming from the perceived fallibility of medical knowledges, incomplete or incoherent procedures, and an appreciation that medical practice is haunted by statistics and margins of error. What if they misdiagnosed? But this sinking feeling is more than just reflective questioning. As Connolly notes, thought and affect are involved in complex feedback loops to create '"affectively imbued thoughts" and "thought imbued intensities"' (Connolly cited in Anderson 2006, 737). This sudden disruption and reorganisation of the field of potential might itself generate a sense of disbelief. Where one's habitual, everyday orientations within and towards the world are torn asunder, how to improvise with confidence? The sufficiency of one's habits and routine modes of embodiment are rendered inadequate. Furthermore, and casting our eyes back to the opening paragraph, it becomes increasingly apparent that uncertainty as an embodied, involuntary and precognitive affective condition in and of itself is key to understanding the experience of the body in pain. Through pain, the ungraspability, indecipherability and incoherence of the body is brought into sharper focus.

Whilst habit and repetition might serve to frame existence by providing an illusion of consistency, affect as an emergent relation is never fixed or certain. In contrast to more traditional geographical approaches that have tended to privilege human intention and consciousness, geographers are increasingly interested in how tending to affect opens up the excessive dimensions of the world. Thinking through affect is, in part, an invitation to consider the excessive dimensions of being in the world that can never be predicted or determined in advance. As such, uncertainty features as a hallmark of such thinking where it becomes a condition of ethical orientation in itself. McCormack for example suggests that the development of an ethics of sense demands an 'openness to the uncertain affective potentiality of the eventful encounter as that from which new ways of going on in the world might emerge' (2003, 503). This valorisation of the uncertain as 'an ethos ... which welcomes such unknowing and uncertainty as the ground from which something worthwhile might emerge' (Gordon 1999 cited in McCormack, 2003, 503), is explicit in many geographies of affect. How to cultivate a disposition and an openness to difference, the multiplicity of life and, crucially, *uncertainty*, is one of the important tasks explored through the McCormack's (2002) experiments with movement. Similarly, Wylie's (2005) corporeal pedestrian movements demonstrate how a multiplicity of entanglements of the body within landscape can give rise to different capacities for affecting and being affected. Here, what a body *can do*, in different timespaces is both provisional and unpredictable. Massumi equates the unpredictable and excessive nature of affect to a promise of change and a possibility of future, arguing that affect 'is nothing less than the perception of one's own vitality, one's sense of aliveness, of changeability' (2002, 41). Here, uncertainty becomes an affirmation of life, characterised by an implicit sense of vitalism. Similarly for Anderson, hopefulness as a disposition to the world is enacted through a particular 'feeling of possibility' that things might be different, which constitutes a 'dynamic imperative to action in that it enables bodies to go on' (Anderson 2006, 744). Where for McCormack it is comportment and movement that generate particular dispositions to affect and be affected, Anderson's catalysts emerge through variety of materialities. This draws attention to the complex micro-topologies of everyday life where the multiple relations between the body and the presence or absence of particular materialities, or the enactment of particular movements, have the capacity to generate particular affects. Similar to Bennett's (2001) notion of 'enchantment', the unpredictability of possible becomings becomes an affirmation of the liveliness of life itself.

Yet in thinking through the body in pain, the affective possibilities of such uncertainty can be detrimental, closing down possibilities to be affected. Here, the excess of possible becomings, often described in a positive vein within geographical research on affect, has the capacity to generate more negative anxieties, tensions, fears and frustrations which can curtail the possibility to affect and be affected. Indeed such an excess contains within it multiple affectual pathways. Anderson's (2006, 748) rendering of hope, for example relies on a 'sense of the tragic', and of melancholia which is always immanent to hope and through which hope emerges from. Far from liberatory, a condition of uncertainty can work to undermine the circulation of positive affects, preventing hope from taking place, narrowing horizons. Put simply, for Deleuze, positive affects *enhance* and negative affects *reduce* our capacity to act. Whilst Deleuze (1988) singles out joy and sadness as being exemplary positive and negative affects respectively, it is of course difficult to narrate and attend to affects in ways that do not qualitatively evaluate their impact on the body. Yet what uncertainty does to the body might be rather more complex than simply bringing about joy *or* sadness. Unlike the naming of other affects such as 'fear' and 'anger' or 'happiness' and 'excitement', each of which emerge through particular skeins of signification, their positive or negative ramifications already mapped onto an ideal body, uncertainty is a more fickle beast. Indeed the affective power of 'uncertainty' in and of itself does not so easily take refuge into a sedimented set of discourses. Echoing Haraway's (1997) process of the 'hardening of the categories', whereby over time particular meanings and significations accrete around specific vocabularies, we might think of 'joy' and 'sadness' as being particularly intense 'firework affects', to borrow Thrift's (2008, 241) term, that are potentially stronger or more durable than other affective modalities.

Whilst for some, the excessive nature of uncertainty might be exciting and liberating, generating a hopefulness that things might be different, for others, uncertainty might serve to engender a series of rather more negative affectual responses. Even though such situations might expand our experiential horizons, financial uncertainty or employment uncertainty are hardly desirable modes of being in the world. Indeed here, is the *excess* of possible futures that generates such negative affects perhaps characterised by a fear that the relatively durable materialities, structures, presences and routines that we rely on to get by every day are always at risk of falling apart, always at risk of being shattered by the unexpected and unforeseen. What is perhaps important here is that ontological uncertainty generates particular orientations, anticipations and dispositions. Indeed some orientations might be particularly debilitating the body. For example, a person suffering from types of obsessive compulsive disorder, far from embracing uncertainty, might spend every waking moment battling *against* uncertainty. Here, a denial of uncertainty might create a range of negative dispositions such as powerlessness, fatalism or pessimism. Where 'certainty' as an impossibility but necessary conceptual device for poststructuralists might be broadly equated with reductionist tendencies that negate excess and close down the creative processes of becoming, a degree of certainty and consistency can be highly attractive. Yet what is of crucial importance here is how these affective dispositions have the capacity to *change* through encounters and experiences in particular places. Put simply, getting to grips with the spatialities of these affective relations, where they unfold and take place is central to understanding these tensions between certainty and uncertainty and how bodies respond to excess. Dispositions towards excess condense, cohere and are sustained through specific places, performances and materialities. So what next for our body in pain?

***

After years of doubt, the writer finally returns to a doctor's room. Hippocrates looks on. The atmosphere hums with a reassuring sense of gravity. A gravity conditioned by centuries of medical knowledge accumulated in the revered figure facing him. A gravity that he fails to remember is always misplaced. This is the place where pain is always tamed, must be boxed and ticked-off. Ideally in 10 minutes. Perhaps that is why he likes it here: a comfort in clinical and medicalised certitude, the scalar and the bounded. This is the place of DNA, of exciting chemical compounds, of synapses and of magical green pieces of paper. Through these objects, this place promises hope. In this place, pain is mapped out onto the body politic. Here the flesh becomes transparent and his pain is made objective. This is the place where the vision and science go hand in hand, where the scopic rules for sure. X-rays, imaging techniques, stethoscope, ophthalmoscope, endoscope, auroscope all enlisted to visualise and reveal pain. It is a pain that is thoroughly located: a pain in the head. Through imaging, the flesh is penetrated and the non-representational is rendered visible. Intensity is condensed: singular and measurable. Cartographies of pain are created through expert maps and pictures; authoritative diagrams and flowcharts. It all looks so manageable. Sensation is codified. The McGill Pain Questionnaire asks "Where is the pain? What does it feel like? Does it move around? Is it constant? What influences its severity? When was the worst pain ever? Why did it happen at that time? Which situations make you feel it might return as badly again? An inventory of pain is created. Every time it is different. As O'Neill (1999) suggests, this must be the place where the war is really waged. Descartes' war of reason against evil: medicine *against* disease. This is the place where Foucault's deviant tissues are isolated and excised. His body has once again become a site of medical warfare, subjected to suppressants to defend against attackers. Sensation here is articulated as a malignant and clinical object-assemblage within and confined by the outer limits of the physical body. Chronic pain. It is made up of faulty neural channels in a specific location. This is sensation exposed to industrial management and bureaucratised discourses to fit into the rationalising, increasingly neo-liberal grammar of the health system. But the experience of pain is absent from these narrations. Where has the body gone? For here it is just the site of clinical symptoms. All else is eroded, downtrodden and neglected: affect closed down and annulled. This space has nothing left to give now. Exhausted.

***

In the doctor's room, uncertainties are categorised. Here, the naming of a set of sensations, percepts and affects as 'chronic pain' invites us to think through the complexities of affect. Particularly owing to its unwilled genesis, it differs from many of the other iterations of research that attends to affect which focuses on the performative body in action and how particular intentional bodily movements have the capacity to engineer particular affective capacities. Chronic pain provides an extremely effective example since, unlike many other neuropathic conditions, it is notoriously difficult to diagnose, let alone treat. The important point to take from this is that non-malignant neuropathic chronic pain is not associated with injury or tissue damage. Yet as I have described elsewhere (Bissell 2009), the experience of chronic pain, particularly that which is stubbornly treatment-resistant, has the capacity to generate a series of negative affects that can potentially engulf and deplete the liveliness of the body.

Chronic pain might be characterised, in part, by its obdurate *certitude*. The certainty that at every moment the pain will be present. In such circumstances, it is difficult to get excited about new becomings. Whilst of course a body is always more than just the dulling effect of pain, other affective intensities may become circumscribed and mediated by the pain. The effect of pain can close down possibility so much that it becomes increasingly difficult to anticipate or bare *any* sort of future. The future of the chronic pain suffer might seem utterly contracted. Whilst many of these negative affects might stem from the discomfort of the pain sensation itself, these are often exacerbated by a series of uncertainties relating to what the pain actually is or might do to the body. Put simply, this is a concern for how pain changes or might change the capacity of the body to affect and be affected. For many, identifying the cause of chronic pain becomes an imperative; a search for meaning and signification which might provide some legitimisation for painful sensations. Yet chronic pain exceeds normalised conventions of rationality. Indeed we could think of it as pure involuntary intensity; pure affect. Newton-John and Geddes argue that chronic pain is such a difficult condition to pin down since multiple diverse pain states, possible emergent from 'neuroendocrine changes, immunological responses, cortical reorganisation, dorsal horn "wind-up" and central sensitisation, and alterations to the hypothalamus–pituitary–adrenal axis' (2008, 199), are masked under the blanket naming of 'chronic pain'. In the face of such ambivalence, chronic pain presents the body with a series of profound uncertainties which can rarely be assuaged by any definitive answers. Why did it emerge? How long it will it last? What might relieve it? What does a pain that has no cause *mean*? Relating-to, and living-with the uncertainty that is chronic pain can be hugely challenging. This absence of vindication in the doctor's room might drive the body in pain to other places.

***

The writer is sitting in a chair. Mikao Usui looks on. There is nothing clinical about this space. An uneasy sense of betrayal looms. Memories of the pain questionnaire, the responsibility of testimony in *that* room slide to unfinished business. There are no instruments or indices here. As the session begins, there is not an invitation to narrate sensation. His voice, his anticipatory dialogue and his intent to narrate have been muted. And then, a blissful sense of relief descends as he realises that he does not have to articulate the pain. Again. Relieved of the burden of responsibility for in this space, pain is not articulated by the sufferer, it is sensed through the healer. There is no imagining and there are no descriptions. Through the silence, the healer stares: eyes following hands, tracing around the shape of his body. Here pain does not have a location. It is not situated within the body and it is not an object to be represented. The shape of sensation shifts, here it is not in his head, but is flowing, rippling, vibrating. These are not metaphors, but vibrations that can be sensed, vibrations that are revealed in colours, vibrations of energy that enclose and gather and collect around the body: a halo. This is a body without organs for real, as the healer draws out these virtual potentials. The healer tells him that his aura is dented. Instead of continuous colour, his aura is dulled and fragmented and inconsistent. Remember, there are 'knots of arborescence in rhizomes' (Deleuze and Guattari 1988, 20). This is the body reimagined, the sensation of pain deterritorialised through vortices of metaphysical energy: chakra points that are not confined to the flesh. As the healer places his hands over the body, the writer does not know how to respond. Think spiritual, think energy. Think new-age, solar plexus, crown. Just try to withdraw. But all he can think of is the comedic effect of Jennifer Saunders' Buddhist chanting in the TV comedy *Absolutely Fabulous*. Why does the ethereal seem so comical suddenly? A post-colonial appropriation of eastern philosophy? Coerced withdrawal is not an option here. Where is the verifiability? This surely iterates and rarefies something of the doctor's room representations, desperate to develop some connections. Unlike the medical narration of sensation, this encounter is not linear, hierarchical or stratified. It is marked by a dynamism of lines and trajectories and contains multiple points of entry. The painful sensation is attended to and narrated by the multiple energy flows between bodies. Sensation is revealed and presented through the hands. Perhaps this is the circulation and transmission of affective intensity made legible? How would Deleuze respond to this? This is the self and the pain decentred, not independent but relational and folded. This reminds him of Gadamer (1996) who insists that doctors only have the capacity to produce 'effects' within bodies such that they reestablish the ill health of a body. Accounting for and narrating a decentred pain sensation as afflicting the soul instead emphasises the transcendental and progressive practice of healing though universal energy. Do not judge; do not dismiss: *you cannot deconstruct healing*. But what of those crystals? Through these objects, this place promises hope. But how to reconcile these two accounts of pain? Desperate to scrutinise but imperative to withdraw, it is time to move to a new place.

***

For the body diagnosed with chronic pain, the assistance and assurances of a doctor can only go so far. Working within a limited set of knowledges, chronic pain confounds and exceeds their parameters. This lack of relief or resolution can therefore intensify the condition of uncertainty, prompting the body to seek out new places beyond the circumscribed world of western medical practice. In the absence of widespread sophisticated pain imaging devices in western medical practice, as hope fades, the body is forced to find other places which attend to pain in different ways. But in seeking consolation in other places, one version of chronic pain is placed against another. The grammars and vocabularies used to attend to pain in different places articulate and animate distinct sets of affective relations, each of which attempt to engineer affective uncertainty in particular, and often unexpected, ways. What is important about these movements between different places is a demand to consider how trust is invested in particular places. This is important since immanent to each investment there is a 'radical loss of the totality of possibilities which we call a world' (Dastur 2000, 185). Investing trust in a particular place in part relies on a negation of the potentiality of other places. It therefore compels us to consider what happens to that trust when a different set of affective relations is animated in a different place. These places are often unexceptional – just rooms, no architectural singularity or importance – but the set of affects precipitated by events in each space make each of these spaces remarkable, special and possibly even hopeful. Yet they are also highly *exceptional* in that they punctuate the relatively stable flow of ordinary experience; a stability that in part gives rise to the exigency of puncture.

***

The writer sits in a chair. Freud looks on. This is where transgressive problems are treated. Painful sensations are brought here to be labelled, judged. Head still buzzing. No crystals here, or body-maps; no computer screen and no blinking cursor. Just a low table, a sad white vase holding some dusty artificial flowers. Next to it, a box of tissues, half empty. The top one is protruding through the card aperture: an invitation to tears. A redundant yet powerful act, is an expression of suffering necessary to narrate pain, to enhance the capacity to be relate or be believed? What if tears don't come? The body stagnated, reterritorialised after years of medication drying up any hope of expressing raw affection. This is not repression but suppression of intensities that were never allowed to emerge. A lady sits opposite: pen poised and clipboard angled in order for rigour to be appropriately exercised. Through these objects, this place promises hope. Here language, structure and Oedipus rule for sure. A Lacanian unearthing, a search for an essence of unconscious form. He knows that with that pen she is searching for hidden significant combinations, a symbolic overcoding of utterances as Deleuze might put it. Yet he is uncomfortable with this archaeology. Is this ontological betrayal? There is no room for pre-personal desire in this clinical enunciation. She wants to know what the pain feels like. How does it make you feel? Rusty, hot barbed wire. But how to provide testimony here? The answers are already formed, there is nothing new here. Trauma of this sensation presents a suspension of language, a blocking of meaning as Barthes puts it. The guilt presses heavily, as if waiting to be found out. This guilt was not present in the place of crystals!

Ricoeur's hermeneutics of suspicion are revealing themselves in full force. In this confessional she holds the key through her masque of certainty. If this is the space of language, he wants to speak of how pain acts on the body and what pain has the capacity to do. For it is here that he is most freely invited to express pain affects: how it has the capacity to affect routine and encounters. But all she wants to do is press structures by weaving a pain that is already pre-formed in her mind: expressiveness reduced to semiotics. When did the pain start? And what was happening to you in your life at that time? Cue Freud's return. A sudden reluctance to describe anything is overwhelming. It seems that in this space, pain is a historically-sedimented and accepted set of feelings that flow from the books on the desk. As no tears come, he envisages how this lack of desired 'authenticity' has reduced his position in her league-table of patient problems. Goodness, how these linear self-evaluations are reminiscent of the pain questionnaire! We're out of time here.

***

Different places; different narrations; different bodies. In each place, chronic pain as an ongoing and benign affliction is made animate in different ways through its calling forth. Indeed narrating affect is paramount to deterritorialisation. In the absence of communication, both the body in pain and multiple potential caregivers are ossified. Each of these places invite pain to be actualised and made conscious in particular ways. It would be easy to comment on the motivation for these different narrations: the rationale in each place fulfils different functional roles. We could contrast the imperative of doctors and neurologists to *diagnose* pain; the imperative of healers to *ease* pain; and the imperative of psychoanalysts to *evaluate* pain. But in each of these places, through different styles of narration, pain becomes something different. Echoing Mol's (2002) ethnography that exposes the multiple ontologies of atherosclerosis, chronic pain is not a singular or simple ailment. Whilst the physical intensity of the pain itself might be relatively obdurate and unwavering, through these performances in different places, chronic pain is *enacted* in different ways such that the affects condensing through and emergent from the pain are really very changeable. Crucially, the form of narration, the way that the non-representational is approached, changes what pain does in relation to the body in that the body becomes something else through each presentation. In the hospital and the doctor's room, the body is a cartography of flesh and organs and is the passive subject of the clinician's gaze. Here the Cartesian body is simplified, made transparent where pain as sensation is readable, rationalised and objectivised through visualising technologies and medical procedure. In the healer's room, the body is a set of energetic vibrations not bounded by flesh but focused around points of intense energy. Here pain as sensation is made legible through the body of the healer, where sensation is felt, tactile and haptic. In the therapist's room, the body is an accumulation of habits, memories and desires that can be expressed through speech acts. Here, pain as sensation is actualised through dialogue and tears in response to particular prompts.

But in encountering each place in turn, and over a sustained period of time, what has this body in pain become? Whilst these are distinct styles of narration with different people at different times, they are interacting narrations in that they coalesce around the same figure over time. Each of these spaces has a clear mandate of what painful sensations are and how they should be attended to. But, crucially, how should he respond to and negotiate these different narrations? What are the cumulative effects of submitting oneself to these different places? As the writer surrenders himself to each place, he is given advice on how to live by those particular logics. Each place offers sanctuary to the body in pain, acting as a point of anchor. Each place offers a set of different practical techniques and performances to take *beyond* that place with which to deal with the quotidian tapestries of everyday life. But in order to be effective, such logics require investment, commitment and loyalty; and are choked by analysis, suspicion and uncertainty. Indeed on consistency and the development of embodied habits, Deleuze and Guattari point out that 'staying stratified, organised, signified, subjected is not the worst thing that can happen; the worst thing that can happen is to throw the strata into demented suicidal collapse' (1988, 161). The jarring and disorientation experienced as the fragile body in pain moves between accounts, where habitual modes of being are rendered inadequate, might be highly detrimental: a 'reluctance to engage, arising partly out of this corporeal vulnerability' (Thrift 2008, 242). Reassurance in each place can only take place through cultivating a disposition of submissiveness which suppresses resistance. For the writer, the confidence of each narration, the assurance that circulates within each space – and that is crucial to their effective functioning – wanes as the number of narrations grows. The intense immediacy of each different style of narration, the conviction that this particular narration has the capacity to alleviate the suffering casts a layer of uncertainty and doubt on the efficacy of the last. Uncertainty multiplies, for it is contagious.

Are these incompatible ontologies? What place do the chakras have under the MRI scanner? And where is Lacan as he meditates with the help of amethyst? The memory of each narration folds into the next as the body twists and weaves through time and space. Each narration wills him to invest belief. But where is his allegiance? And how to reconcile each of these actualisations of pain? Or perhaps put more simply, why should the power and conviction of each narration appear so farcical when it is out of place? When taken together, the affectual half-life of each narration seems to be pretty short. This concurrent uncertainty through multiplicity, so important to non-representational theory, seems at odds with the cathartic role that each individual narration of pain assures – cathartic in the sense that the experience of pain moves beyond a pathology of fear to one of serenity (Kearney 2007), where pain sometimes finds some release. By the very definition, only through their boundedness, through their different and mutually-exclusive internal logics does each narration have the capacity to *minimise* uncertainty. Only by introducing another way of actualising pain does uncertainty about what it is and therefore how it can be attended to increase.

Whilst multiple narrations potentially open up new configurations of possibility, he is given no advice on how to coordinate, manage or negotiate the divergent logics of each different enactment of chronic pain. Indeed it is through the cumulative multiplication of different ontologies that absence, division, discord and uncertainty intensifies. Or at least this is how things might appear.

With pain there is no sense of arrival, no moment of triumph that occurs when all uncertainty has been smoothed and ironed out. Experiential uncertainty is at the heart not only of idiopathic pain, but also of narrating sensation. Unknown sources. Imperfect and incomplete knowledges. 'Whatever pain achieves, it achieves in part through its unsharability, and it ensures this unsharability through its resistance to language ... to have great pain is to have certainty; to hear that another person has pain is to have doubt' (Scarry 1985, 4). There really is no way to feel someone or something else's pain. Sensation itself is untranslatable. But where does this leave the writer? And what hope does this give to those who suffer from pain who must narrate and attend to their sensation on an everyday basis?

Whilst it is important to acknowledge that non-representational approaches within geographical research are certainly not uniform, what they share is a concern for the sensate and (post-)phenomenological dimensions of existence. The prioritisation of becoming over being and process over fixity has enabled geographers to consider how the strata of signification, discourse and meaning emerge first and foremost through practice and performance. Attending to affective life specifically emphasises the excessive dimension of existence; the excesses and surpluses; the 'pressing crowd of incipiencies and tendencies' (Massumi 2002) that is the realm of the potential. Positive connotations are often attributed to this affective excess where the potentiality of life is taken to be a 'pure gift' (see Anderson 2006). Attuning the body through experimental performances (McCormack 2003) might provide the conditions for something new to emerge. Multiplicity and uncertainty are taken to be affirmative of the liveliness of existence. Yet multiplicity and uncertainty can affect bodies in very different ways that are far from enabling or affirmative. This is certainly not to say that geographies of affect have been poor at appreciating the diverse affective capacities and capabilities of *different* bodies. Indeed an asset of this work is the way that it has expanded the ways in which diversity can be apprehended in ways that transcend the normative conventions of signification. Rather, it is important to keep in mind how negative dimensions of uncertainty can potentially close down the possibilities of particular bodies – such as a body in pain – to affect and be affected.

As illuminated in the movement between the different places traversed in this chapter, the uncertainty of excess might, in itself, be disabling. Incoherence and uncertainty might be debilitating; even traumatic. Whilst the medical room or hospital is often the first place of call for the sufferer, chronic pain frequently confounds even the most experienced medical practitioner. This is a body that has reached the extremities of 'medical knowledge': a deviant body that the system is ill-equipped to handle. But movement away from this place dramatically alters the body's field of potential in ways that may be both enabling and disabling.

On one hand we could consider how the affects emergent through the limitations and dissatisfactions encountered in each place provide the drive for searching for other places; other logics; other ontologies of pain. Perhaps, as Thrift suggests, such a drive is generated though a 'thoroughly healthy anxiety about losing the future' (2008, 235). Multiple narrations give rise to points of connection and points of divergence. Indeed these narrations, these expressions of pain take place at the intersection of different bodies and objects. Narration is decentred through the enlistment of a variety of materialities. When taken together, these multiple narratives form what Frank refers to as a 'restitution story' (1995), an account of conflict and collaboration with myriad clinicians and caregivers. They are truly rhizomatic. Perhaps it is not about achieving harmony between accounts, but rather to embrace the uncertainty that emerges from these multiple narrations of pain. This prevents one narration, one simulation, one form of testimony, from gaining ultimate authority. The painful body is not blocked and reterritorialised by the strata and structure of particular individual narrations but is open to the potentiality that multiple narrations offer. It is this ethic of unfinishedness that might provide a sense of hope.

Yet on the other hand, it is this uncertainty through multiplicity that is also disabling. Each place encountered that promises to reshape the body, but does not alleviate the pain reduces the possibility that the pain will ever shift. Through each broken promise, hope fades. Hope has limits, but pain knows no bounds. 'Suffering does not have a limit; like an event which does not concern you it continues regardless of the point where you can no longer go on', (Harrison 2007, 594). Furthermore, movements between places, between systems, between ontologies which dramatically alter the field of potential for the body in pain might shock the body into submission. Now, desire might wane; life might diminish, particularly if such movements between happen with such frequency and in such close proximity. Indeed Deleuze and Guattari warn against such violence; against such potentially jarring movements. Far more cautiously, they advise that you must 'lodge yourself on a stratum, experiment with the opportunities it offers, find an advantageous place on it, find potential movements of deterritorialisation, possible lines of flight, experience them, produce flow conjunctions here and there, try out continuums of intensities *segment by segment*, have a *small* plot of new land at all times' (1988, 161, emphases added).

***

But the writer has grown tired of searching. Even though he should know better, the writer is now drawn to hospitals. After all, sense is not necessarily sensible. The condition of uncertainty can be fatiguing as the shifting kaleidoscope of 'what might be otherwise' generates an endless succession of cycles that oscillate through hope and distress. In the dark of night and in the midst of a city in slumber, the hospital lives. The hospital promises to relieve the burden of uncertainty, and the weight of decision, reducing the volatility of the body to an object under the clinician's gaze. The desire to be 'overdetermined' is satisfied in the sanctuary of the hospital. Its structures and knowledges promise to organise and discipline the unruly body in pain, for here life appears ordered. Objects and intrusions are metallic and sterile. The desire to be monitored and attended to is temporarily satiated. The rhythmic pulsation of coloured lights offering the reassurance of regularity. The pots of medication pledge to reassemble the body into something less volatile. The body is blanketed and its unwieldy agency quiesced. He knows that this is incongruous, but once the threshold is crossed the body does not have to make choices. The responsibility to make decisions is suspended. In yielding the body is never alone. An illusion? Possibly. But this place promises escape.

References

Anderson, B. (2006), 'Becoming and being hopeful: towards a theory of affect', *Environment and Planning D: Society and Space* 24(5), 733-752.

Anderson, K. and Smith, S. (2001), 'Emotional geographies', *Transactions of the Institute of British Geographers* 26(1), 7-10.

Bennett, J. (2001), *The Enchantment of Modern Life: Attachments, Crossings and Ethics* (Princeton, NJ: Princeton University Press).

Bissell, D. (2009), 'Obdurate pains, transient intensities: affect and the chronically-pained body', *Environment and Planning A* 41(4), 911-928.

Bondi, L. (2005), 'The place of emotions in research: from partitioning emotion and reason to the emotional dynamics of research relationships', in Davidson, J., Bondi, L. and Smith, S. (eds), *Emotional Geographies* (Aldershot: Ashgate), 231-246.

Burgess, C., Hunter, M.S. and Ramirez, A.J. (2001), 'A qualitative study of delay among women reporting symptoms of breast cancer', *British Journal of General Practice* 51(473), 967-971.

Chapple, A., Ziebland, S. and McPherson, A. (2004), 'Qualitative study of men's perceptions of why treatment delays occur in the UK for those with testicular cancer', *British Journal of General Practice* 54(498), 25-32.

Clough, P. (2007), 'The affective turn: introduction', in Clough, P. and Halley, J. (eds), *The Affective Turn: Theorising the Social* (Durham, NC: Duke University Press).

Dastur, F. (2000), 'Phenomenology of the event: waiting and surprise', *Hypatia* 15(4), 178-189.

Davidson, J., Bondi, L. and Smith, M. (2005), *Emotional Geographies* (Aldershot: Ashgate).

Deleuze, G. and Guattari, F. (1988), *A Thousand Plateaus: Capitalism and Schizophrenia*. Trans. B. Massumi (London: Continuum).

Deleuze, G. (1988), *Spinoza: Practical Philosophy* (San Francisco: City Lights Books).

Frank, A.W. (1995), *The Wounded Storyteller* (Chicago: University of Chicago Press).

Gadamer, H-G. (1996), *The Enigma of Health: The Art of Healing in a Scientific Age*. Trans. J. Gaiger and N. Walker (Stanford, CA: Stanford University Press).

Greco, M. and Stenner, P. (2008), 'Introduction: emotion and social science', in M. Greco and P. Stenner (eds), *Emotions: A Social Science Reader* (London: Routledge).

Haraway, D. (1997), *Modest_Witness@Second_Millennium.FemaleMan©Meets_ OncoMouse™* (London: Routledge).

Harrison, P. (2000), 'Making sense: embodiment and the sensibilities of the everyday', *Environment and Planning D: Society and Space* 18(4), 497-517.

—— (2007) '"How shall I say it ... ?" Relating the nonrelational', *Environment and Planning A* 39(3), 590-608.

Kearney, R. (2007), 'Narrating pain: the power of catharsis', *Paragraph* 30(1), 51-66.

Kwan, M-P. (2008), 'From oral histories to visual narratives: Re-presenting the post-September 11 experiences of the Muslim women in the United States', *Social and Cultural Geography* 9(6), 653-699.

McCormack, D. (2002), 'A paper with an interest in rhythm', *Geoforum* 33(4), 469-485.

—— (2003), 'An event of geographical ethics in spaces of affect', *Transactions of the Institute of British Geographers*, 28(4), 488-507.

Massumi, B. (2002), *Parables for the Virtual: Movement, Affect, Sensation* (Durham, NC: Duke University Press).

Mol. A. (2002), *The Body Multiple: Ontology in Medical Practice* (Durham, NC: Duke University Press).

Newton-John, T.R. and Geddes, J. (2008), 'The non-specific effects of group-based cognitive–behavioural treatment of chronic pain', *Chronic Illness* 4(3), 199-208.

Pain, R. and Smith, S.J. (2008), *Fear: Critical Geopolitics and Everyday Life* (Aldershot: Ashgate).

Relph, E. (1976), *Place and Placelessness* (London: Pion).

Rose, M. and Wylie, J. (2006), 'Animating landscape', *Environment and Planning D: Society and Space* 24(4), 475-479.

Scarry, E. (1985), *The Body in Pain: The Making and Unmaking of the World* (Oxford: Oxford University Press).

Sedgwick, E. (2003), *Touching Feeling: Affect, Pedagogy, Performativity* (Durham, NC: Duke University Press).

Thrift, N. and Dewsbury, J-D (2000), 'Dead geographies and how to make them live', *Environment and Planning D: Society and Space* 18(4), 411-432.

Thrift, N. (2004), 'Intensities of feeling: towards a spatial politics of affect', *Geografiska Annaler B* 86(1), 57-78.

—— (2008), *Non-Representational Theory: Space, Politics, Affect* (London: Routledge).

Tuan, Y-F. (1977), *Space and Place: The Perspective of Experience* (Minneapolis: University of Minnesota Press).

Wylie, J. (2005), 'A single day's walking: narrating self and landscape on the South West Coast Path', *Transactions of the Institute of British Geographers* 30(4), 234-247.

Chapter 5

Non-Representational Subjects?

John Wylie

Introduction

Two quotes, to start with:

> Non-representational theory is resolutely anti-biographical and pre-individual. It trades in modes of perception which are not subject based (Thrift, 2008, 7).

> All of this said, I do want to retain a certain minimal humanism ... dropping the human subject entirely seems to me to be a step too far (Thrift, 2008, 13).

Placed beside each other, these citations from Thrift's recent *Non-Representational Theory: Space, Politics, Affect,* serve to indicate the topic, or more accurately the dilemma, of this chapter. When I first began to write this piece, it was going to be *about* writing, and the place of writing within non-representational geographies, with especial reference to notions of 'creativity' and experimentation. In one way, all I ever wanted was to be taken seriously as a writer ... and so I thought perhaps to open with the declarative thought that geographers should engage freely with the techniques and presentational formats of the creative arts. Personally, I'd like to write descriptive prose, travel narrative, fiction, poetry even. I'd like to paint pictures, take photographs, maybe make videos, produce site-specific pieces – and so on.

That *was* my aim. But, in the course of thinking and writing, things have changed. A feature of the opening paragraph above is its profusion of 'I's: *I'd like to ... I want to*. This would be innocuous, perhaps, if it weren't the case that the status of the 'I' – the gazing subject, the writing subject, the body-subject – has been one of the standout problematics for several generations of critical inquiry in the social sciences and humanities. More than almost anything else, the humanist notion that creativity, agency and inspiration are qualities rooted in, and in some sense defining, the individual artist, writer and so on has been exhaustively critiqued and deconstructed.

And this anti-humanism, in particular anti-subjectivism, is something that non-representational theories clearly share in common with many other traditions of inquiry, as the first citation above from Thrift indicates. Hence the question-mark in my title – an unqualified 'non-representational subject' would perhaps be a contradiction in terms. And hence also, from the very start (fretting, fingers hovering over the keyboard), my anxiety that this sort of thing – 'creative geographies' – would be a stupid thing to argue for. Even if it came hedged with disclaimers pointing to the criticality of so much contemporary artistic practice/commentary, including that by geographers themselves (see, for instance, Yusoff 2007; Butler and Miller 2005; Quoniam 1988), wouldn't an argument for creative geographies just make me look frivolous? At some dark level wouldn't the argument be haunted by concerns, spoken or not, about relevance, rigor, pretentiousness, propriety, sobriety and so on?

Well, no – be reassured that at least there's no question here of retreating into some indulgent solipsism, or of endorsing a specious notion of individualistic creativity somehow set outside of all location, duration and circumstance. After all, no art form or creative product – no novel, poem or painting, no film – has *ever* stood outside of these critical contexts.

These are givens, perhaps, but they are worth stressing here, in the context of this book on geography and non-representational theory, because the suggestion has been made at times that, if taken in the wrong directions, non-representational approaches would involve abjuring some of the agendas of critical geographies, and might indeed instead lead us down the paths of a certain romanticism, partly by drifting into a baroque realm of conceptual abstraction, and partly again *via* a näif celebration of events and performances that said little about the politics and positions of academic theorising, let alone the glaring iniquities of the economic, political and environmental *status quo* (e.g. Nash 2000; Cresswell 2002; Castree and Macmillan 2004; Tolia-Kelly 2007). However, I have always found these arguments difficult to accept. There may have been a move away from a certain type of critical identity politics, and a certain politics of representation, but it seems to me that the ubiquity of 'the political' – I mean assumptions regarding the essentially political nature of academic practice, and of its objects of inquiry – is just as evident within non-representational theories as it is anywhere else. Thus, over the past few years probably the single most pressing issue for geographers and others aligning with non-representational theories has been precisely the elaboration of a new, specifically critical and political *milieu* – a 'politics of affect', a 'non-representational politics' and so on. Or, to put this another way, the task for some has been to show that non-representational theories were and are *always already* 'political' – that a critical/political moment was never supplementary, never something to factor in, as it were, after the event. For the most part, as is evident in several noted publications (Dewsbury 2003; McCormack 2003; Thrift 2004; Anderson 2006), and as is moreover quite clear in the structure and content of this very book, questions of affective politics, of ethics, of witnessing and hoping, are there from the very start in the delineation of non-representational agendas.

I'll go further – I'd even suggest that, *contra* their common association with conceptual and methodological experimentation, (e.g. Dewsbury, Harrison, Rose and Wylie 2002; Lorimer 2007; Davies and Dwyer 2007), and *contra* also their derivation, at least in part, from the vitalist, 'life' philosophy of Gilles Deleuze, non-representational theories have actually had relatively little to say about creativity *per se*. Contrary to what might seem to be the case, the critical/political nature of non-representational theories is a given, whereas the potential for those theories and agendas to inform senses of 'creativity' in geographical practice still awaits, I think, a more specific consideration, even after Thrift's (2008) reassertion of the centrality here of notions of experimentation.

So in sum we have, a) a residual sense, in the face of much evidence to the contrary, that non-representational theories might involve a retreat of sorts towards more individualistic, subjective geographies; b) a feeling on my part that notions of creativity, and more widely the *subject* of creativity, have in actuality been undercooked *within* non-representational geographies, and lastly; c) a broader and deeper background across the social sciences and the humanities in which the very concept of the 'creative subject' has been extensively problematised. And more widely, as Osborne (2003) cogently argues, the very concept of 'creativity' has been extensively devalued and domesticated through the advent of new governmental contexts, in academia and beyond, in which the staking of claims regarding creativity and innovation in research processes and outputs has become almost mandatory. Straightaway, in other words, any direct petition on behalf of 'the creative' within geography becomes difficult to sustain, becomes pernicious even. And yet still, all I ever wanted was to be taken seriously as a writer …

Therefore the subject of this chapter *is* subjectivity, and more particularly the variegated, sometimes tenuous articulation of subjectivity within current non-representational and poststructural geographies. Specifically in this regard, the chapter argues that non-representational theory may be understood as part of poststructuralism in the broadest sense, and in this context supplies an overview of the Deleuzian and Derridean idioms that have circulated at least as widely as any others in writing in this area. The subject, I will argue, is of course in no simple sense either eliminated or re-asserted by non-representational approaches, rather it continues to haunt contemporary geographies in a way that *is* potentially creative and productive.

Non-Representational Subjects

Non-representational theory – or, poststructuralism

One way in which the question of subjectivity within non-representational approaches has come to the fore for me personally has been through the experience of trying to engage undergraduate geography students with aspects of the topic. For example, in 'methods' classes I have suggested to students that non-representational approaches involve an ethos of the affective, the emergent and the experimental, in which a certain premium is placed upon 'creativity'. If human geographers are expected to be conversant with social-scientific methods, from statistical analysis through to in-depth interviews and focus groups, and if they have equally embraced the interpretative and discursive approaches traditionally associated with disciplines such as art history and literary criticism, then why should it not also be possible to engage with the approaches and aspirations of work in the creative and performing arts – with creative writing, dance, site-specific art and performance and so on? I felt (and to an extent still feel) that this was a plausible way of introducing and positioning at least some of the impulses of non-representational work for students coming wholly new to the topic. However, I have also found that student responses to these sorts of suggestions highlight a key problem – the problem of the subject, and more specifically the persistence of an undisturbed humanism. I mean by this the persistence of beliefs in the inviolate, coherent and given existence of a free-standing 'creative' subject – an undisturbed 'I' who feels, speaks, expresses and so on. Presented with the argument that non-representational geographies take us towards creative writing and performance, a common response has been that these geographies are therefore all about expressing personal beliefs and feelings. This is the unintended outcome of speaking, perhaps too soon, about non-representational theories and 'creativity'.

What lessons might be drawn from such experiences? Of course I am the one who has cause to reflect – in delivering a particular account I had unwittingly guided students down a particular path. It certainly didn't expose some essential lacunae in non-representational approaches, it was perhaps mostly a technical question of changing the form and content of my teaching. But this has also made me see, again, how classes on the determination of subjectivity within geographies of power, inequality and exclusion in everyday life are a necessary corollary to any other work.

However, to leave teaching issues behind at this point, the wider consequence has been a slowly-dawning sense that the status of the *subject* with non-representational approaches deserves, maybe even demands, specific attention (just as in other contexts, for example in work on material assemblages and networks, concepts of structuration and determination have been re-thought (e.g. Latour 2005; De Landa 2002). As Dewsbury (2007, 444) comments, 'surprisingly … focus on the subject as such has slipped the direct scope of post-structurally minded geographers who follow non-representational endeavours'. In the light of this, where and how is the subject variously placed in the array of non-representational theories? What nuances and differences are there in the work of particular authors here? Or are non-representational theories essentially anti-humanist and anti-subjective in some way? On one hand, non-representational theories, as I will note in more detail below, may be understood in terms of a much broader post-structural dislocation in which notions of subjectivity, agency and presence are untethered from their humanist anchorage within, and as the essence of, human individuality. On the other, even so, such notions do not cease to be problematic, howsoever they are questioned, dispersed and multiplied – for instance into the relational, into objects, into various subjects we might not traditionally have thought of *as* subjects (animals, machines, networks, 'natural phenomena') – and in this sense they cannot be finally resolved or somehow made to disappear completely.

A further irony here is that, as previously noted, non-representational theories have been criticised on the grounds of a perceived over-emphasis upon individual agency and singular events. This, commentators have continued to argue (e.g. Nash 2000; Castree and Macmillan 2004; Lorimer 2007), runs the risk of obscuring or even denying the central role of wider power relations and socio-economic structures in the determination of individual subjectivities. Difficult though it is for me to see how such a conclusion could be gleaned from reading the various papers and editorials in which petitions of behalf of non-representational theories have been issued, a sense persists that a critical, ideological or structural analysis of cultural politics is in some way threatened by the advent of such theories. Indeed, my own work (e.g. Wylie 2002, 2005) has itself been judged wanting in terms of an excessive subjectivism (e.g. Blacksell 2005; Massey 2006; Tolia-Kelly 2007).

So there would seem to be a need for more explicit consideration of the forms of subjectivity at work within non-representational endeavours. Partly this is because it is important, in the light of critique, to highlight the complex, distinctive and differential understandings of subject that are now emerging in this area (see most recently Dubow 2004; Rose 2006; Dewsbury 2007; Harrison 2007). Partly also it is because even any very implicit plea on behalf of creative writing, as here, would have to grapple with the question of the *subject* of creativity, necessarily in the light of post-structural dislocation.

I am speaking broadly here, in order to set up the point I wish to make. In the context of 'mapping subjectivities', one account of the advent of non-representational research in geography is that it heralds a move away from, perhaps even a rejection of, the discursive cultural politics of identity and representation that characterised cultural geography in the early 1990s, and concomitantly a move towards, perhaps even a return to, a less-overtly politicised phenomenology of body, sense and world. Another account, from a broader cultural studies perspective, might contextualise non-representational theories within a wider 'performative' agenda, in which notions of the discursive construction of subjectivities have segued and morphed into notions of the ongoing practice and performance of subjectivities in everyday life. Another might stress the links between non-representational theories and the advent of a new sociological and anthropological *lingua franca* of relations, networks and assemblages. What all these accounts fail to clarify, however, is an in some ways more basic, albeit more parochial framework – one in which non-representational approaches are the ongoing and evolving articulation of various branches of poststructural thinking and writing within Human Geography – an articulation that began, in geography, back in the late 1980s and early 1990s, with the cultural turn's recognition of the crisis of representation, and its initial foregrounding of discursive and critical-deconstructive epistemologies.

Thinking in this way is useful, firstly, because it foregrounds continuity and evolution rather than rupture and division. This works against what I think is the unhelpful impression, generated by some debates, of a cultural geography dividing up into opposing and conflicting representational and non-representational 'camps' (e.g. see Thien 2005, and responses by Anderson and Harrison 2006;

McCormack 2006). Secondly, to place non-representational theory, from the outset, *within* poststructuralism, would certainly help to avoid any identification of this 'approach' with a humanistic focus upon personal feelings, attachments, senses of place and so on. Lastly, and most pointedly, I think that conceiving non-representational approaches *as* poststructuralism redux also offers a path to clarifying some nascent distinctions and tensions *within* these approaches – approaches which have of course from the start only ever presented themselves as heterogenous and non-programmatic.

Deleuze and the emergent subject

As with aspects of poststructural theory in the academy at large, non-representational geographies may be understood in terms of a broad theoretical division between, on the one hand, work written within a Deleuzian ontological idiom of force, vitality, materiality and relationality, and, on the other, work emphasising issues of being, language, ethics, writing and perception; here drawing upon a post-Heideggerian and post-phenomenological current of thought most closely associated with authors such as Derrida, Levinas, Nancy and Blanchot. This distinction is of course quite heuristic in some ways, although it does reflect a 'Deleuze/Derrida' threading that stitches and unstitches contemporary cultural theory (see Patton and Protevi 2003). However, in terms of cultural geography, the salience of the distinction, and its potential implications, have perhaps been obscured by the fact that the initial energies of non-representational theory were decidedly Deleuzian in tone and argument (e.g. see Dewsbury 2000; Thrift and Dewsbury 2000; McCormack 2002). Therefore, up until the time of writing, the visage of non-representational theory has usually been painted for geographers in a Deleuzian hue (e.g. see Lorimer, 2007), and it is partly for this reason, I would argue, that these theories have often been apprehended as a rupture – a breaking-away from the concerns and epistemologies of 1990s new cultural geographies, and in particular a breaking-away from an historicist, discursive and in the broadest sense Foucauldian analysis of power, space and identity.

Why this should be so is obvious from a brief consideration of Deleuzian geographies' treatment of the intertwined topics of the subject and creativity. Deleuze of course emphasises creativity, and cognate notions of invention, experimentation, connectivity and vitality; these are some of the key motifs of a philosophy of becoming and transformation, of rhizomes, assemblages and folds, which, on first encounter, appears quite alien to the critical analyses of much social science. But, this is creativity *without* subjectivity – creativity without either a subject-who-creates, or a subject-created as an end-point or culminating moment. Deleuze notably eschews most of the standard 'philosophy of the subject', from Descartes through the phenomenologies of Heidegger and Merleau-Ponty to the more recent critiques of being and presence advanced by Derrida and Levinas. Instead, *via* the elaboration of an alternative baroque lineage, his philosophy, and his work in collaboration with Félix Guattari, paints a picture of a world of incessant non-personal, pre-personal and trans-personal relations of becoming, currents of intensity and affectivities – a world which, in its ongoing creative evolution, refuses to ever really settle down into more familiar patterns of subject and object, animate and inanimate, cause and effect. One influential response by geographers has been more wholeheartedly anti-humanist, segueing Deleuzian motifs with the relational materialism of actor-network theory, and producing thereof a topological and connectivist world-picture in which creative agency is only ever an effect, never a cause, and certainly not an attribute possessed by individual human beings, these being themselves but the contingent and occasional substantiation of a ramifying and *a priori* inexaustible tangle of folds and flows which, in its endless inclusivity, scrambles conventional distinctions and demarcations of culture and nature, human and non-human (e.g see Whatmore 2002; Harrison, Thrift and Pile 2004). Another response, different in degree but not in kind, and influenced also by Deleuzian variants of cultural theory (e.g. Bennett 2001; Massumi 2002), has been to locate creativity and subjectivity firmly within the circulation of affect – that is, within bodily movement and sensation, and within the emotional atmosphere and tonality of particular situations and relations (e.g. see McCormack 2002; Anderson 2006; Bissell 2008). The agenda underpinning this response is clarified in Dewsbury et al.'s (2002, 439) statement that 'affects … are that through which subjects and objects emerge and become possible'. And Lorimer (2007, 96), in summarising the outcomes of this focus upon affective subjectivites, also clarifies the sense of a rupture from a politics of representation: 'to more traditional signifiers of identity and difference (class, gender, ethnicity, age, sexuality, disability), have been added another order of abstract descriptors: instincts, events, auras, rhythms, cycles, flows and codes'.

As these abstract descriptors imply, what is at work within these varied but primarily Deleuzian geographies is a notion of the subject as *emergent*. The subject, in other words, is no longer presumed as a locus of thought and action; nor, however does it emerge 'once and for all', to reflect back, from a now-detached perspective, upon the affective swirl from which it arose. Instead, a creative subject – a subject who senses and responds, perceives and paints – is posited as a changeable possibility, at once arising within and folding back into a processual affectivity indexed *as* creative by a Deleuzian insistence on the necessary production of the new, the experimental forging of new transformations. The articulation of such a view of the subject, beyond the 'gridlock' (Massumi 2002) imposed by structuralist thinking, was, in fact, the goal and the conclusion of my own work on walking and narrating the South West Coast Path (Wylie 2005, 245):

> Of course it is *I* who have chosen to assemble the paper in this particular way; it was *me* who experienced these things, but not as an unaffected, unaffecting atom. I am equally assembled and dispersed in this pathfinding process, I precipitate amid tones, topographies, theoretical discourses. This is a *credo* of sorts. As Rose argues, engaging post-humanist geographies will require that we 'recognise not

only the movement of deconstruction but also the movement of what Derrida (1976) calls our "dreams of presence": our dreams of being a subject'.

Derrida and the ghost of the subject

The invocation of Derrida in the above quote provides an entrance point to what might be termed a second filigree of thinking within non-representational geographies; one that I think has been, until recently, relatively hidden beneath, more dominant Deleuzian chords. This filigree could be called post-phenomenological, and more precisely post-Heideggerian, insofar as it gains direction and impetus from a nexus of writers, notably including Derrida, Levinas, Nancy and Blanchot, each of whose work is characterised by a desire to query and move away from Heidegger's conception of subjectivity in terms of being-in-the-world, while remaining, at the same time, haunted by that conception. Here, given the more nascent character of this work, I want to move somewhat from summation of extant writing to speculation regarding future directions.

Of course, to labour the point again, I am not trying to say that there is some straightforward choice to made between, broadly, Deleuzian and Derridean idioms – the citation from my own work above is perhaps indicative of the ways in which the two are often segued and shuffled together. Clark (2003) also supplies a pertinent account of the ways in which Deleuze and Derrida share a general commitment to expressing 'the play of the world'. Most recently, Dewsbury's (2007) essay on 'undoing subjectivity' in the light of Alain Badiou's philosophy could be read as another attempt to hold together a more general and all-embracing non-representational account of poststructuralism. Here, the motif of 'undecidability' invokes both Deleuze and Derrida in the cause of a geography-beyond-representation:

> As extended into the realm of the subject and the question of subjectivity, undecidability cuts into the simplistic idea within cultural geography that the subject is, in our post-colonial, post-everything, times delimited and defined by cultural 'othering' (Dewsbury 2007, 449).

This vision of a subject constituted by, and faithful to, epochal truths and events, precisely places the subject on the outside of systems of representation and signification – outside, that is, of the 'subject positions' already mapped out in advance by those systems, such that 'a subject constitutes a rupture in knowledge, in that which is already said, named, positioned and represented' (ibid., 451). However, the accent here remains, in conclusion, upon a performative and decidedly emergent subject:

> The subject is not then inherently fixed, with a predetermined array of political concerns, for performativity accounts for the ways in which subjects may resignify social practices in the very realisation of those cultural constructs and

structural determinations that make their subjecthood possible in the first place. And what I want to argue quite strongly is that there is a growing agenda to go much further than this towards a subject formation that is much more molecular, evental and material (Dewsbury 2007, 453).

In partial contrast to these 'evental', emergent tonalities, other work by non-representational geographers seems to be moving in a different direction; or, at least, seems to be working towards a quite different register for the subject, one which emphasises motifs of absence, passivity and responsibility. Harrison's (2007) critique of the concept of the relational, and its uptake in contemporary human geographies summarises the sense which underlies this move, in noting that, 'in the proliferation of biophilosophy, the unstoppable materialisation of actor networks and constructivist totalisations of the social or the cultural, few have been asking about breaks and gaps, interruptions and intervals, caesuras and tears' (Harrison 2007, 592). The irony of relational thinking, he goes on to note, is that it actually harbours within itself an irreducible non-relationality – a non-relation in no way equating with a humanist or traditionally phenomenological subject, but one reducible neither to a 'relational effect': 'the somewhat counterintuitive result of this thought is not so much (or not only) another "decentring of the subject", the dissolving of our selves and responsibilities into so many flickering networks, or the ecstatic evacuation of ipseity along lines of flight but rather (or also) a rebounding intensification of severance and responsibility' (ibid.).

Severance, responsibility, absence, passivity. Harrison's paper draws mostly on Levinas for inspiration, but here, to try to draw out some more summary themes from these nascent (emergent!) inquiries, I will focus upon Derrida (and see also Rose 2006, 2007; Wylie 2007). Like Deleuze, Derrida is committed to a re-thinking of subjectivity, and especially to a critique of Heidegger's conception of 'being'. Rather than supplying an alternate account in terms of an incessant 'becoming', however, Derrida, as is well known, places the category of 'being' under erasure: ~~being~~. In other words, there is no being 'as such', no complete, coherent and self-present subject who speaks, acts and senses. The aim here is to undercut and unsettle all philosophies – from Descartes' *cogito* to Heidegger's *Dasein* – which take as their beginning and end a given and present subjectivity. A particular target for Derrida is the notion of a subject present-to-itself; that is, an originary speaking subject, enclosed in itself, and *coinciding* with itself, with no break or opening to interrupt its voice – and no gap either on the 'inside', between its voice and its intentions – and many of his earlier analyses were notably concerned to deconstructively demonstrate how the trace of the outside, or the other, always and necessarily haunts all the attempts at self-definition that we find encapsulated in dualities such as self/other, speech/writing, culture/nature and so on.

The subject, then, is 'erased'. Subjectivity, for Derrida, *is* loss – our sense of 'ourselves' is experienced as a sort of mourning for something constitutively absent – or as a sort of yearning for something that will never arrive, a 'dream of presence', to adopt Rose's (2006) term. Yet the more crucial and poignant point is that this erasure is equally never complete. Just as there can be no full presence or self-coincidence, then so there is no pure absence, no void. Erasure leaves a mark, or trace – an absence of presence. This is why, in the broadest sense, places and landscapes are constitutively haunted by absent presences – this is why displacing and distancing are what places and landscapes *are*, in a certain way (see Dubow 2004; Wylie 2007).

I want to draw out two points from these arguments. Firstly, notions of absence/presence, haunting and erasure are becoming quite widespread, not just within 'non-representational' work, but also more generally in geographies of landscape, memory and materiality (e.g Till 2005; Edensor 2005; De Silvey 2006). What is less evident, however, is explicit discussion of the subject *as* ghost, or trace, and of the narratological, ethical and epistemological possibilities this may afford. It is not enough, in other words, to demonstrate how 'we', our places and landscapes, are haunted in various ways. The further task is to explore the implications of Derrida's suggestion that *we are the ghosts*, already and necessarily, insofar as firstly our dwelling-in-the-world is from the start displaced from itself and haunted (the stranger, the visitor, is always already inside the house), and secondly, relatedly, because our subjectivity is precisely neither pure presence nor pure absence.

The latter point I wish to make relates to recent work by geographers that addresses the 'later' Derrida's concerns with themes of ethics, friendship, hospitality and so on (e.g. Popke 2003; Barnett 2005). This work quite rightly attends to how Derrida, alongside and in dialogue and critique with Levinas, develops a certain notion of subjectivity as indelibly invested within responsibility to others, and indeed also respect for the 'otherness' of others. Therefore Derridean subjectivity is introduced *via* an ethos of hospitality-to-others, and is discussed through terms such as acknowledgement, inclusion, and cosmopolitanism. And so we find here the embryo of a quite different but still recognisably 'relational' subject, and potentially a quite distinctive cultural politics, to that proposed within Deleuzian geographies. But, working to an extent against this understanding of the later Derrida, Hillis-Miller (2007) draws a detailed distinction between his writing on the *self-other relation* (from whence come notions of unconditional hospitality and the idea that 'every other is altogether other'), and his writings on *community* (which are, Hillis-Miller claims, consistently deeply critical of the concept). *Via* a rejection of any thought of community there emerges, he argues, a Derridean subject phrased more starkly in terms such as solitude, separation, isolation and incommunicability. This is even in a way a necessary consequence of Derrida's thought: just as every self is haunted by others, so the other must remain apart, unknowable in any absolute sense, leaving us as much marooned as connected by responsibility. And what I find interesting here is the sense of isolation, distance and separation that thus characterises the self – you and me – even if it is also true that there is no singular presence, no being-as-such, no pure self-consciousness. Each of us, all of us, in some way, *always alone*. There is no world, there are only islands:

> Neither animals of different species, nor men of different cultures, nor any individual, animal or human, inhabits the same world as another, however close and similar these living individuals may be, *and the difference from one world to the other will remain forever uncrossable*, the community of the world being always constructed, simulated by a group of stabilizing positings, more or less stable, therefore also never natural, language in the broad sense, codes of traces being destined, with all the living, to construct a unity of the world always deconstructible and nowhere and never given in nature. Between my world, the 'my world'; what I call 'my world', and there is no other for me, every other world making up part of it, *between my world and every other world there is initially the space and the time of an infinite difference*, of an interruption incommensurable with all the attempts at passage, of bridge, of isthmus, of communication, of translation, of trope, and of transfer which the desire for a world and the sickness of the world, the being in sickness of the world, will attempt to pose, to impose, to propose, to stabilize. *There is no world, there are only islands* (Derrida 2003, unpublished, cited in Hillis-Miller 2007, 265-266, my emphasis).

There is a strangeness and even, as Hillis-Miller notes, a 'wildness' to this passage. I have little doubt that it will seem problematic to many, and perhaps especially so to geographers and others whose work and aspirations are premised upon connectivity or relatedness or interdependency as the preconditions of critical and political thought. But of course Derrida is not staking a claim here for solipsism, or unfettered subjectivism. The largely anti-Heideggerian rhetoric of the passage above serves instead as a sort of nagging reminder of a constitutive loss, a necessary failure-to-connect that, in its refusal of any move towards totality or being-in-the-world, is just as much part of an ethos for the subject as all of our equally necessary gestures in the other direction, towards hospitality and community.

It may be that we are on the cusp of a more sustained exploration of these sorts of ideas by geographers inspired by non-representational and post-structural theories. These collected Derridean tropes – of responsibility and hospitality, but simultaneously *also* of distance and apartness – might possibly, and alongside equally post-Heideggerian notions of being, language and subjectivity from writers such as Nancy and Blanchot, inform a new array of narratives and subjectivities within cultural geography. This would clearly be on one level a more sober and even somber non-representational subject, attentive more to critical issues around incommunicability and the limits of representation. In the concluding section which follows, however, I want to return, more pragmatically, to the ideas of writing and creativity with which I began.

Conclusion

All I ever wanted was to be taken seriously as a writer ... and so I have to acknowledge that I often feel envious of those who have, in my eyes at least, succeeded in this aspiration, by publishing work that, while remaining academic and scholarly in inspiration, reaches and speaks to other disciplines and wider reading publics. In my own domain of writing on landscape, culture, nature and biography, I am thinking especially here of a swathe of recently-published, closely-connected books on these themes in the British Isles, for example Robert Macfarlane's (2007) *The Wild Places*, Tim Robinson's (2006) *Connemara: Listening to the Wind*, and Richard Mabey's (2005) *Nature Cure*.

Each of these books is an intense literary evocation of a specific landscape (the 'wild' peaks, shores, moors and islands of Britain for Macfarlane (2007); the famine-haunted landscape of Connemara for Robinson (2006); avian East Anglia for Mabey (2005)). Each further seeks to communicate more holistic and transcendant messages about our relationship with land, with other creatures, and with our own vexed histories. And they are each also characterised by a distinctive authorial 'presence' and voice: personal experience and biography is the tiller which steers them through explorations of diverse topics. But what I have also found intriguing about these books is a sort of dissonance within their own subjectivities and narratives. On one level all these authors take what can be called a 'critical' and informed approach – they are all haunted by their awareness of the constructed, contextual and contingent nature of both our ideas of ourselves and of nature, for example. They all know that history is in some ways a story of exclusion and erasure, they all know that nature and 'the wild' are cultural ideas: this awareness is what in fact drives and animates these narratives: these are passionate books. But at the same time (although I believe Robinson's work is the most nuanced and aware of the issue), there is relatively little questioning of the writing subject here; in each case the narrative is very much that of a given self, still there in conclusion. In each case, in truth, a certain solitary romanticism of individual experience works so as to stitch the text together.

So in the end, in a way, the initial problem rears its head again. I began this chapter with a plea for 'creative geographies', and a note of my initial desire to make, here, an argument on their behalf from a perspective informed by non-representational geographies. But just as soon as this was posited, so difficulties arose – the problem we'll never solve: how to invoke 'creative writing' beyond forms of subjectivism? And so the topic of the chapter broadened commensurately, to reflect on concepts of the subject more generally. Subsequently the chapter sought to outline the post-humanist and poststructuralist conceptions of subjectivity at work within current non-representational writing, and as may be evident, my sense at present is that the post-Heideggerian philosophies of writers such as Derrida have much to offer geographers in search of a subjectivity without subjectivism, a way of creatively and critically engaging that does not lapse back into voluntarism or romanticism.

What do we want from non-representational theories? I know what I want, what I've always wanted: a set of intellectual and practical resources through which I could inhabit – that is, *ghost* – certain spaces between the critical and the creative, the 'academic' and the 'literary'. I must admit I've only ever thought of these approaches in terms of the creative or experimental licence they afford. To put this another way, I seem to always find myself in the space between the two citations from Thrift at the start of the chapter – the space between 'modes of perception which are not subject based', and a 'certain minimal humanism'. I'd like to think that this is a plausible and potentially productive space to be in, perhaps precisely because its so tense and irresolute. On the one hand, I think that to have argued for the extension of agency, affectivity and sensibility into all manner of non-human and trans-human materialities, processes and emergences is a significant achievement, (and its certainly not one accomplished wholly or even chiefly by non-representational work). On the other, I still wonder about and want to explore the minimal remainder of the self within this array of perceptions and sensibilities – even if this self is something necessarily already lost, bereft, astray, estranged, and haunted.

Its only quite recently, though, that I've begun to seriously think about what these positions might imply in terms of trying to reach out to writers and practitioners in other disciplines, and to other wider publics. When I read works such as those mentioned above by Macfarlane, Robinson and Mabey, I always wonder about the cultural geographies I also enjoy reading. Much discussion within geography on connecting with other audiences has taken place in a different register, in terms of more inclusive and engaged research practices: I am thinking especially here of recent moves towards forms of public and participatory geographies (e.g. mrs kinpaisby 2008; Fuller and Askins 2007). My abortive plea on behalf of 'creative and affective geographies', resurrected here, would perhaps have in mind more issues of publication and dissemination – the dissemination of the new forms of place-writing and landscape-writing that non-representational and poststructural theories afford.

And strange as it may seem I do believe that the ideas of the subject I've outlined here – the emergent subject, the affective subject, the ghostly subject, hospitable and lonely – provide new ways of creatively and critically writing through self, landscape, nature, history. They may seem baroque, the chapter may seem to delve into territories ever more abstract and attenuated, but it is strangely from here, and not through the resurrection of a more comfortable humanistic style, that I think we stand the best chance of writing work that compels.

Acknowledgements

Many thanks to Ben and Paul for their advice and criticism, which has helped me avoid not only interpretative errors, but also, I think, possible embarrassment with this chapter. Thanks to Ian Cook in particular for support and encouragement, and also to the members of the Historical and Cultural Geography Research Group at Exeter for reading and commenting on an earlier draft.

References

Anderson, B. (2006), 'Becoming and being hopeful: towards a theory of affect', *Environment and Planning D: Society and Space* 24(5), 733-752.

Anderson, B. and Harrison, P. (2006), 'Commentary: Questioning affect and emotion', *Area* 38(3), 333-335.

Barnett, C. (2005), 'Ways of relating: hospitality and the acknowledgement of otherness', *Progress in Human Geography* 29, 1-17.

Barnes, A. (2007), 'Geo/graphic mapping', *Cultural Geographies* 4, 139-147.

Bennett, J. (2001), *The Enchantment of Modern Life: Attachments, Crossings and Ethics* (Princeton: Princeton University Press).

Bissell, D. (2008), 'Comfortable bodies: sedentary affects', *Environment and Planning A* 40(7), 1697-1712.

Blacksell, M. (2005), 'A walk on the South West Coast Path: a view from the Other side', *Transactions of the Institute of British Geographers* 30(4), 518-520.

Butler, T. and Miller, G. (2005), 'Linked: a landmark in sound, a public walk of art', *Cultural Geographies* 12, 77-8.

Castree, N. and Macmillan, T. (2004), 'Old news: representation and academic novelty', *Environment and Planning A* 36(3), 469-480.

Clark, N. (2003), 'The play of the world' in Pryke, M. et al. *Using Social Theory: Thinking Through Research* (London: Sage), 28-46.

Cresswell, T. (2002), 'Bourdieu's geographies: in memoriam', *Environment and Planning D: Society and Space* 20, 379-383.

Davies, G. and Dwyer, C. (2007), 'Qualitative methods: are you enchanted or are you alienated?', *Progress in Human Geography* 31, 257-266.

DeLanda, M. (2002), *Intensive Science and Virtual Philosophy* (New York: Continuum).

DeSilvey, C. (2007), 'Salvage memory: constellating material histories on a hardscrabble homestead', *Cultural Geographies* 14, 401-424.

Dewsbury, J-D (2000), 'Performativity and the event: enacting a philosophy of Difference', *Environment and Planning D: Society and Space* 18, 473-496.

—— (2003), 'Witnessing Space: knowledge without contemplation', *Environment and Planning A* 35(11), 1907-1933.

—— (2007), 'Unthinking subjects: Alain Badiou and the event of thought in thinking politics', *Transactions of the Institute of British Geographers* 32(4), 443-459.

Dewsbury, J-D, Wylie, J., Harrison, P., and Rose, M., (2002), 'Enacting Geographies', *Geoforum* 32, 437-441.

Dubow, J. (2004), 'The mobility of thought: Reflections on Blanchot and Benjamin', *Interventions: The International Journal of Postcolonial Studies* 6(2), 216-228.

Edensor, T. (2005), *Industrial Ruins: Space, Aesthetics and Materiality* (Oxford: Berg).

Harrison, P. (2007), 'How shall I say it...? Relating the non-relational', *Environment and Planning A* 39(3), 590-608.

Harrison, S., Pile, S., and Thrift, N. (eds) (2004), *Patterned Ground: Entanglements of Nature and Culture* (London: Reaktion Books).

Hillis-Miller, J. (2007), 'Derrida enisled', *Critical Inquiry* 33(Winter), 248-276.

Latour, B. (2005), *Reassembling the Social: An Introduction to Actor-Network Theory* (Oxford: Clarendon).

Lorimer, H. (2005), 'Cultural geography: the busyness of being "more-than-representational"', *Progress in Human Geography* 29(1), 83-94.

—— (2006), 'Herding memories of humans and animals', *Environment and Planning D: Society and Space* 24, 497-518.

—— (2007), 'Cultural geography: worldly shapes, differently arranged', *Progress in Human Geography* 31(1), 89-100.

Mabey, R. (2005), *Nature Cure* (London: Chatto & Windus).

Macfarlane, R. (2007), *The Wild Places* (London: Granta).

Massey, D. (2006), 'Landscape as a provocation: Reflections on moving mountains', *Journal of Material Culture* 11, 33-48.

Massumi, B. (2002), *Parables for the Virtual* (Durham: Duke University Press).

McCormack, D. (2002), 'A paper with an interest in rhythm', *Geoforum* 33, 469-485.

—— (2003), 'An event of geographical ethics in spaces of affect', *Transactions of the Institute of British Geographers* 28, 458-508.

—— (2004), 'Drawing out the lines of the event', *Cultural Geographies* 11, 211-220.

—— (2006), 'For the love of pipes and cables: a response to Deborah Thien', *Area* 38(3), 330-332.

Nash, C. (2000), 'Performativity in practice: some recent work in cultural geography', *Progress in Human Geography* 24(4), 653-664.

Osborne, T. (2003), 'Against Creativity: a philisitine rant', *Economy and Society* 32(4), http://www.informaworld.com/smpp/title~content=t713685159~db=all ~tab=issueslist~branches=32 – v32, 507-525.

Patton, P. and Protevi, J. (eds) (2003), *Between Deleuze and Derrida* (London: Continuum).

Popke, E.J. (2003), 'Poststructuralist ethics: subjectivity, responsibility and the space of community', *Progress in Human Geography* 27, 298-316.

Quoniam, S. (1988), 'A painter, geographer of Arizona', *Environment and Planning D: Society and Space* 6(1), 3-14.

Robinson, T. (2006), *Connemara: Listening to the Wind* (London: Penguin).

Rose, M. (2004), 'Re-embracing metaphysics', *Environment and Planning A* 36, 461-468.

—— (2006), 'Gathering "dreams of presence": a project for the cultural landscape', *Environment and Planning D: Society and Space* 24, 537-554.

—— (2007), 'The problem of power and the politics of landscape: stopping the Greater Cairo ring road', *Transactions of the Institute of British Geographers* 32(4), 460-476.

Rose, M. and Wylie, J. (2006), 'Animating landscape', *Environment and Planning D: Society and Space* 24, 475-479.

Thien, D. (2005), 'After or beyond feeling? A consideration of affect and emotion in geography', *Area* 37, 450-454.

Thrift, N. (2004), 'Intensities of feeling: the spatial politics of affect', *Geografiska Annaler Series B* 86, 57-78.

—— (2008), *Non-Representational Theory: Space, Politics, Affect* (Routledge: London).

Thrift, N. and Dewsbury, J-D (2000), 'Dead Geographies – and how to make them live', *Environment and Planning D. Society and Space* 18, 411-432.

Till, K. (2005), *The New Berlin: Memory, Politics, Place* (Minneapolis: University of Minnesota Press).

Tolia-Kelly, D. (2007), 'Fear in Paradise: the affective registers of the English Lake District landscape revisited', *Senses and Society* 2, 329-352.

Whatmore, S. (2002), *Hybrid Geographies: Natures, Cultures, Spaces* (London: Sage).

Wylie, J. (2002), 'An essay on ascending Glastonbury Tor', *Geoforum* 33, 441-454.

—— (2005), 'A single day's walking: narrating self and landscape on the South West Coast Path', *Transactions of the Institute of British Geographers* 30, 234-247.

—— (2007), 'The spectral geographies of W.G. Sebald', *Cultural Geographies* 14, 171-188.

Yusoff, K. (2007), 'Antarctic exposure: archives of the feeling body', *Cultural Geographies* 14, 211-233.

PART II
Representation

Chapter 6

Representation and Difference

Marcus A. Doel

Difference is not and cannot be thought in itself, so long as it is subject to the requirements of representation.

Gilles Deleuze, *Difference and Repetition*

The evil demon of language resides in its capacity to become object, where one expects a subject and meaning.

Jean Baudrillard, *The Ecstasy of Communication*

Ordinarily, representation is bound to a specific form of repetition: the repetition of the same. Through representation, what has already been given will come to have been given again. Such is its fidelity: to give again, and again, what has always already been given, without deviation or departure. Such is its fidelity to an original that is fated to return through a profusion of dutiful copies; an original whose identity is secured and re-secured through a perpetual return of the same, and whose identity is threatened by the inherent capacity of the copy to be a deviant or degraded repetition, a repetition that may introduce an illicit differentiation in the place ostensibly reserved for an identification.

The problematic of representation is constrained to keep its repetitions, reproductions, and copies in order: to ensure that they do nothing more than return originals, identities, and givens. When all is said and done, one rightly expects representation to re-present the eternal identity of the same. Ideally, re-presentation should give back what has already been given. Representation should resemble rather than dissemble. There are, however, two fatal flaws with this despotic characterization of representation. First, to repeat is to differ and defer. The same that returns ineluctably returns otherwise. Representation is inevitably transformation and differentiation, even when it is a transposition of the semblance of one medium into that of another. By necessity, it brings forth more than the same. Representation is always in excess of itself (Lyotard 1990, 1998). Second, given that originals, identities, and the same are repeatable, and can only be secured and affirmed through repetition, they are always already marked by difference and differentiation. A signature, for example, is the quintessential mark of originality, identity, singularity, and the same, and yet it is of the essence of a signature that it be repeatable, a repeatability that distributes its originality, identity, singularity, and similitude across a multitude of times, spaces, and contexts. Without repetition and difference, identity could not be established. The same, then, is not what holds together through representation, it is what is halved together – differed and deferred

– through repetition, and in so being it is placed under serial erasure (Derrida 1988, 1991). This is the great lesson of structuralism and poststructuralism. No irrelative position exists. All re-presentation is differentiation. Representation, even when it is ostensibly devoted to a return of the same, is transformation (Doel and Clarke 2007).

Non-representational theory should not, then, be understood as a refusal of representation *per se*. It is a refusal of representation yoked to the problematic of a repetition of the same. Herman Melville's short story, 'Bartleby, the Scrivener', is a wonderful illustration of what is at stake in declining this problematic. A scrivener is a professional writer, a scribe or copyist, a clerk, secretary or amanuensis, or a notary. Bartleby is employed within a law firm solely for the purpose of copying legal documents, which he does with all due diligence. Imagine his horror, then, when his employer requests him to participate not in copying, but in verifying – and thereby valorizing – the fidelity of that which has been transcribed. As a comparative method, verification ensures that what would otherwise be singular (a copy that may or may not accord with an original) is rendered commensurable, exchangeable, and substitutable (a faithful reproduction of the same). Verification is therefore not only authorization; it is valorization. Verification underwrites the proper identity of a piece of writing and thereby authorizes its issuance as an exchange-value and a use-value. A copy may enter into licit circulation only after it has been verified and valorized.

> Being much hurried to complete a small affair I had in hand, I abruptly called to Bartleby. In my haste and natural expectancy of instant compliance, I sat with my head bent over the original on my desk, and my right hand sideways, and somewhat nervously extended with the copy, so that, immediately upon emerging from his retreat, Bartleby might snatch it and proceed to business without the least delay.
>
> In this very attitude did I sit when I called to him, rapidly stating what it was I wanted him to do – namely, to examine a small paper with me. Imagine my surprise, nay, my consternation, when, without moving from his privacy, Bartleby, in a singularly mild, firm voice, replied, 'I would prefer not to' (Melville 1968, 11).

As soon as the lawyer invites Bartleby to participate in verification and valorization ('business') all entreaties for him 'to be a little reasonable' (Melville 1968, 25) will elicit the same indefinite and dysfunctional reply: 'I would prefer not to'. 'You *will* not?' enquires the lawyer. 'I *prefer* not' retorts Bartleby (Melville 1968, 17). The advance withdrawal of Bartleby's refrain 'hollows out an ever expanding zone of indiscernibility or indetermination between some nonpreferred activities and a preferable activity. All particularity, all reference is abolished. The formula annihilates "copying"' (Deleuze 1997, 71). Having eschewed the problematic of the repetition of the same, the only thing that remains is suspense. Bartleby has found a way of working without labouring, of copying without resembling or dissembling, of engendering sign-values without exchange-values or use-values (Baudrillard 1981). For the lawyer, the real of the copy is taken to be hyperreal: '*that of which it is possible to give an equivalent reproduction*' (Baudrillard 1983, 146). For Bartleby the scrivener, the real of the copy is simulacral: 'different relates to different *by means of* difference itself. ... no *prior identity*, no *internal resemblance*' (Deleuze 1994, 299).

Like Bartleby's preference 'not to', the phrase 'non-representational theory' is an inelegant rallying cry for those who wish to have done with a certain kind of representation (Thrift 2004). Representation obviously casts a very long shadow over everything that we do. Representation is second nature. Almost everyone expects one medium to rehearse what has already been given in some other medium. A face is painted. A sunset is photographed. A room is described. What has already taken place in one medium passes over into what takes place in another medium. A presentation is re-presented, and so there is an original and a copy, and the relationship between the one and the other lends itself to an evaluation in terms of the degree of similarity and resemblance. Consequently, to be spellbound by representation is to be bound by a certain duty: representation has an obligation to *give back* a true semblance of that which it re-presents – and woe betide any representation that falls short in this duty. But let us be clear on the following point. It is not necessary for the original presentation, the original medium, to be real or worldly. A great deal of unnecessary confusion has been caused by people too hastily equating presentation with reality and re-presentation with language, etc. Re-presentation is simply a relationship of semblance between media within which the same returns. One medium repeats another. One medium is transposed into another medium. But which medium will serve as the original and which the copy is entirely arbitrary and contingent. Now, it should go without saying that all of this is laughable. There are media, to be sure, each with a specificity all of its own, but the transposition of one medium into another is not duty bound to be governed by resemblance. The face that is painted is not the face that one faces. The sunset that is photographed is not the sun that sets. The room that is described is not the room that is inhabited. Each is estranged from the other. Each takes place according to a trajectory that is essentially oblivious to the other. Re-presentation, if there is such a thing, is a differentiation: a bifurcation in the order of things – a bifurcation that is ramified in exact proportion to the number of media that are mobilized. Accordingly, that inelegant phrase 'non-representational' does not so much signify a stance that is opposed to the enslavement of representation to the duty of resemblance, but rather an affirmation of the pulverization of re-presentation and the proliferation of media, each of which is liberated – yes, *liberated* – to follow its own path (cf. Deleuze 2003; Latour and Weibel 2002). Faces face. Suns set. Rooms room. Painters paint. Photographers photograph. Writers write. And nothing obliges each to re-present the imperatives of any other. Nevertheless, each medium may accommodate itself to the traits of another medium, but in so doing each always accomplishes this in its own terms. When one paints a face, the paint takes command. When one photographs a sunset, the camera takes charge. And when one describes a room, the words dictate what takes place.

Simplifying to the extreme, the phrase 'non-representational' announces the bypassing of the subservient relationship between one medium and another. Hence the fact that 'non-representational' theoretical practices have been advanced by those who consider themselves to be radical: those who refuse to be burdened by re-presenting that which is always already given by another medium and those who seek to unleash the disruptive potential of all manner of media. Faces. Sunsets. Rooms. Paintings. Photographs. Writings. Et cetera. And never the twain shall meet. This is what it means to bypass the re-presentation of the same.

Non-representational styles of thought treat everything usually regarded as representational (e.g. words, concepts, ideas, perceptions, and images) as events in their own right. They 'transform the relations of representation *against* representing, against the universalizing conditions of exchange; representation held to use (a definition of Brechtian distanciation), that is, to division, disunity, disturbance of the (social) contract' of faithful reproduction (Heath 1981, 242). Simplifying to the extreme, non-representational styles of thought collapse the longstanding separation of the world, which is reputedly over there, somewhere in the Real, from its re-presentation, supposedly over here, somewhere in the imaginary and the symbolic, exemplified by speech and writing. By refusing to yield to the onto-theological dogma of re-presentational second comings (i.e. the dutiful copying in words, concepts, and pictures of revered originals, such as being, identity, intention, reality, sense, truth, and value), non-representational styles of thought foreground the *eventfulness* of 'a moment-ary world … which must be acted into', and 'not a contemplative world' that should be held at a reverential or critical distance (Thrift 2000, 217). Every actuality is accompanied by the virtual rather than pre-empted by the possible (Deleuze 1994). Actuality does not emerge from the future and come to rest in the past (possibility → actuality → memory). It precipitates out of the differential relations of creative encounters as one medium yields to another. Past, present, and future happen once and for all, as one of the tribes inhabiting Jorges Luis Borges' fictive world of Tlön appreciates: '[T]he present is undefined and indefinite, the future has no reality except as present hope, and the past has no reality except as present recollection' (Borges 1999, 74). Little wonder, then, that non-representational styles of thought should have moved so rapidly with their 'simulacra' and 'pure means' onto the processual terrain of actions, situations, events, praxis, performance, and *phrōnesis* – 'a commitment to opening up the moment' through '*effectivity* rather than representation' (Thrift 2003, 2023; and Thrift 2000, 216, respectively).

What does this eventfulness mean in practice? It means that the world is not given in advance. It is not always already suspended in reserve as a set of countless possibilities or eternal and ethereal Platonic forms, which simply await their successive realization in the course of everything that happens. The world does not take place as the serial realization of possibilities and forms, which would make of the world and its occurrence nothing but an impotent repetition of the same and a dutiful re-presentation of the identical, such that the world would amount to little more than the fleeting and ephemeral passage of a succession of degraded realizations, materializations, and manifestations of what is always already given and accounted for in a higher dimension. The world that takes place is not simply the addition of reality to a prefigured possibility, an immaterial possibility that would be realized by momentarily dressing it in the garb of materiality. The world that returns is never the same world. What returns with the taking place of the world is neither the same, nor the identical, nor the possible – but the event. Indeed, 'the event (is what) deconstructs' (Derrida 1988, 109). The world is not suspended in the cleft between the possible and its realization, or the same and its reproduction, but in the repetition of reconstruction and remaking. What returns with the event of the world is difference. Non-representational theory attends to this difference.

Ordinarily, difference is conditioned by and derived from identity, either in terms of the calibrated dissimilarity between identities (the diverse) or the negation of identities (the negative). 'In other words, we never think difference in itself' (Deleuze 2006, 42). Yet precisely because no irrelative position exists, identity cannot ground difference. Identity can only be established after the fact, by a twist of re-affirmation that will have ruined it in advance. Identity will not have taken place. We must therefore seek a ground for difference in difference itself: 'Not to maintain together the disparate, but to put ourselves there where the disparate itself *holds together*, without wounding the dis-jointure, the dispersion, or the difference, without effacing the heterogeneity of the other', as Jacques Derrida (1994, 29) once put it. One of the great lessons of poststructuralism is that difference, knowing nothing of identity, is 'inexplicable'. Difference is always displaced from itself, and only ever sensed through its effects. 'In thinking it *as such*, in recognizing it, one misses it', writes Derrida (1982, xi-xii). 'One reappropriates it for oneself, one disposes of it, one misses it, or rather one misses (the) missing (of) it, which … always amounts to the same'. Difference is always already otherwise: displaced, disfigured, disguised, and deconstructed. 'Difference is not diversity', notes Gilles Deleuze (1994, 222). 'Diversity is given, but difference is that by which the given is given, that by which the given is given as diversity'. The world therefore has two asymmetrical and incommensurable halves: *diversity* on the one hand (*n*), and *difference* on the other hand (*n*-1). The two halves always sum up to the more or the less (i.e. the dissimilar and the unequal), but never to a whole or a one (i.e. the identical and the equalized).

> Difference is explicated, but in systems in which it tends to be cancelled; this means only that difference is essentialy implicated, that its being is implication. For difference, to be explicated is to be cancelled or to dispel the inequality which constitutes it. ... We cannot conclude from this that difference is cancelled out, or at least that it is cancelled in itself. It is cancelled in so far as it is drawn outside itself, *in* extensity and *in* the quality which fills that extensity. However, difference creates both this extensity and this quality. ... Difference of intensity

is cancelled or tends to be cancelled in this system, but it creates this system by explicating itself (Deleuze 1994, 228).

In place of the dialectical play of identity and difference under the auspices of the repetition of the same and the negative realization of the possible (e.g. being, non-being, and becoming or position, negation, and negation of the negation), Deleuze offers the asymmetrical play of difference and diversity under the auspices of the repetition of the divergent and the affirmative actualization of the virtual (Deleuze and Parnet 2002). While realization adds nothing to the possible, save for its ephemeral (re)appearance in the world through limitation and selection, actualization creates something new through integration and resolution. So, rather than posit identities correlated to differences, Deleuze posits two forms of difference: *difference in itself* as the generative power of differential repetition and creative evolution, and *explicated difference* as the resulting disparateness that holds together by being halved together. Difference in itself is intensive, unqualified, and implicated – a virtual power of differen*t*iation. Explicated difference is extensive, qualified, and unfolded – an actual power of differen*c*iation through which difference in itself is distributed, equalized, cancelled, and resolved. The two, unequal halves of difference (differen*t/c*iation) relate to one another without resembling one another, and each is fully real: the virtual is real without being actual, and the actual is real without being virtual.

> The reality of the virtual consists of the differential elements and relations along with the singular points which correspond to them. The reality of the virtual is structure. We must avoid giving the elements and relations which form a structure an actuality which they do not have, and withdrawing from them a reality which they have (Deleuze 1994, 209).

The halving together of virtual–actual, differen*t*iation–differen*c*iation, implication–explication, and intensity–extensity does not re-present a static system that is fated to return nothing but the same. Instead, this halving together affirms a dynamic system whose dissimilation opens up to a return of the different. 'Difference and repetition in the virtual ground the movement of actualisation, of differenciation as creation' (Deleuze 1994, 212). Hereinafter, the repetition of the same and the identical gives way to the repetition of the different and the divergent, 'distributing the disparities in a multiplicity' (Deleuze 1994, 50). The interminable play of difference and repetition, of differing and deferring, which Derrida (1982) famously called *différance*, opens a fissure of dissimilation and dissemination that puts everything that happens under erasure (Derrida 1981).

> One of these repetitions is of the same, having no difference but that which is subtracted or drawn off; the other is of the Different, and includes difference. One has fixed terms and places; the other essentially includes displacements and disguise. One is negative and by default; the other is positive and by excess. ...

One involves succession in fact, the other coexistence in principle. One is static; the other dynamic. One is extensive, the other intensive. ... One is a repetition of equality and symmetry *in the effect*; the other is a repetition of inequality as though it were a repetition of asymmetry *in the cause* (Deleuze, 1994, 287).

When difference no longer issues from identity as a secondary effect or derivative it is spared from negative determination (no longer is difference not). 'The negative appears neither in the process of [virtual] differen*t*iation nor in the process of [actual] differen*c*iation' (Deleuze 1994, 207). This is why difference is *affirmed* as inexplicable cause (virtual differen*t*iation) and explicated effect (actual differen*c*iation), and why difference *returns*, eternally, since actual differen*c*iation never exhausts virtual differen*t*iation. 'The Negative does not return. The Identical does not return. The Same and the Similar, the Analogous and the Opposed, do not return. Only affirmation returns – in other words, the Different, the Dissimilar' (Deleuze 1994, 299). In short, the world remains open. 'The powers of repetition include displacement and disguise, just as difference includes [the] power of divergence and decentring' (Deleuze 1994, 288). Through differen*t*iation, the virtual renders the real problematic: evental. Through differen*c*iation, the actual renders the real qua problematic solvable: eventful. Yet there is no final solution that would exhaust the world's problematic. Through differen*t*/*c*iation, difference is always called back into play, displacing, redistributing, and transforming everything that has been given. This is why the world takes place as an event, why the event is perpetually unsettling and unsettled, and why the world qua event has the character of the eternal return.

The eternal return affirms difference, it affirms dissemblance and disparateness, chance, multiplicity and becoming. ... The eternal return eliminates precisely all those instances which strangle difference and prevent its transport by subjecting it to the quadruple yoke of representation. ... namely, the Same and the Similar, the Analogue and the Negative (Deleuze 1994, 300).

The eternal return is a deconstructive, differential repetition: repeating/altering, dividing/displacing, extracting/grafting, and differing/deferring in accordance with an iterability and supplementarity that 'ties repetition to alterity' (Derrida 1988, 44). 'Eternal return is tied not to a repetition of the Same, but to a transmutation', writes Deleuze (2006, 207). Through the eternal return, the two-fold repetition of difference qua virtual–actual differen*t*/*c*iation returns the repetition of the same whence it came: the simulacrum.

Simulacra are those systems in which different relates to different *by means of* difference itself. What is essential is that we find in these systems no *prior identity*, no *internal resemblance*. It is all a matter of difference in the series, and of differences of difference in the communication between series. What is

> displaced and disguised in the series cannot and must not be identified, but exists
> and acts as the differentiator of difference (Deleuze 1994, 299-300).

This characterization of the simulacrum may be aligned with Jean Baudrillard's definition of the real in the era of simulation – 'The very definition of the real becomes: *that of which it is possible to give an equivalent reproduction.* ... At the limit of this process of reproducibility, the real is not only what can be reproduced, but *that which is always already reproduced.* The hyperreal' (Baudrillard 1983, 146) – provided that one understands that simulation has abolished the distinction between originals and copies.

> The simulacrum is not a degraded copy. It harbors a positive power which denies
> *the original and the copy, the model and the reproduction.* At least two divergent
> series are internalized in the simulacrum – neither can be assigned as the original,
> neither as the copy. ... Resemblance subsists, but it is produced as the external
> effect of the simulacrum, inasmuch as it is built upon divergent series and makes
> them resonate. Identity subsists, but it is produced as the law which complicates
> all the series and makes them all return to each one in the course of the forced
> movement. ... The same and the similar no longer have an essence except as
> *simulated*, that is as expressing the functioning of the simulacrum. There is
> no longer any possible selection [amongst reputedly well-founded claimants
> and supposedly unfounded pretenders to participation in the original]. ... Far
> from being a new foundation, it engulfs all foundations, it assures a universal
> breakdown (*effondrement*), but as a joyful and positive event, as an un-founding
> (*effondement*) (Deleuze 1990, 262-263).

Virtual–actual differen*t/c*iation, eternal return, and simulacra all affirm differential repetitions, divergent series, and decentred circles within the eventual taking place of the world.

> Only the divergent series, insofar as they are divergent, return: that is, each
> series insofar as it displaces its difference along with all the others, and all series
> insofar as they complicate their difference within the chaos which is without
> beginning or end. The circle of the eternal return is a circle which is always ex-
> centric in relation to an always decentered centre. ... [W]hat is excluded, what
> is *made not to* return, is that which presupposes the Same and the Similar, that
> which pretends to correct divergence, to recenter the circles or order the chaos,
> and to provide a model or make a copy (Deleuze 1990, 264-265).

Accordingly, structuralism supplements the age-old dialectic of the *real* and the *imaginary* with a third order, that of the *symbolic*. The advent of the symbolic ruins representation, and decisively envelopes the play of the real and the imaginary within the simulacrum. '[S]tructure is at least triadic, without which it would not "circulate" – a third at once unreal, and yet not imaginable' (Deleuze 2004, 172).

Structure is virtual differen*t*iation. '[I]t is more a combinatory formula supporting formal elements which by themselves have neither form, nor signification, nor representation, nor content, nor given empirical reality, nor hypothetical functional model, nor intelligibility behind appearances' (Deleuze 2004, 173). The symbolic is a structured space of differential relations, positional elements, and singular points, whose transcendental topology determines whatever comes to occupy and traverse it. Deleuze draws several consequences. Value, meaning, and sense arise as *positional effects* derived from the combination and permutation of elements that are themselves bereft of value, meaning, and sense. Precisely because they are structural effects, there is always an *excess*, overproduction, and over-determination of value, meaning, and sense. Sense and nonsense are always in *play*, and this play *takes place* amid the different/ciation of manifold series: 'every structure is serial, multi-serial, and would not function without this condition', says Deleuze. 'Indeed, the terms of each series are in themselves inseparable from the slippages [*décalages*] or displacements that they undergo in relation to the terms of the other. They are thus inseparable from the variation of differential relations' (Deleuze 2004, 182-183).

Play, slippage, and displacement are essential characteristics of structural and symbolic space. Little wonder, then, that metaphor and metonymy should have such a prominent role in structuralism, since 'they express the two degrees of freedom of displacement, from one series to another and within the same series' (Deleuze 2004, 184). Play, slippage, and displacement occur because of the existence of a differential element that is 'always displaced in relation to itself' (Deleuze 2004, 185). This element is the 'empty square' that is perpetually on the move within the structure, and without which nothing would circulate and take place. Its errancy traverses and transforms the structured space of differential relations, positional elements, and singular points. By way of the play of the 'empty square' that lends it consistency, every structure takes place as a 'destabilization on the move' (Derrida 1988, 147). When all is said and done, structure cannot be pinned or penned down: 'difference is behind everything, but behind difference there is nothing. Each difference passes through all the others; it must "will" itself or find itself through all the others' (Deleuze 1994, 57). Difference cannot be pinned/penned down because it is forever taking place. Structure is never static. It is a transformer. In other words, in a structured space 'the total milieu ... is constantly being reinscribed and thrown back into play' (Derrida 1981, 339. Cf. Deleuze and Guattari 1986, 1988). So, different/ciation is not what happens *to* a structure. Different/ciation *is* structure. Structure is always already the immanence of complication–explication, virtualization–actualization, and differential repetition.

Now, when it comes to 'the strictest possible determination of the figures of play, of oscillation, of undecidability, which is to say, of the *différantial* conditions of' that which takes place (Derrida 1988, 145), there is a tendency to turn to signs. Conventional approaches to representation assume that the sign can be pinned down: to a realm of meaning, sense, and intention on the one hand and a domain of matter, substance, and things on the other hand. As sign users, human beings are essentially semiotic creatures: not only because they make their way in the world through signs, without which they would be both ontologically and epistemologically lost, but also because they are a special (spectral) effect of the sign. To be human is to take (one's) place amid the always already structured play of signs. Bereft of signs, human beings and their worlds would cease to exist. Without signs there would be no-thing, no-one, and no-body, only non-sense. What, then, is a sign? A sign is not simply a vehicle for the conveyance of sense. It is not simply a material embodiment of meaning and reference. Whenever and wherever a sign signifies, it necessarily opens an abyss. This is the abyss between the signifier (S) on the one hand (e.g. a dribble of ink or an emission of sound) and the signified (s) on the other hand (e.g. a meaning that one perceives in the ink or a sense that one hears in the sound); the abyss between the sensible matter of meaning and the intelligible meaning of matter. The sign is the exemplary empty square, whose displacement, errancy, and ex-centricity in relation to itself enables the two divergent series of signifiers and signifieds to take place: not through realization and reproduction, but through actualization and repetition.

Three key aspects of the sign need to be emphasized. First, signifiers and signifieds are mutually constituted: 'no S without an s, no s without an S' (Olsson 2007, 106), and yet they necessarily fail to coincide. Signifiers and signifieds are arrayed into two series that always sum up to the more or the less (i.e. the dissimilar and the unequal), but never to a whole or a one (i.e. the identical and the equalized). The two series are fated to slip past one another, like tectonic plates (Deleuze 1988). This is graphically illustrated by the infamous Lacanian algorithm: S/s. Signifier and signified take place as a differential relation. Second, precisely because of this asymptotic non-coincidence, signifiers and signifieds are enchained in a series of displacements and substitutions without origin or end. This is illustrated by the Derridean notion of *différance*, a neologism that foregrounds the perpetual differing and deferring of meaning, reference, and sense (Derrida 1978). Third, signifieds (and the referents that they evoke) slip beneath the play of signifiers and in so doing they are destined to be placed under erasure: signifiers do not re-present signifieds, since signifieds are the spectral effect of the occult play of signifiers. This is illustrated by the Saussurean Bar: '————'.

Drawing upon his long-standing penchant for spatial science, which employs mathematics for theoretical affect, Gunnar Olsson takes the Saussurean Bar, '————', this 'dividing/unifying divisor between Signifier and signified' (Olsson 2007, 240), and forces it to approach the limit conditions of zero and infinity, becoming 'minimally thin' and 'maximally thick' in the process. In the case of the former, the wafer-thin Bar between the signifiers and signifieds ('————') appears to erase itself in the congruence of the speech act (e.g. saying as doing, doing as saying) and the repetition of the same. The non-coincidence, spectral play, and slippage of signification is artificially arrested, so that signifiers and signifieds appear to merge seamlessly with one another – and all appears well with the wor(l)d. In the case of the latter, however, the thickened Bar ('■') appears to ossify, so that signifiers and signifieds are forcibly barred from coming into contact with one another – whereupon signification falters and stalls.

Accordingly, the thinning and thickening of the Bar appears to disclose two, heterogeneous limits: one of which marks the boundary between the Bar and the play of signifiers; while the other marks the boundary between the Bar and the play of signifieds. Yet this disclosure is an illusion because each of the series can only be put into play by way of the empty square. Signifiers and signifieds are not the limits of the Bar (a.k.a. the limits of language, the prison-house of language, and the limits of one's world). They are the faces of the empty square. The blankness of the Bar cannot but fail to make sense, and in so doing the Bar dramatizes the baselessness from which sense is withdrawn and into which sense continuously plunges. Signification traverses the abysmal structure of the Bar, and in so doing sense takes place: it is disjoined and dislocated. Sense takes place entirely within the upper and lower limits of the Saussurean Bar '——', between its cavity wall '=', and around its central abyss, which is the ex-centric empty square '□', whose errancy de/structures and de/constructs that which takes place. When all is said and done, it is because of the empty square that a world comes to take place in the manifold, simulacral play of heterogeneous series: not as a return of the same, but as a return of difference, divergence, and differen*t/c*iation.

> Difference must become the element, the ultimate unity; it must therefore refer to other differences which never identify it but rather differenciate it. Each term of a series, being already a difference, must be put into a variable relation with other terms, thereby constituting other series devoid of centre and convergence. Divergence and decentring must be affirmed in the series itself. Every object, every thing, must see its own identity swallowed in difference, each being no more than a difference between differences. Differences must be shown *differing* (Deleuze, 1994, 56).

The world takes place in difference. In difference, the world never forms a whole, an identity or a self-same. The One is always subtracted from the in difference of the world (n-1). This is why the notion of multiplicity has become pivotal to poststructuralism (Badiou 2000, 2005, 2009; Deleuze and Guattari 1986, 1988). Multiplicity is not the disparate that is *held together*, since this conception would still retain a supplementary dimension within which the multiplicity is summed up as a whole (transcendence). Multiplicity is the in different that *holds together* without reference to something other than itself (immanence). It is therefore not enough for non-representational theory to affirm the taking place of the world as an eventful simulacrum (becoming). It must also affirm the taking place of the world as an in different multiplicity (rhizome). We should therefore be extremely cautious about the growing popularity of actor-network theory (Latour 2005), since its ontology of association remains enamoured by the disparate that is *held* together through addition ($n + 1$), rather than by the in different ontology of multiplicity that *holds* together through subtraction ($n - 1$) (Doel 2009; Cf. Badiou 2009).

As a parting gesture, it would be tempting to mix metaphors and suggest that non-representational theory has two sides, two faces, and two cutting edges: actor-network theory on the one hand (the 'more' of association: being-multiple), and poststructuralism on the other hand (the 'less' of subtraction: becoming-other). This would be a grave error, however, since in the round the more and the less form a single side, a single face, and a single edge on a Möbius strip of complication–explication, different/ciation, and differential repetition. Our world is given, fully – as an open hole (in which nothing, not even nothing, is lacking). What takes place is the taking place, and what becomes of place is in the taking (Doel 2008). The world is indeed a double take, but not in terms of originals and copies: '*everything divides, but into itself*' (Deleuze and Guattari 1984, 76). It is the eternal return of the event-full. No less. No more. In different. When all is said and done, then, non-representational theory, like the world, must be suspended in the void: not as a Platonic representation of the same – *Once and Again* – but as a Kafkaesque repetition of the in different – *And, But*. Non-representational theory needs 'a hinge-logic, a hinge-style' (Lyotard 1990, 123; Cf. Deleuze and Guattari 1986, 1988; Derrida 1981; Doel 1999). This is why it should align itself with poststructuralism.

References

Badiou, A. (2000), *Deleuze: The Clamour of Being*. Trans. L. Burchill (Minneapolis: University of Minnesota Press).

—— (2005), *Being and Event*. Trans. O. Feltham (London: Continuum).

—— (2008), *Number and Numbers*. Trans. R. Mackay (Cambridge: Polity).

—— (2009), *Logics of Worlds: Being and Event 2*. Trans. A. Toscano (Continuum: London).

Baudrillard, J. (1981), *For a Critique of the Political Economy of the Sign*. Trans. C. Levin (St. Louis: Telos).

—— (1983), *Simulations*. Trans. P. Foss, P. Patton, P. Beitchman (New York: Semiotext(e)).

—— (1988), *The Ecstasy of Communication*. Trans. B. Schutze, C. Schutze (New York: Semiotext(e)).

Borges, J.L. (1999), *Collected Fictions*. Trans. A. Hurley (Harmondsworth: Penguin).

Deleuze, G. (1988), *Foucault*. Trans. S. Hand (Minneapolis: University of Minnesota Press).

—— (1990), *The Logic of Sense*. Trans. M. Lester, C. Stivale (London: Athlone).

—— (1994), *Difference and Repetition*. Trans. P. Patton (London: Athlone).

—— (1994), *Francis Bacon: The Logic of Sensation*. Trans. D.W. Smith (London: Continuum).

—— (1997), *Essays Critical and Clinical*. Trans. D.W. Smith, M.A. Greco (Minneapolis: University of Minnesota Press).

—— (2004), *Desert Islands and Other Texts: 1953-1974*. Trans. M Taormina (New York: Semiotext(e)).

—— (2006), *Two Regimes of Madness: Texts and Interviews 1975-1995*. Trans. A. Hodges, M. Taormina (New York: Semiotext(e)).

Deleuze, G. and Parnet, C. (2002), *Dialogues II*. Trans. H. Tomlinson, B. Habberjam, E.R. Albert (London: Continuum).

Deleuze, G. and Guattari, F. (1984), *Anti-Oedipus: Capitalism and Schizophrenia*. Trans. R. Hurley, M. Seem, H.R. Lane (London: Athlone).

—— (1986), *Kafka: Toward a Minor Literature*. Trans. D. Polan (Minneapolis, MN: University of Minnesota Press).

—— (1988), *A Thousand Plateaus: Capitalism and Schizophrenia*. Trans. B Massumi (London: Athlone).

Derrida, J. (1978), *Writing and Difference*. Trans. A. Bass (Chicago: University of Chicago Press).

—— (1981), *Dissemination*. Trans. B. Johnson (Chicago: University of Chicago Press).

—— (1982), *Margins of Philosophy*. Trans. A. Bass (Chicago: University of Chicago Press).

—— (1988), *Limited Inc.* (Evanston: Northwestern University Press).

—— (1991), *A Derrida Reader: Between the Blinds*. P. Kamuf (ed.) (Hemel Hempstead: Harvester Wheatsheaf).

—— (1994), *Spectres of Marx: The State of the Debt, the Work of Mourning, and the New International*. Trans. P. Kamuf (London: Routledge).

Doel, M.A. (1999), *Poststructuralist Geographies: The Diabolical Art of Spatial Science* (Edinburgh: Edinburgh University Press).

—— (2008), 'Dialectics revisited. Reality discharged', *Environment and Planning A* 40, 2631-2640.

—— (2009), 'Miserly thinking/excessful geography: from restricted economy to global financial crisis', *Environment and Planning D: Society and Space* 27(6), 1054-1073.

Doel, M.A. and Clarke, D.B. (2007), 'Afterimages', *Environment and Planning D: Society and Space* 25(5), 890-910.

Heath, S. (1981), *Questions of Cinema* (London: Macmillan).

Latour, B. (2005), *Reassembling the Social: An Introduction to Actor-Network Theory* (Oxford: Oxford University Press).

Latour, B. and Weibel, P. (eds) (2002), *Iconoclash: Beyond the Image Wars in Science, Religion, and Art* (Cambridge, MA: MIT Press).

Lyotard, J-F. (1990), *Duchamp's* TRANS/*formers*. Trans. I McLeod (Venice, CA: Lapis).

—— (1998), *The Assassination of Experience by Painting–Monory*. Trans. R. Bowlby (London: Black Dog).

Melville, H. (1968), *Bartleby and the Lightning-Rod Man* (Harmondsworth: Penguin).

Olsson, G. (2007), *Abysmal: A Critique of Cartographical Reason* (Chicago: University of Chicago Press).

Thrift, N. (2000), 'Afterwords', *Environment and Planning D: Society and Space* 18, 213-256.

—— (2003), 'Performance and ...', *Environment and Planning A* 35, 2019-2024.

—— (2004), 'Summoning life', in Cloke, P., Crang, P. and Goodwin, M. (eds), *Envisioning Human Geographies* (London: Arnold), 81-103.

Chapter 7

Representation and Everyday Use: How to Feel Things with Words

Eric Laurier

1.

A person X says to person Y 'I'm bored' or 'You annoy me'. Common enough things for someone to say to someone else, and common enough expressions for both to understand, yet professional analysts of language are troubled by what 'I'm bored' or 'You annoy me' *means*, it seems of quite a different order to 'this is a tree' or 'if you do not eat meat then you are a vegetarian'. It would not be uncommon for certain logicians or linguists to stay with the words themselves. In staying with the words themselves, cutting away what class, gender or age of person said such words to what other category of person. Cutting away which place, what time period, in which culture and various other elements. Trimming away, then, most of the context and dealing with the words as if their meaning was internal to themselves.

There are two things I should mention about 'I'm bored' or 'You annoy me'. Firstly, they are examples used to teach, explore and analyse what *perlocutionary acts* in language are. These are words which are related to action but do not perform that action in saying them. Words which, like indexicals, such as 'it', 'this' and 'you', cause endless troubles for formal logic and for translation software. Secondly, 'I'm bored' and 'You annoy me', while not bizarre instances, in fact recognisably and acceptably ordinary, are made-up examples. Made-up by Stanley Cavell (2005) as cases of locutionary acts aimed at having effects on the feelings, thoughts or actions of others. Cavell put 'I'm bored' and 'You annoy me' to use in order to extend and gently critique Austin's (1962) theory of performative actions in his renowned collection of lectures 'How to do things with words'. Pertinent to this collection, Austin's theory of the performative dimensions of language is an unexcavated cornerstone of non-representational theory (Thrift and Dewsbury 2000), ANT (Pels, Hetherington, and Vandenberghe 2002) and ethnomethodology (Garfinkel and Sacks 1970). For each of the latter there is an emphasis on the always ongoing accomplishment of joint action, the insufficiency of the discursive (or representational) and a rethinking of what makes social order possible. Sometimes the emphasis in non-representational theory appears to be as decontextualised and universalist as that of the linguist mentioned above (Laurier and Philo 2006; Tolia-Kelly 2006). What both actor-network theory and ethnomethodology offer to return to certain non-representational theory analyses is a tone of worldliness. They bring context back to speech through contextualising, though as I will hope to show in what follows not external social scientific contextualisation, rather, contextualising as internal to ordinary conversations. My parallel ambitions in what follows, then, are to re-examine speech in context and context in speech and to use this to return, by the close of this chapter, to Cavell's thoughts on how passionate utterances are related to performative ones.

As a first step in an ethnomethodological direction I would like to shift our attention away from Cavell's examples provided for thinking with, to some words actually said, come upon in looking for something else. As Thrift (1994) puts it in his seminal introduction to his *Spatial Formations* collection that in some ways launched non-representational theory over a decade ago, this is 'an ambition to move away from doing theory by conducting abstract thought experiments towards a style which attends to the knowledge we already have, and does not assume a common background when this is precisely what is at stake' (Thrift 1994, 3). In a similar manner, Harvey Sacks (1992 a and b) throughout his studies of conversation analysis warned his students (and those other colleagues in receipt of his lectures) to avoid beginning with a theory and then, either inventing a suitable example or, looking for a quote from a transcript to pull out to illustrate it. For the former what any member of a given research community views as reasonable provides the limit on suitable examples and, for the latter, why bother with ordinary conversation at all?

In describing to his students why they are looking at a round of introductions in a therapy session Harvey Sacks offers his reasons for labouring over conversations that appear to have no 'lay interest'.

> People often ask, 'Why do you choose the particular data you choose? Is it some problem that you have in mind that caused you to pick out this group therapy session?' And I'm very insistent that I just happened to have it, somebody had found this segment, it became fascinating, and I spent some time at it. Furthermore, it's not that I attack it by virtue of some problem I could bring to it (Sacks 1992a, 292).

Sacks goes on in the same lecture to say that he has developed a 'counter-strategy' to the concept of 'interesting' data and picks deliberately uninteresting materials. In that way he is avoiding exploiting material that is already assumed to be exciting, important or salacious.

In the quote I will begin with, the speakers here *are* beginning saying the kind of thing that might be interesting enough to catch the eye of a social scientist with, if not coding in mind, then at least topic:

the vast majority of retailers in Britain

On the basis of such a generalisation it might appear as if someone is about to state their belief or opinion about shops in the UK: 'the vast majority of retailers in Britain are encouraging us to overspend'. If that were the ending of the quote, while it might be taken as an opinion or statement of belief, it raises a number of questions. The statement still has not shaken off its indexicality nor, indeed, will it ever, nevertheless those persons present when it was uttered 'manage to make adequate sense and adequate reference with the linguistic and other devices at hand' (Lynch 1993, 22). Quite what it could mean will surely require a few more salient details. An early and ongoing solution in cultural geography to dealing with this problem of indexicality was to place the statement in a context of what category of person said the statement. To examine whether it was a man or a woman or child, the Chancellor of the Exchequer or Nigel Thrift, would help us secure the stability and certainty of what X could have meant in saying 'the vast majority of retailers in Britain'.

The prevailing tendency in doing research projects with more ordinary members of society than the Chancellor of Exchequer would be to allocate this person according to one of the social categories which are stock-in-trade of the social sciences: their gender, their class, their race, their age. With the last category we would begin to be more certain about what the phrase means if a 10-year-old says this, a teenager, 26-year-old, or a 70-year-old. So what kind of person said this? It was a man somewhere in his forties, white, middle class and middle management. If we pause for a moment, while a 10-year-old could have uttered our first ready-made example 'I'm bored', by contrast, 'the vast majority of retailers in Britain' is not the kind of thing we imagine 10-year-olds saying at all. With the social categories in hand it suddenly sounds like the kind of opinion that someone occupying those categories could say with no need for special explanation. The point about this is that we start to come upon how, in examining a number of statements, they predicate particular categories of person. A classic example here being 'I sentence you to 10 years in Pentonville Prison'. It is not the free-for-all that an example like 'I'm bored' might seem to imply.

Even though we have the social science categories of this person available to us now, the statement remains pruned of its branches so we do not yet know whether it is opinion or what else it might be. A little more of what follows this speech in the transcript of the conversation it was uttered within will help us make greater sense of what is going on:

A: As with the vast majority of retailers in Britain, I'm afraid.

'I'm afraid' in this context is not the equivalent of 'I'm bored' of course. It is not an expression of a feeling nor an attempt to have someone act to remedy the dullness of the situation. One thing it does is simply mark A's turn in the conversation as completed. Here it notes a preceding complaint, so that, for instance were someone to have said 'I'm bored', then B Could reply 'There's nothing I can do about that, I'm afraid'. So, just as it looked like we were getting somewhere in terms of the linking together of speech and speaker settling disputes over what this phrase means, we find that we cannot make sense of what A, the man in question, is saying because he is not prefacing a statement of his own, he is replying to a previous statement without which we cannot identify what he believes about British retailers. When presented with a statement like this from a respondent in our research, practical solutions during interview situations have then been to, either transcribe what the interviewer had said that A is replying to, summarise it in one way or another, or, indeed, instead of transcribing A simply summarise the whole thing as his opinion. Perhaps, then the combination of social science categories of the speaker and the speech's replacement in its dialogical context can settle its meaning. Indeed, as it turns out, this move has profound consequence for how we understand context in that each part of a pair in a dialogue provides the context for the other.

> B: Your stuff's shit. Better fucking correct it. And your customer service is
>   pish as well
> A: As with the vast majority of retailers in Britain, I'm afraid

In fact the statement, now that we can see it as a response in a dialogue between two speakers, becomes all the more intriguing and puzzling. Settling an individual's speech into conversation displaces the importance of A being white, male and middle class. The talk, in this case, is not generated from a more or less formal interview, it is, to adapt a phrase of Ed Hutchins (1995), 'talk in the wild'. As such the repetitive standardisation of the interviewer-interviewee disappears, to be replaced by a multitude of possible dialogical pairings: doctor-patient, parent-child, teacher-student, (on the phone) caller-called, teaser-teased, driver-passenger. The shift from orphaned statements to unfolding conversations is a further step in an ethnomethodological investigation of the social ordering at source in our everyday talk. Common to both ethnomethodology and conversation analysis is that the problem of meaning which fascinates cultural studies is subsumed by the problem of doing. From what people are doing emerge potential roles, characterisations, responsibilities, motive and, indeed, meaning, for them. To grasp what is being done during any action or interaction gives us, and participants in the original situation, resources for settling on who is doing it. As such conversation analysts ask themselves 'what is this word doing? What is this preface doing? What is this response doing?' and so on, often before they check to see what type of person was saying it. Trying to express meaning, or, indeed, repair misunderstandings (e.g. 'what I meant to say was ...'), is one amongst a range of possibilities. As likely, there are more practical purposes afoot: complaining and responding to a complaint.

From the two halves of this dialogue – a complaint and its response – it would appear that they predicate the members of a category-collected pair (Hester and Eglin 1997): buyer and seller. Or, if we use the categories at source: customer and retailer. B, as a customer, is making a brightly coloured complaint about the items on sale and the customer service. He sounds angry as hell. Is A joking with him by saying that we, the retailers of the UK, are almost all like this? Actual dialogue is full of puzzles like this. We, and A, have to make sense of what B is saying is happening by reference to what is happening – [complaining]. A's response could be taken to accept that, yes, their products are of poor quality, as is their customer service. However that is not what a complaint with the force and directness of A's would expectedly require. Such a charged complaint as one half of a pair of conversational parts, would surely solicit an apology and an explanation:

> B: Your stuff's shit. Better fucking correct it. And your customer service is
>   pish as well
> A: I am terribly sorry that you have had such a bad experience with our
>   company. We can replace your item or offer you a full refund.

That would be the training-manual response by A to B's complaint which quickly accepts the complaint as legitimate and offers a standard way for a retailer to right their wrong. Responses to complaints as they are actually produced show a number of ways of handling a complaint: defences, denials or acceptance with attribution of the fault elsewhere (Dersley and Wootton 2001; Edwards 2005; Sacks 1992b). A's acceptance with its humour might further enrage B if he fails to or refuses to enjoy A's wit. Indeed not only is A witty, he aligns himself with B in that his response identifies a common awful situation that they will have to endure together. There is no inevitability in how we respond to complaints, indeed the meaning, consequence and force of B's complaint is open to local adjustment by A (Latour 1986). In what A says as the recipient of the complaint, by his wit he can try and show that while he accepts the complaint, the fault lies with a more general problem with UK retailing and that the 'your' which is the basis of the complaint is not 'ours', it is a misdirected complaint. If A is a retail manager then speaking so seems a curious way of righting the wrong that is the basis of the complaint (as was the case in the training-manual response).

Other well-worn social science categories of who B might be are perhaps of assistance here. He is also white, male, middle class, if a few years younger than A and this mutually recognisable match between them might provide the underpinnings for trying out a witty response. And yet, A is not in the business of social science theorising and if we try and pick out his remark to support an argument we would like to make about his opinions it misses what he is doing in saying what 'as with the vast majority of retailers in Britain, I'm afraid', for a start he is not offering it as his opinion nor anyone else's (e.g. by ending his response with 'according to the Daily Telegraph'). His generalisation would be part of deflecting the complaint so that rather than customer/retailer we are two men of the world who appreciate the steady decline of UK retailing over the last few decades. And his deflection could be ignored, questioned, challenged or taken as provocation by B. He might then show his understanding of its tone by saying 'don't patronise me!'

2.

Even with two halves of a pair in the dialogue we are still not all that much closer to what A could be meaning with his 'as with the vast majority of retailers in the UK, I'm afraid'. As Bruno Latour puts it when reviewing the close analysis of laboratory scientists' conversations 'one has the same feeling as reading a newspaper with a microscope' (Latour 1986, 545). The solution surely is to zoom backwards and sideways and take in the preceding newspaper column inches. We can look at how this conversation has produced a preceding and emerging context for this moment of confrontation (for a similar move see also Schegloff 1992):

> B: Out on my, my bike last night. Another puncture
> A: Same tyre
> B: Nah. Front this time
> A: You'll need to get the same done for the front then ((laughing))

> >>   B: Nahh, so I'm taking it back tonight and just giving it over
> A: A bit of feedback
> B: A whole load, yeah. Aye, a whole pile of feedback
> A: Yeah?
> B: Yeah
> A: Didn't spend 350 quid for bla bla bla bla bla
> B: Your stuff's shit. Better fucking correct it. And your customer service is pish as well
> A: As with the vast majority of retailers in Britain I'm afraid
> B: Ts, yeah I was most unhappy.

With a wider angle of perspective on the conversation, everything seems to change. As we read down the transcript, with now the beginnings of the upcoming topic of this conversation, a bicycle tyre puncture and its consequences, it becomes apparent that A and B are not seller and buyer. A's joke is not what we had thought it was nor is B's angry complaint. In fact, the shift in perspective on B's complaint is reminiscent of a classic narrative device in film where we discover we are hearing a dry-run of a line rather than the line's delivery to its recipient. A's 'as with the vast majority …' is not a witty response to try and defuse an angry customer. B's complaint seems to be an angry expansion upon both the suggestion of and rehearsal by A (e.g. 'didn't spend 350 quid for bla bla bla bla'). The whole description of what is going has been turned upside down. Wait a moment though, not as much changes as we might at first imagine: there remains a complaint from B in what is happening and A is still its recipient and his response is still a little puzzling.

We learn that A has been party to previous puncture reports by his saying 'same tyre'. Had A responded by saying 'what a pain' he would have been sympathetic but not registered that he remembered that B had had a puncture before. In one sense, this quick response shows that A's mind is with B (Sacks 1992a), while at the same time it can be heard as the beginning of a diagnostic sequence. The diagnosis being offered in the line before '>>' where, while chuckling, A offers that whatever fixed the puncture on the rear can be done to the front. At the marked line, B tells A of his planned response to 'another puncture' which is that he will not be repairing it. By his use of 'taking it back', rather than 'taking it to', B primes A that the party that will receive the puncture has an ongoing connection to the bike. From the 'retailers' remark from A we know that it is a retailer that will be getting the bike back.

Even though zooming out and back puts A's speech in a new context, in this longer run of the conversation the context does not stabilise, quite the contrary we begin to get a feel for context in flight as it is ongoingly being achieved by the parties to the conversation. The episode begins with the preliminaries of what is not the first (e.g. it is 'another') and might be a longer stretch of *troubles*, 'another puncture' by B, which presents A with a problem of how to appreciate this recurrence of trouble, with sympathy or not? A common feature of descriptions of troubles such as punctures is to provide an assessment as the thing emerges (e.g. 'another bloody puncture' or by laughing while saying 'another puncture' (Goodwin 1992)). That a trouble, in whatever form (punctures or divorce or a stock market crash), requires an appreciation of what stance to take on it, is all the more marked because A laughs while offering the diagnosis of what to do about a puncture in the other tyre. His initial treatment is that the recurrence of punctures is one of those annoying, though potentially humorous, misfortunes of riding a bike. Punctures being laughable in ways in which the bike being stolen, for instance, would not be. B's prefatory 'nah' makes clear A's error and he goes on to show a departure from dealing with punctures by repairing them himself, the implications of this puncture are not to be a basin of water and a puncture repair-kit, it will be taken back and given over to the retailers to fix.

As Edwards (2005) notes the word 'complain' or 'complaint' is seldom used when a person makes a complaint. One reason being that if speakers are not making a complaint they can then try and characterise what they are doing as reporting in a neutral manner on observations they have made. A second related reason being that they care about their dispositions in various ways, not least in terms of their character for others (Edwards 2006). In any particular episode that could be found by others to be a complaint, the public character of the person so doing, is at risk. They are open to what they are aggrieved over being attributed to their character as someone who is 'always' complaining about this or that, is difficult or unreasonable in their affairs. To avoid having what one is doing being straightforwardly taken as a complaint is one way of handling how one's actions are appreciated. So it is, then, in making available his revised appreciation of 'another puncture', A not only correctly anticipates what B will be doing in returning his bike he formulates it as 'feedback', rather than a complaint. Feedback being what businesses specifically ask for and, as such, A's selection of 'feedback' rather than 'complaint' plays up firstly, the positive aspects of what B is doing in that he will be helping the business improve, secondly, that 'feedback' is not seen as self-interested or motivated by other personal problems in ways in which a complaint is. A's delivery is yet more artful than that, he uses the diminutive 'a bit of'. In keeping this minor key he allows B to then respond by either staying with this business-like tone or more satisfyingly, as he does, inflating it significantly:

A: Didn't spend 350 quid for bla bla bla bla bla

B: Your stuff's shit. Better fucking correct it. And your customer service is pish
    as well

A leaves B to provide the details of what is wrong with the bike by saying 'bla
bla bla bla'. A's rehearsal of the line to be delivered on handing the bike over is
responded to with a second upgrading by B with his angrier, blunter and more
confrontational set of assessments. What B accomplishes in his outraged 'feedback'
is both producing speech hearable as of a more general nature 'stuff' thus bringing
the fault not on this particular bike. Were it the particular bike the fault might lie
with the manufacturer and not the shop that sells it. Moreover rather than asking
for some form of recompense as would be the case with a complaint, he tells the
retailer to correct their 'stuff'. 'Correct' predicating a mistake or an error rather
than a broken or defective object. This is not the speech of a bleating sheep or
grumpy old man. What is not available from the transcript is the calm tone with
which B delivers his line to A, one which rather than sounding outraged as one
might expect, is controlled. If it were more exaggerated A might have heard it as
ironic in some way (Edwards 2000).

What we can see in the two pairs above is a produced similarity in structure
where, in each, A is allowing B to pump himself up (if you'll pardon the pun) for
the return of the bike that night. The planned line that emerges from this inflation
sequence is highly unlikely it will actually be delivered. Were B to walk in and
deliver that line to a sales assistant their first response might justifiably be 'calm
down sir, what is actually wrong with your bike?' While A has helped B to get
pumped up and, in doing so, express his genuine annoyance with the agency
that sold him the bike he has also taken him to the highest step in this step-wise
progression. A suggested small complaint begets a final huge complaint. The
expectations of spending that sum of money are elevated to problems that beset
the whole company. What would B do after that or as the consequence of that?
Shut down the company, punch the shop assistance, or, in ultimate desperation,
write a letter to the Daily Mail?

3.

Just when all the contextualising work of the conversants is beginning to display its
ongoing sense and sense-making I want to use the Latourian macroscope against
us. We will move it again and add a perplexing visual element to our close reading
of the transcript.

A: As with [Figure 7.1] the vast majority of retailers in Britain, I'm afraid

A and B are in a car when A says 'As with the vast majority …'. That seems a
pretty fundamental absence. One of the uses of supplying the missing context

Figure 7.1   Image as part of conversation transcript

is the murder-mystery moment where, by supplying the missing information, suddenly the speech makes sense (Schegloff 1992). To reveal that A and B are in the car surely changes everything, though in the opposite way, the speech makes less sense. The shift in perspective is disorienting. In a geographical *denoument* we could argue that the space of the car is central to our understanding of what is happening. Well, is it?

Certainly we have a new set of categories to bring into consideration, alongside the usual suspects of the social sciences which may or may not be relevant to what is happening, the locally produced complaint-maker and indirect recipient of the complaint now we have 'driver' and 'passenger'. Equally alongside these categories we have the activities that generate them: [complaining] and the parallel activity of [driving]. Why though should the context not serve as a stable background? Is the activity 'driving'? It could also be characterised as travelling, journeying, racing and commuting. It can be broken down into an array of skilled practices: cornering, reversing, overtaking, hill-starts, dodging potholes and so on.

In fact to close the microscope back in on the action:

> B: Your stuff's shit. Better fucking correct it. And your customer service is
>   pish as well
>
> ((B drives along stretch of road with a gentle curve requiring small turning of steering wheel))
>
> A: As with the vast majority of ((looks slightly to passenger side, then
>   returns to looking ahead out of front window)) retailers in Britain, I'm afraid
>   ((B puts on indicator))
>
> B: Ts, yeah I was most unhappy.

While there are times when unfolding events on the road lead to more or less significant re-arrangements of the organisation of conversation, here there are only a few events worth remarking on. One is A turning his head slightly before going on to say 'retailers'. The second is that the turning on of the indicator appears to offer a way for B to return to 'another puncture' and finally delivering his, by this stage, unsurprising stance on it. Both of these potentially driving-related functions require one more element that has been an ongoing entity of interest for conversation analysis and ethnomethodology. An entity that turns many of us toward the transcendental or at the very least seems as if it marks the limits of language. Silence.

Let's put the silences back in (in brackets in seconds):

A: Didn't spend 350 quid for bla bla bla bla bla
B: Your stuff's shit. Better fucking correct it. And your customer service is
    pish as well
    (3.0)
A: As with the vast majority of (2.0)

retailers in Britain I'm afraid
(6.0)
B: Tsk yeah I was most unhappy

These apparent absences of speech from the transcript in fact are one more part of what we need to supply to any conversation to make sense of it. Silences are not the limits of language rather they are at the heart of our speaking. The silences play out in language along with pauses, serving all manner of purposes: silences that speak volumes, calm silences, studied silences, dramatic pauses, marks of seriousness, poesis, displays of understanding, displays of misunderstanding (Lynch 1999). We are missing the myriad uses of silence when we think of silence in opposition to speech, or between speech acts. Sometimes 'the occasions of silence are extremely dangerous to all persons present' (Sacks 1992a) and sometimes, as in the car or out fishing, they are not. Where silences are dangerous or could be taken the wrong way it may be the speaker's task to mark out a pause with an 'uh' before leaving a gap in speech (Sacks 1992b, 547). If we return to the complaint, we have a pause of some length between B's 'is pish as well' and A's response 'as with the vast'. There is a remarkably long pause between 'I'm afraid' and 'tsk yeah'. And with A's speech around which this chapter has revolved there is a pause mid-way through.

In the consideration of what pauses are doing and indeed their very analysability by those talking together the car returns as a particular setting for speech (Brown et al. 2008; Laurier 2002). The car journey is almost the opposite of talk-radio where a silence is 'dead air', in the car there is always the other activity as safety net – driving in the car together. Pauses and silences are less noticeable, or better, less threatening to the talk itself. So pauses and silences can safely be put to use in the car, so that the last pause above of six seconds between 'I'm afraid' and 'tsk yeah'.

The first pause is after B's rehearsal of what he will say when he takes the bike back with the puncture. As we have noted already A is the indirect recipient of B's complaint, and the pause of several seconds after the pumped-up complaint before 'As with the vast majority of' serves to give some distance between that last voicing of the complaint and a further remark. While in an earlier analysis of the conversation Barry Brown and myself felt that A was missing the point of B's complaint. In a later examination with Ignaz Strebel we came to the sense that having taken A to the highest step in his beef with the shop that sold him the bike, A offers B a way out. He picks up on the generalisation made by A in 'stuff' and 'customer service' and takes that a step higher into the national sphere. A sphere which is clearly beyond the remit of even competent managers like A and B to deal with. In closing his generalisation with 'I'm afraid' he marks out the excusability for such poor service and quality of goods. In other words he is sorry to have to be the one to remind B of the low quality of the UK retail sector and thereby blunt, but not ironise, B's exaggerated feedback (e.g. Edwards 2000, 365 onwards). In doing so we are taken back to our earlier point that appeared to have been rendered irrelevant when we considered the longer sequences of the conversation. The earlier point noting the shift toward generalisation of the complaint by A does indeed deflect the anger being directed at the particular shop. A *is* contextualising B's complaint where the contextualisation is part of getting them both to a point of agreement on the basis of their general world views. In other words, this generalisation would be where their conversation touches upon what they care about and are responsible for. As managers, they reproduce and reshape organisations and indeed spend a great part of their time discussing how their respective sectors (private and public) function and malfunction. We begin to see that not only are we contextualising A's speech, the conversation itself has been contextualising 'another puncture' throughout in a journey that has taken us from a puncture to the state of one economic sector of the UK.

What I have been trying to bring us towards here is not the application of 'context' rather it is the ongoing work of contextualising. This form of, and occurrence of, contextualising is not the analyst's privilege rather it is a common resource for analyst and member and analyst as member. Equally quite how much context is required is related to whatever activity is underway because 'actions and their contexts are inter-articulated' (Coulter 1994, 694). As such we begin to see what Thrift might have been gesturing toward in his brief remarks:

> These four obsessions [time-space, practice, subject, agency] are coded in nearly all of my work as concern for the *context* of the situation ... By 'context' I most decidedly do not mean passive backdrop to the situated human activity. Rather, I take context to be a necessary constitutive element of interaction, something active, differentially extensive and able to problematise and work on the bounds of subjectivity (Thrift 1994, 3).

4.

In this final section, I want to return to J.L. Austin's (1962) idea of performative speech, an idea which, as noted at the outset, has lead via de Certeau (1984) and Butler (1990, 1996) amongst others into cultural geography's current interest in performativity. Yet Austin's idea has been at the same time the target of an unexpectedly impatient reading by Derrida (Coulter 1994; Cavell 1995; Derrida 1977), for its reliance on analyses of 'a context exhaustively determined' (quoted in Coulter 1994, 693). There is in Derrida's critique a failure to realise that context need not be exhaustively determined in Austin's idea of performatives. So what are performatives?

Austin used examples such the bride saying 'I do' at the correct point in a marriage ceremony, or 'I name this ship' by the appropriate person on the launch of a vessel to argue that there are forms of speech which do not *represent* anything in their utterance, they *do* the thing. That is, they marry you to another person, or, they name the ship. The target of Austin's argument was a branch of philosophy that saw numerous parts of our ordinary language as compromised in their logic and meaningfulness because they could not be demonstrated to be either truths or falsehoods. Austin showed that performatives were essential parts of our speech that were certainly not nonsense and underlay the very possibilities of proving things true or false. Rather than truth, or falsity, performatives' conditions of success or failure were found in what Austin called their felicity or infelicity. Austin (1962) went on to specify a number of conditions that had to be met for a happy performance, most of which rested on convention, such as 'there must exist an accepted conventional procedure having a certain conventional effect, the procedure to include the uttering of certain words by certain persons in certain circumstances' (26). In looking at how we do things with words Austin's aim was to 'lift the non-descriptive or non-assertional or non-constative gestures of speech to renewed philosophical interest and respectability' (Cavell 2005, 159). While Austin delighted in reminding philosophers of the details of what would allow 'I name this ship' to successfully name a ship, this did not entail that a pre-given list of contextual details need be assembled or checked-off before a ship could be named. In an Austinian spirit we can examine, and have examined here, when we do offer contextualisations for an event (or our actions).

Austin's raising up of performatives hopefully strikes the right chord in this collection on the non-representational in Human Geography. I wanted to retain his work as a touchstone in this otherwise empirically-guided chapter for two reasons. The first being that Austin's work opens up the 'what are these words doing' approach to speaking, acting and convention that is taken up by ethnomethodology and conversation analysis (e.g. Sacks 1992a, 343). In the conversation:

B: Your stuff's shit. Better fucking correct it. And your customer service is pish
    as well
A: As with the vast majority of retailers in Britain, I'm afraid

B is 'doing' a complaint. To be a little more precise he is rehearsing a complaint. Or, as we built up to earlier, he is expressing his anger with the shop that sold him his bike. The second reason for introducing Austin is to leave us all too briefly with Stanley Cavell's response and extension rather than Derrida's critique and departure. Cavell is a former student of Austin's and a current philosopher of, not only ordinary language but also, moral perfectionism (Cavell 1990, 1998). Cavell has written and reflected extensively on his relationship with Austin's ideas, in terms of his conversion to serious inquiry, his influence and of how one elicits conviction in ordinary language. Austin's anxiousness over emotion in his study of performative utterances has lead Cavell to extend Austin's theories into the study of passionate utterances (Cavell 2005, Chapter 7), or from the illocutionary to the perlocutionary. To make this a little easier to grasp, we are shifting from the *doing* of complaining *in* saying 'Your stuff's shit' to what is *done by* saying 'Your stuff's shit' which is not so straightforward since it could be intimidating, upsetting, annoying, riling and a number of other possible effects on and responses by the other.

What Cavell picks out for us is, that when B expresses his annoyance in 'your stuff's shit' he is doing complaining, we could not say whether he is satisfying, amusing, unsettling or boring A. Unlike 'I bet you', to say 'I bore you' requires disclaimers such as 'do I bore you?' or 'I seem to bore you'. To try and bore someone by saying 'I bore you' could only work were I a talented hypnotist. Quite how you will respond to my doing something to you by my utterance lacks the conditions of felicity or infelicity listed by Austin. Instead Cavell draws out a contrasting set of conditions, an important one being that, with the passionate utterance *there is no conventional procedure involved* that will produce the desired effect for the speaker. To produce our desired affectual effects, imagination and virtuosity are required. Nor are there pre-specified persons (such as bride and groom) that go with passionate acts as there would be with Austin's performatives. Here I can only massively summarise Cavell's remarks and leave you to reconsider their relevance to the dialogue between A and B. Without these pre-given roles the speaker must offer their standing with you and at the same time 'single you out' (Cavell 2005, 181). Moreover, when I speak from my emotion I must be suffering that feeling and thereby demand a response from you which you will be moved to offer (Cavell 2005, 182). Finally, and crucially, Cavell adds a further asymmetry: you may contest any, some, or all, of those elements of my passionate speech.

It is here where Cavell's extension of Austin's 'How to do things with words', returns us to Harvey Sacks and conversation analysis. Beginning as conversation analysis did with all manner of dialogues, it was attuned to the defeasibility and fragility of even a compliment (e.g. Lecture 29 'Weak and safe compliments, Sacks 1992a) let alone a complaint as made by one to another. In its pursuit of conversation as a joint social action it has traced out ongoingly *how* it is that actions get done, alongside, *what* and *who* gets done by them. Sacks described the asymmetries of expression and response that are allied with joy, boredom and suffering as they are expressed by others and responded to by others. How, in the case we have examined, a complaint is assembled jointly by A and B. How B, after a false start, helps set up the space for A to express his justified passion over yet 'another puncture' in his bike. How B ends up providing a social explanation to calm those passions, one that makes their source an object that can be dealt with by managers. The conventions that underpin many of the methodical ways in which we act are constantly being re-pinned as the affective force of our actions shakes them loose. Ethnomethodology and conversation analysis are well known for sharing in Austin's procedural focus and his celebration of the feats of ordinary language. What is perhaps less often appreciated is their sense of ordinary language's constant crumbling, intermittent eruptions and ongoing repair by those who put it to use in their everyday affairs. Ethnomethodology and conversation analysis help us see not only 'how to do things with words', they help remind us how to feel sometimes common, and sometimes exceptional, things in the words of ourselves and of others. As such we can also appreciate why it is that affect has become such a topic of concern in both ethnomethodology and non-representational theory.

References

Austin, J.L. (1962), *How To Do Things With Words* (Oxford: Clarendon Press).

Brown, B.A.T., Laurier, E., Lorimer, H., Jones, O., Juhlin, O., Noble, A., et al. (2008), 'Driving and passengering: notes on the natural organization of ordinary car travel and talk', *Mobilities* 3(1), 1-23.

Cavell, S. (1990), *Conditions Handsome and Unhandsome: The Constitution of Emersonian Perfectionism* (Chicago: University of Chicago Press).

—— (1998 (original 1979)), *The Claim of Reason, Wittgenstein, Skepticism, Morality and Tragedy (New Edition)* (Oxford: Oxford University Press).

—— (2005), *Philosophy the Day After Tomorrow* (Cambridge, MA: Harvard University Press).

Coulter, J. (1994), 'Is contextualising necessarily interpretive?', *Journal of Pragmatics* 21, 689-698.

Derrida, J. (1977), 'Signature, event, context', *Glyph* 1, 172-197.

Dersley, I. and Wootton, A.J. (2001), 'In the heat of the sequence: Interactional features preceding walkouts from argumentative talk', *Language in Society* 30, 611-638.

Edwards, D. (2000), 'Extreme case formulations: Softeners, investment, and doing nonliteral', *Research on Language and Social Interaction* 33(4), 347-373.

—— (2005), 'Moaning, whinging and laughing: the subjective side of complaints', *Discourse Studies* 7(1), 5-29.

—— (2006), 'Managing subjectivity in talk'. In Hepburn, A. and Wiggins, S. (eds), *Discursive Research in Practice: New Approaches to Psychology and Interaction* (Cambridge: Cambridge University Press), 31-49.

Garfinkel, H. and Sacks, H. (1970), 'On formal structures of practical actions'. In McKinney, J. and Tiryakian, E. (eds), *Theoretical Sociology: Perspectives and Developments* (New York: Appleton-Century-Crofts), 337-366.

Goodwin, C. and Goodwin, M.H. (1992), 'Assessments and the construction of context'. In Duranti, A. and Goodwin, C. (eds), *Rethinking Context: Language as an Interactive Phenomenon* (Cambridge: Cambridge University Press), 147-190.

Hester, S. and Eglin, P. (eds) (1997), *Culture in Action: Studies in Membership Categorization Analysis* (Washington DC: International Institute for Ethnomethodology and Conversation Analysis and University Press of America).

Hutchins, E. (1995), *Cognition in the Wild* (London: MIT Press).

Latour, B. (1986), 'Will the last person to leave the social studies of science turn on the tape recorder', *Social Studies of Science* 16, 541-548.

Laurier, E. (2002), 'Notes on dividing the attention of a driver'. *Team Ethno Online, http://www.teamethno-online.org/*.

Laurier, E. and Philo, C. (2006). 'Possible geographies: a passing encounter in a café', *Area* 38(4), 353-363.

Lynch, M. (1993), *Scientific Practice and Ordinary Action: Ethnomethodology and Social Studies of Science* (Cambridge: Cambridge University Press).

Lynch, M. (1999), 'Silence in context: Ethnomethodology and social theory', *Human Studies* 22(2), 211-233.

Pels, D., Hetherington, K. and Vandenberghe, F. (2002). 'The status of the object. Performances, mediations, and techniques', *Theory, Culture and Society* 19(5/6), 1-21.

Sacks, H. (1992a), *Lectures on Conversation, Vol. 1* (Oxford: Blackwell).

—— (1992b), *Lectures on Conversation, Vol. 2* (Oxford: Blackwell).

Schegloff, E.A. (1992), 'In another context'. In Duranti, A. and Goodwin, C. (eds), *Rethinking Context: Language as an Interactive Phenomenon.* Cambridge: Cambridge University Press, 191-228.

Thrift, N. and Dewsbury, J-D (2000), 'Dead geographies – and how to make them live', *Environment and Planning: D, Society and Space* 18, 411-432.

Tolia-Kelly, D. (2006), 'Affect – an ethnocentric encounter? Exploring the "universalist" imperative of emotional/affectual geographies', *Area* 38(2), 213-217.

Chapter 8

Language and the Event: The Unthought of Appearing Worlds

J-D Dewsbury

Introduction: Geography Unbound

> We need to free ourselves of the sacralisation of the social as the only instance of the real and stop regarding that essential element in human life and human relations – I mean thought – as so much wind. Thought does exist, both beyond and before systems and edifices of discourse. It is something that is often hidden but always drives everyday behaviours. There is always a little thought occurring even in the most stupid institutions; there is always thought even in silent habits (Foucault 1998, 456).

One attractive feature of non-representational theory, underpinning its emergence as a productive presence within geography's social scientific endeavours in the late 1990s and early 2000s, is its experimental and expansive engagement with different philosophical texts, the perspectives therein on art and science, and the grounded soundings that act within the world. This chapter is then a short treatise towards non-representational thinking which approaches the non-representational from the standpoint of the event, and understands this through the task, and the politics, of theorizing or at least thinking the singular (see Hallward (1998) for a critical overview of such an approach in Badiou). Part of this project is to play up the importance of philosophy in our social scientific engagements as we scan back and forth to the humanities and sciences, always remembering that as a geographer we cannot forget that this pivots off from a topographic, phenomenological position of being in the world – philosophy doesn't take the higher ground, rather it is put to work on the ground. Non-representational theory is then about an ontology of sense that is also materialist and concrete and that pivots off the belief that prior to the distinction between ideality and materiality sense comes about as a bodily event, 'as an opening up of meaningful spaces and a meaningful world' (James, 2006: 107). Ideal distinctions, of identity for example, are thus only thinkable as such after such constitutive openings. In terms of philosophy the emphasis here is in thinking through the ontological, of having a sense of what being as such is and that this is different from our actual existence in the world. Explicitly here I am pushing for the understanding that philosophical thought (thinking) and lived empirics (being) are mutually constitutive. Our experience is often given meaning and orientation through the representational logic of language and signification, but we run the danger of forgetting the world itself; too often word and world get segued together. So the unenviable undertaking of non-representational theory is to affirm life in an intelligible manner in a way that thinks matter *and* movement despite having 'no language to express what is in becoming' (Klossowki 2005, 38). Thus the arguments in this chapter claim that non-representational theory has only just begun and that the agendas, passions, and interventions crafted out of the genealogy of concepts, technological practices, and fleshy performatives underwriting the work done under the name have a vibrant and vital future.

One burgeoning entry point to this thinking of the singular performance of being and being-there, world and word, comes in Alan Badiou's axiomatically precise gaze over three models for considering the sense in which thinking and being are identical (2000, 79): (1) that this identity comes from what happens, that we reference it to events and the outcomes they give rise to; (2) that there is 'a structural articulation of being and thought via the mediation of language and linguistic criteria of coherence and construction' (Hallward 2003, 4); (3) that this identity is grounded on an inarticulable, ultimately mysterious, first principle – a transcendence. Now the stage for conversation and debate in light of this positions non-representational theorists as those wanting to think upon the first (not necessarily buying it hook, line and sinker); understands that geography as a whole tends towards thinking through the second gaining insight and politics from the intellectual social scientific debates of the 1970s (Marxist geography), 1980s (Feminist geography and identity politics), 1990s (cultural turn and linguistic perspectives); and that academia more broadly is secularist now and fears being accused of the third even though this need not be, and perhaps is far from being, the worst option of the three.

One of the issues about representation then, and that non-representational theory stands against, is of this attachment to the mediate (in the context of this chapter this is about words standing in for the world; more generally it signals the way in which the world is orientated to us via all sorts of meaning-making machines: codes, signs, rules, technologies etc). Therefore to talk of the non-representational for me is to question but not abandon 'a commitment to the realm of language and meaning in either the semantic/historical or the religious/ineffable sense' (Hallward 2003, 15). As such mediations are fine if taken as performative, however it is easy to fix these mediations in place. It is easy to take the replacement of the metaphysics of truth with a belief in language as that which heralds and holds the plurality of meaning that is both ethical and contemporaneous to our times, but language is not an open and innocent arena. Non-representational theory offers an alternative: what if there is a regime of the thinkable that is irreducible to denotative language (after Badiou 2006, 321)? It is not that a lot of geography and social science doesn't think this; it is that most of the time we don't consider it important enough to attend to it, preferring instead to get things done immediately with the language, economy of communication, and knowledge we have. But we cannot escape the fact that our findings are then bound by the parameters given to them and that this in turn forecloses certain vistas. To reiterate, we are always at a loss: there is always more going on, and what is going on is always more than we can apprehend. As such, and to mark the closure of this introductory section, perhaps, as Deleuze thought, our:

> thought is not abstract enough, that it pitches its level of abstraction too low and consequently forfeits the virtues of abstraction. That which is more abstract is that which incorporates a larger, that is, multiple, sample, considering not only human but also animal, inorganic and other inhuman becomings (Mularkey 2005, 16).

What follows is in two parts. In Part 1 I attend to the aspect of non-representational theory that questions the relation of meaning and truth as precisely that gap between world and word. What I take here as part of the challenge of non-representational theory is the task of understanding how meaning and truth are also bridged by 'modes of perception which are not subject based' (Thrift 2007, 7). This addresses the phenomenality of the world, the ways in which through different systems and technologies of thought the world can be made present and thus different, instead of accepting the implicitly universal phenomena of the world as perceived by subjects. This pays attention to the habits, traditions, cultures and politics of 'the "how" of manifestation over the "what"' (Hart 2007, 39) where the 'how' of manifestation is precisely this question of phenomenality: how do we register the world, how does the world afford registrations to us, and in combination how do both bring the world into being. And herein we should be cautious about objectifying this registering. This is why the 20th Century underpinnings of the genealogy of non-representational conceptualizations are significant, and it is perhaps this that marks the difference to a lot of the other more representational modes of understanding in geography:

> Recently however, the teachings of Nietzsche, Bergson, and the life philosophers set the standard for this claim concerning the objectifying character of all thinking and speaking. To the extent that, in speaking, we say 'is' everywhere, whether expressly or not, yet being means presence, which in modern times has been interpreted as objectivity – to that extent thinking as re-presenting and speaking as vocalization have inevitably entailed a solidifying of the intrinsic flow of the 'life-stream', and thus a falsifying thereof (Heidegger 1998, 57).

The section closes by turning directly to the concept of the event and how language apprehends and gives a reality to the ongoing appearance of worlds in the aftermath of an event's singularity (quite precisely a non-objectifiable moment in that it cannot be predicted in advance). In Part 2 I set the stage for non-representational theory's resolute experimentalism insofar as the arguments presented in Part 1 raise the spectre that the presence of the world is untouchable, always necessarily in doubt, and that a fully graspable representation of the world in its eventhood is impossible, and that this uncertainty sets up the desire to know in the first place. Here we will look to the necessary uncertainty in the art of non-representational theory, showing, via Beckett, how this ethos of experiment and failure allows thought itself to be staged, and thus sites a space in thought for intervening in how the world appears. The chapter closes by reflecting briefly on a certain minimal humanism in relation to the event.

Part 1: Meaning and Truth/Thinking and Being

How we bring the world into being through our modes of registration and representational communication is the principle research question for non-representational theorists. This may seem esoteric and abstract but non-representational theory deals with the question of world-forming (often meaning world-framing), that is it focuses upon the everyday performative practices across the sciences and the arts, and thus across technology and the meanings we make to inform our actions and values. At the moment constructions of affective ecologies are foremost in this endeavour (see Thrift 2007, 171-197 and 220-254). More expansively and consistently however is the push towards the thinking of spaces of human relation, of spaces of meaning-in-action held in common, propagating potential spaces of signification contra (after Nancy (2007)) a globalization of ideas that enclose, in an 'undifferentiated sphere of unitotality' (ibid., 27), a thinking of the world 'that is perfectly accessible and transparent for a *"mastery-of*!" without remainder' (Raffoul and Pettigrew 2007, 2; my emphasis and alteration). The point is, there is always a remainder – an excess (Marion 2002), a supplement (Derrida, 1997 (especially p. 144)), an inconsistent multiple (Badiou 2006), an outside (Blanchot 1989; Foucault 1990), a virtual open whole (Deleuze 1994) – that is constitutive of the reality of the world. This excess is eventful that is it is a surprise, secret, and always unknowable as such in advance. It presents the 'empiric' that there is always something non-representational given that 'this something' comes before it has been made manifest and apprehensible in representation, not because it is transcendent and prior, but because in the actualization of the manifestation of the representation of the world, that gives it its reality, the world is changed in the process. All of which is to say that the world 'is a possibility before being a reality' (Nancy 2007, 65): there are a crowd of pretenders to actualization that never get actualized and a multitude of micro manoeuvres and potential bifurcations that never happen. We are always thinking a world into being in terms of creation 'as an unceasing activity and actuality of this world in its singularity' (ibid.). In sum, the world in the present tense is always other than its representation, of what we know of it; it is always in excess and outside of representation and all horizons of calculability. This is the structure of the Derridean event: that we have to think the possibility of the impossible, namely 'that which happens outside the conditions of possibility offered in advance by a subject representation' (Raffoul and Pettigrew 2007, 9) such that we have to think 'an experience removed from the conditions of possibility of a finite knowledge, and which is nevertheless an experience' (Nancy 2007, 65). I believe that this as yet impossible thought, this unthought, is an *experience* of this world; I believe that this is what makes thought what it is and why thought thinks us as much we it; and I believe that non-representational theory acknowledges this, risks its grounding, and tasks us with the imperative and political vigilance to think towards this experience. This imperative is non-representational theory: to think an experience and to present thinking in an experience so as to think as experience. This means that our experience of the world is of an experience that we don't have the means to signify as such. This type of thinking is precisely the act of regarding thought itself and thus addresses Foucault's solicitation represented by his words that opened this chapter. In this it equips us to think the event: an unforetold experience as the constitutive realm of thought as molecules, energies and imaginations refigure themselves in response.

In this regard, thinking the event drives us to consider that the meaning of the world 'cannot signify the sense of the world as objective genitive, an encompassing of the world as totality on the basis of an external overview', but rather suggests 'a subjective genitive, produced from the internal references of the world' (Raffoul and Pettigrew 2007, 5-6). For me the provocation of non-representational theory is to think that part of the world that does not exist prior to itself. It is to think upon or towards that which exists singularly and always as 'a gap [the non-representational] from itself, without ground or against the background of nothing' (ibid.). This is where the work of Jean-Luc Nancy is worth the effort: for him, when we speak of meaning or representational signification, we

> do not intend by this term the same thing as 'signification', in the sense of an accomplished given meaning, but rather the opening of the possibility of the production of significance. Meaning is not given, it is to be invented, to be created, that is to say, as we will see, out of nothing, *ex nihilo* … (ibid., 6).

This is where the problem of and for non-representation lies: how do we, the human, endure if meaning comes from nothing? Surely, we rely on durable forms and identities that are in some way prefigured and have a consistency that affords us a consistent immutable presence? And we (actual, human, individual, subject) have different specifics (situations, capacities, opportunities, empowerments) and contexts (presents shaped by the past, coded by genetics, framed by biography, structured by degrees of inequality). All this matters immensely. But what also matters is always coming about and I believe we need to think on that constitutive and abstract basis as well so as to reverse the trend that makes the world secondary to the concept of a world 'view' (see Nancy's excellent essay *Urbi et Orbi* 2007, 33-55).

The philosophical conceptualizations of phenomenology are thus significant because they involve the account of subjectivity with its intimate bond with the space and spatiality of the world itself, the central connections of the body, and the temporalizing movement of sense in a way that is situated but eventful. The world thus can be taken to be always a place of possible habitation where 'what takes place takes place in a world and by way of that world. A world is the common place of a totality of places: of presences and dispositions for possible events' (Nancy 2007, 42). Whilst the world does not exist prior to itself (that is as an essential thing), it exists in relation to itself: there is a being-world of the world and this is often referred to as the existence of the world considered in its immanence. We are in the world through our situatedness, immanent through the experience of one's self through an affectivity that speaks of 'the indivisible identity of that which affects and that which is affected' (Hanlon 2003, 98) in the sheer simple fact of being inscribed there. Thus:

> Whilst the other sciences study specific phenomena – physical, chemical, biological, juridical, social, economic, etc. – phenomenology explores what allows a phenomenon to be a phenomenon. Phenomenology investigates pure phenomenality as such. One can confer various names upon this pure phenomenality: pure manifestation, showing, unveiling, uncovering, appearing, revelation, or even a more traditional word: truth. As soon as the object of phenomenology is understood in its difference from the object of other sciences, a further distinction seems to impose itself: that of the phenomenon considered on the one hand in its particular content, and on the other hand in its phenomenality. Such is the distinction between that which shows itself, that which appears, and the fact of appearing, pure appearing as such (Henry 2003, 100).

In its focus on the present and on intelligence-in-action rendered as a kind of know-how in manipulating material presence, and thus in its interest in the empirics of practice and the theorization of performativity, non-representational theory is often accused of falling foul of going straight to the things themselves, a naïve empirical realism whereby our body somehow holds the answers. This is why it is important to think through how phenomena come into being through modes of phenomenality. And I think that non-representational theory does do this, and as such circumnavigates such critiques, but it tends to it only implicitly. First, how is a phenomenon given? How do we know phenomena, and hence the world? Crudely, so the starting point of Henry's argument goes, we begin with Husserl and the traditional phenomenological notion that the phenomenon comes about through consciousness and a sense of interiority, and thus intention, whereby the appearing is 'a setting at a distance'. Thus, 'the possibility of vision resides in this setting at a distance of that which is placed in front of the seeing, and is thereby seen by it' (Henry 2003, 101). The question then is how this consciousness, which shows or makes the world appear, appears or reveals itself to itself? When you think, don't you think in words? Don't you talk to yourself? Without these words, without this phenomenality, this mode of making appear, would you exist? Can you 'be' without language? In other words, does 'another mode of revelation exist other than the showing of intentionality' (ibid.)? As Henry argues, it is with Heidegger that the appearing of the world itself, a showing without intentionality, is taken to the highest degree by thinking through the relationship between Being and Dasein (being-there). However, I want to remain with Henry, to set out his phenomenology of life itself through what he characterizes as three decisive traits of the *appearing* of the world.

First, the appearing of the world consists in the coming outside of the Outside. By this he means that all that shows itself, shows itself literally outside itself, as other, as different. This is an ek-static structure that is a primordial alterity – a distancing if you want – that nonetheless is identically a Difference (see Henry 2003, 101). It is also affirmation of the 'non' of non-representational thinking: there has to be difference for the world to 'be' given that the world is becoming. We are here, abandoned in Heideggarian terms as being-in-the-world, as a being of the world and nothing more: as Nancy says 'being a subject in general means having to become oneself ...' (2007, 41). Second, 'the appearing which unveils in the Difference of the world does not just render different all that which unveils itself in that fashion, it is in principle totally indifferent to it, it neither loves it nor desires it, and having no affinity with it, it does not protect it in any way' (Henry 2003, 101-102). All that is before us in the world, stands before us in the same way; it is just there; it is just the 'there is'. Third, this indifference of the appearing world is incapable of conferring existence upon it; it is incapable of setting out a reality; Heidegger's unveiling unveils, or opens, but does not create (ibid.). In sum: Difference, indifference, and non-existence. This might be the brute 'truth' of the world in its bare 'factiality' (Meillassoux 2008). This might be the logic of the event as conceived by many post-continental philosophers. But perhaps the humanism, or what makes us human, is the unavoidable stance that there is no world without representation, without language, without the conference of reality in sense; although the project of speculative realism, of which Meillassoux is an attributed protagonist, offers a Nietzschean hammer to this unavoidable. And perhaps too non-representational theory represents a half-way house (a weak humanism) appreciating that the world is different and indifferent to us, and that the world did, does, and will exist without us but is brought into being through us. For me, the powerful empiric of the world is its Difference, indifference, and temporality towards non-existence: as such the world is precisely scripted by singularities, through singular moments. That we can inhabit these singularities comes down to the domain of the event and the role of language in apprehending and making sense of the singular logics of the world; it is then to the concept of the event and its relation to language that we now turn.

Part 2: Event and Language

There are many concepts of the event in the philosophical genealogy behind non-representational theory and continental philosophy more broadly: unpacking these and their comparability is another project. In relation to the appearing of the world, which can advent many worlds, many topoi, the event is the moment of appearing and it is singular. It is a surprise and non-relational, that Difference and indifference of Henry's; but time takes on a particular relation within the event, and as such, and in order to think the singularity of the event, we have to understand it as a realm. Now switching to Deleuze's terms, although speaking in general here, the event is a meanwhile and incorporeal (thus an inbetween space) which means it is not entirely actual; and it is never brought to completion in an instance, those instances registered in the state of affairs of the body that undergoes it. The event is also, however, without relation, being of infinite movement, being of 'a virtual that is no longer chaotic, that has become consistent or real on the plane of immanence that wrests it from the chaos' (Deleuze and Guattari 1994, 156). The event causes something to happen, but 'this cause is nothing outside of its effect' which creates an immanent relation to the open whole of the virtual turning the product, that which it causes to happen, 'into something productive' (Deleuze 1990, 95). It is singular being 'neither particular nor general, universal nor personal', 'entirely independent of both affirmation and negation', and, as already intimated, it cannot be empirically determined thanks to 'intuitions or positions of empirical perception, imagination, memory, understanding, volition' etc, and 'from the point of view of relation, it is not confused within the proposition which expresses it, either with denotation, or with manifestation, or with signification' (Deleuze 1990, 101).

In other words, and thus: if the event is the only concept capable of ousting the verb 'to be', as it is for Deleuze (1994, 141), then where do we find our place in the world that is continually being created in the infinite movement of the event? We are always beginning, in this grammatical logic of thinking our being-in-the-world, in the order of speech, in the fissure of a proposition, in the moment and continual movement of appearing. Thus a fleshy body in a topos, like, as we shall see, a Samuel Beckett body opening its mouth, becomes for Deleuze and for us, the moment when the appearing of the world gains a reality: 'There is always someone who begins to speak. The one who begins to speak is the one who manifests; what one talks about is the denotatum; what one says are the significations' (1990, 181). The 'there is' is just given, and something happens in this given. And so, in sum:

> Pure events ground language because they wait for it as much as they wait for us, and have a pure, singular, impersonal, and pre-individual existence only inside the language which expresses them. It is what is expressed in its independence that grounds language and expression – that is, the metaphysical property that sounds acquire in order to have a sense, and secondarily, to signify, manifest, and denote, rather than to belong to bodies as physical qualities. The most general operation of sense is this: it brings that which expresses it into existence; and from that point on, as pure inherence, it brings itself to exist within that which expresses it. Without it [sense], sounds would fall back on bodies, and propositions themselves would not be 'possible' (Deleuze 1990, 166).

So as we move from questions of ontology (the question of being as such) to the topoi of the worlds in which we find ourselves, part of non-representational theory's legacy is in experimenting with the possibility of other means of representing and thus creating more flexible and inventive analytical frameworks for sighting and intervening in the worlds that come about. In these terms, no analytical framework, nor any mode of discourse or system of signification, can fix the world and make it graspable and knowable once and for all. I caution against sighting this 'outside' as something 'more-than-representational' (Lorimer 2005) as that moniker does, for me at least, do a disservice to the art of representation. The very *raison d'être* of representation is a struggle, and a creation, to put into meaning – through words, images, movements, etc. – the appearing of the world. Representation gives to the appearing world its reality; without it the world is just given and vacuous (see Olsson 2007). The beef of non-representational theory is to deny a too fast foreclosing of what manner this reality might take given it does not believe in a representational system that is productive of an authentic and universally available world. The 'non' of non-representational then becomes very important, and sidesteps the easy slip into any form of transcendent plane (the 'more-than'): I believe that it presents the vital affirmative nihilism of difference itself – in other words, the singular logics of the world in its becoming. Significantly though, and on close inspection, the non-representational frame is always striving or bent towards the artistic more than the social scientific. In this it echoes the aphorism of Nietzsche that 'we have art so as not to go under on account of truth' (quoted in Blanchot 1989, 239). Non-representation is then a mode of foraging for this 'bottomless abyss' (ibid.) that lies outside the domain of truth. There is some concern that as geographers, as social scientists, we are not equipped to be artistic enough. There may not be a geographer on the way to being the next Beckett, Braque, Bergman or Bach, but I do think that there have been a fantastic number of geography articles that exhibit that honest, anxious, and therefore artistic, endeavour to grapple with representation through thought; and part of this achievement is in employing registers of expression that acknowledge stuttering. I think we need to do this because like as not that is precisely what is needed to think the unthought, the singular, and allow different worlds to appear.

The point to remember here is that the work of *doing* art is an event of disclosure,[1] a limit we can approach but cannot extend, can touch but cannot know. The *work* of art 'is what always occurs and continues to occur on the hither side of the world we inhabit' and as such it precisely *inscribes* the rift of Being and beings (Bruns 1995, 535). As Foucault was always at pains to show, 'the breakthrough to a language from which the subject is excluded, the bringing to light of a perhaps irremediable incompatibility between the appearing of language in its being and

---

1   'People are constantly putting up an umbrella that shelters them and on the underside of which they draw a firmament and write their conventions and opinions. But poets, artists, make a slit in the umbrella, they tear open the firmament itself, to let in a bit of free and windy chaos and to frame in a sudden light a vision that appears through the rent – Wordsworth's spring or Cezanne's apple, the silhouettes of Macbeth or Ahab' (Deleuze and Guattari 1994, 203-304).

consciousness of the self in its identity' (Foucault 2006, 15), is the exposure of being, and also the point where the potential of being often gets closed out in false significations of normalization. For the social import of geography it is worth reflecting on the possibility that people don't ask to be known, understood, they ask to be acknowledged. Perhaps then, our politics should be in placing emphasis on this responsibility to acknowledge over that of cognition, that writing 'work' towards what Clark has labelled the 'science of the singular' (1992). This is a perilous zone as it breaks habits and questions intently the 'suffering of being';[2] it is how we are forever tasked to make reality out of the brute phenomenological fact of just 'being there' in the 'there is'.

> Whoever devotes himself to the work is drawn by it toward the point where it undergoes impossibility. This experience is purely nocturnal, it is the very experience of night. In the night, everything has disappeared ... But when everything has disappeared in the night, 'everything has disappeared' appears ... Here the invisible is what one cannot cease to see; it is the incessant making itself seen (Blanchot 1989, 163).

This is then the supposition of the chapter: existence, always from a point of view, rests in the imperative of saying – 'On. Say on. Be said on. Somehow on. Till nohow on' (Beckett 1983, 7)/Capturing the 'onflow' (see Thrift 2008, 4) and the push of life. Then at the same time the imperative of saying emerges out of a cosmology of the pure being of the 'there is'. No one has rendered this 'there is' as powerfully as Beckett, who in literature, script and film, presents it as the void, the dim, and the 'grey-black'. These present the interval being the passage from pure being to being there, from being as being (Being) to topoi (Dasein). Whilst, this is only one way of telling the fiction we tell of the relation/non-relation between the intelligible and the sensible, between thinking and being, between language and event, it is important to accept the that the fiction of the world is not of an all knowing 'I', but nor is it stuck impassively in the ontological and constitutive background of the Difference and indifference of the 'grey black' appearing of worlds. In other words, thinking the singularity of the event, moves us beyond having 'a tormented subject of language, on the one hand, and a non-intentional analysis of the "landscape" of being, on the other' (Power and Toscano 2003, xix). So again, neither absolute meaning nor bare brute empiricism; there is no way of

---

2   'Habit is a compromise effected between the individual and his environment, or between the individual and his own organic eccentricities, the guarantee of dull inviolability, the lightning-conductor of his existence ... The creation of the world did not take place once and for all time, but takes place every day. Habit then is the generic term for the countless subjects that constitute the individual and their countless correlative objects. The periods of transition that separate consecutive adaptations represent the perilous zones in the life of the individual, dangerous, precarious, painful, mysterious and fertile, when for a moment the boredom of living is replaced by the suffering of being' (Beckett 1987, 18-19).

fully knowing, nor is there an essential world to know; there is never 'all' nor is there ever 'nothing else'.

> All of old. Nothing else ever. Ever tried. Ever failed. No matter. Try again. Fail again. Fail better (Beckett 1987, 7).

When it comes to language and the event, the mantra of non-representational theory is to experiment in thought with what constitutes the world: 'Try again. Fail again'. If you don't experiment and 'worsen' inscription you implicitly work on a false economy of perfecting representation towards the illusion of a representational system that can only deaden our senses to the event of the world.

Conclusion

> No symbols where none intended (Beckett 1998, 255)

If we think back to Foucault's indictment to think thought seriously, I think we can agree with the sentiment but it is clear that it is less easy to concentrate upon its demands: to embrace the essential element in human life and human relations, that of thought itself. In other words, to think 'what if thought grasped itself as the thinking of thinking' (see Badiou 2003b, 12)? That thought and language equates to the speaking and writing of Being is a question of the enunciation, and the imperative of the announcement, of appearing: we have to speak or think in words to give a reality of the world (indicative of some kind of language even if via some rudimentary mimicry of others, even animals); try not doing so! But what does it achieve and from whence does it come? But what if these questions have no object, no answer? Rather it is just that the world is achieved – the one that you, in your particularly topos, are operating in now.

> I would like to know what you are searching for.
> - *I too would like to know.*
> - *This not knowing is rather carefree, is it not?*
> - *I'm afraid it may be presumptuous. We are always ready to believe ourselves destined for what we seek by a more intimate, a more significant relation than knowing. Knowledge effaces the one who knows ...*
> - *But we will also lose certitude, a proud assurance ...*
> - *Perhaps ... Uncertainty does not suffice to render modest men's efforts. But I admit that the ignorance in question here is of a particular kind. There are those who seek, looking to find – even knowing they will almost necessarily find something other than what they are searching for. There are others whose research is precisely without an object* (Blanchot 1993, 25).

In Beckett's *Texts for Nothing* the testimony of the texts makes the realization, 'not that there is nothing (Beckett will *never* be a nihilist), but that writing has nothing more to show for itself' (Badiou 2003b, 15). Like Francis Bacon, this is an optimism about nothing; and he 'is not using empty words when he declares that he is cerebrally pessimistic but nervously optimistic, with an optimism that believes only in life' (Deleuze 2003, 43), that here we are in the world, and whilst nothing is certain, there is life and the continual opportunity to create ways of going on. This is thought itself: 'something in the world forces us to think' (Deleuze 1994, 139). In other words, when we write or speak, we are not decoding the world, we are creating worlds; or again, we are not just dealing in interpretations of representations, but also equally presentations for thinking thought and being anew.

It is easy, in the architecture of thought required in thinking non-representationally, to forget the question of humanism even though it is always present. So what of it here? Like Thrift, perhaps we should 'retain a certain minimal humanism' (2008, 13), for, like Badiou, the human, or more precisely humanity, comes about as the 'pure capacity to be affected by the irruption of novelty and to decide upon the event' (Power and Toscano 2003, xxi). That is not to say that this capacity is an essential quality overriding the singular instance of affection in an event, thus defining the human definitively and providing grounds for redemption and removing in the last instance an individual's responsibility for their own life. Rather it performs a pared-down way of thinking the human in its 'atemporal determinants' (ibid., xxii). Non-representational theory is thus akin to Wittgenstein's ladder in writing of a Human Geography where the human appears and disappears in the appearing of the world: thus, the concept, the rung of the ladder, disappears once you have used it to go on in your thought; it does not, then, add-up to a system of thought that can be, innocently and in an hermetically sealed way, translatable to a different empirical site. In other words, it contains in its ethos the fact that the empirical site, or encounter, affects the thought that is thought there: it makes explicit an ethos of attunement to the event of thought itself, to the experience of thinking. The task then is to re-treat representations exactly as they are: presentations of thought in the wake of the event.

Acknowledgements

I want to thank Ben Anderson and Paul Harrison for doing all the hard work in putting this book together, for all their comments, and for allowing me the space to write this particular perspective of and for NRT. Many thanks too to all those conversations with (chronologically!) Nigel Thrift, Paul, John Wylie, Nick Bingham, Claire Pearson, Billy Harris, Derek McCormack, Ben, Mitch Rose, and to my colleagues and postgraduates at Bristol, especially here Charlie Rolfe.

References

Badiou, A. (2003a), *Saint Paul: The Foundation of Universalism* (Stanford: Stanford University Press).

Badiou, A. (2003b), *Dissymetries: On Beckett* (Manchester: Clinamen Press Ltd).

Beckett, S. (1963), *How It Is* (London: John Calder Publishers).

Beckett, S. (1982), *Ill Seen Ill Said* (London: John Calder Publishers).

Beckett, S. (1983), *Worstward Ho* (London: John Calder Publishers).

Beckett, S. (1987), *Proust and Three Dialogues with Georges Duthiut* (London: John Calder Publishers).

Beckett, S. (1998), *Watt* (London: John Calder Publishers).

Blanchot, M. (1989), *The Space of Literature* (Lincoln: University of Nebraska Press).

Blanchot, M. (1993), *The Infinite Conversation* (Minneapolis: University of Minnesota Press).

Deleuze, G. (1990), *The Logic of Sense.* Trans. M. Lester with C. Stivale (London,: The Athlone Press).

Deleuze, G. (1994), *Difference and Repetition.* Trans. P. Patton (London: The Athlone Press).

Deleuze, G. (2003), *The Logic of Sensation.* Trans. D.W Smith (London: Continuum Books).

Deleuze, G. and Guattari, F. (1988), *A Thousand Plateaus.* Trans. B. Massumi (London: The Athlone Press).

Deleuze, G. and Guattari, F. (1994), *What is Philosophy.* Trans. H. Tomlinson and G. Burchell (London: Verso).

Derrida, J. (1997), *Of Grammatology.* Trans. G.C. Spivak (Baltimore: John Hopkins University Press).

Foucault, M. (2001), *The Order of Things.* Trans. A. Sheridan (London: Routledge).

Foucault, M. (2000), *Michel Foucault: Power – The Essential Works 3* (London: Penguin Press).

Foucault, M. (1990), *Maurice Blanchot: The Thought from Outside.* Trans. J. Mehlman and B. Massumi (New York: Zone Books).

Gibson, A. (2006), *Beckett and Badiou: The Pathos of Intermittency* (Oxford: Oxford University Press).

Hallward, P. (1998), 'Generic Sovereignty: The philosophy of Alain Badiou', *Angelaki* 3(3), 87-111.

Hallward, P. (2003), *Badiou: A Subject to Truth* (Minneapolis: University of Minnesota Press).

Hart, K. (2007), 'Phenomenality and Christianity', *Angelaki* 12(1), 37-53.

Heidegger, M. (1998), *Pathmarks* (Cambridge: Cambridge University Press).

Henry, M. (2003), 'Phenomenology of life', *Angelaki* 8:2, 97-110.

Klossowki, P. (2005), *Nietzsche and the Vicious Circle* (London: Continuum Books).

Lorimer, H. (2005), 'Cultural geography: the busyness of being "more-than-representational"', *Progress in Human Geography* 29(1), 83-94.

Marion, J-L. (2002), *In Excess: Studies of Saturated Phenomena.* Trans. R. Horner and V. Berraud (New York: Fordham University Press).

Meillassoux, Q. (2008), *After Finitude: An Essay on the Necessity of Contingency.* Trans. R. Brassier (London: Continuum Books).

Mullarkey, J. (2006), *Post-continental Philosophy: An Outline* (London: Continuum Books).

Nancy, J-L. (1991), *The Inoperative Community* (Minneapolis: University of Minnesota Press).

Nancy, J-L. (2000), *Being Singular Plural.* Trans. L. Dosanto and D. Webb (Stanford: Stanford University Press).

Nancy, J-L. (2007), *The Creation of the World or Globalization.* Trans. F. Raffoul and D. Pettigrew (Albany: State University of New York Press).

Nancy, J-L. (2008), *Noli Me Tangere: On the Raising of the Body.* Trans. S. Clift, P-A. Brault and M. Naas (New York: Fordham University Press).

Olsson, G. (2007), *Abysmal: A Critique of Cartographical Reason* (Chicago: University of Chicago Press).

Power, N. and Toscano, A. (2003), '"Think Pig"! An Introduction to Badiou's Beckett, in Alain Badiou, *Dissymetries: On Beckett* (Manchester: Clinamen Press Ltd), xi-xxxiv.

Raffoul, F. and Pettigrew, D. (2007), 'Translators' introduction' in Jean-Luc Nancy *The Creation of the World or Globalization* (Albany: State University of New York Press), 1-26.

Thrift, N. (2007), *Non-Representational Theory: Space/Politics/Affect* (London: Routledge).

Chapter 9

Testimony and the Truth of the Other

Paul Harrison

Introduction

This chapter is about testimony, and is so in two ways. Firstly, it is an attempt to think through the specificity of testimony as a particular 'act'. Secondly, in working to arrive at such an understanding, the chapter offers a brief genealogy of how testimony has been apprehended and engaged by forms of systematic interpretation, theorisation and calculation.

The initial motivation for this chapter comes from a very familiar situation for many social scientists: in one way or another our informants speak, they signal to us, and in turn we contextualise, code, work-through and interpret. This is, to a large extent, our task and our inheritance. Our inheritance insofar as to place the subject within the context of the social and accepting the 'objectivity of the social', however determined, is, in many respects, *the* constitutive gesture of modern social analysis. As Adorno (2008, 15-17) suggests, social science is, for the most part, established on the basis of a dialectical gesture which asserts the explanatory priority of the social (be it as, for example, the symbolic, the discursive, the economic, the practical, the patriarchal, the libidinal) over the conscious concerns of the individual existent. Crudely put, social science is founded upon the assertion that the latter is an effect of the former, be it is as an epiphenomenon, a symptom or a running fold. This statement is as true if we trace the origins of modern social science to Durkheim's account of 'social facts' *or* to Tarde's of imitation; in both cases the I or the *ipse* of the existent is fused with and understood as a moment within the wider (social) context, the first by (collective) representation the second by (collective) practice (see Abensour 2002).[1] Our

---

1   There is not room here to discuss Marxist approaches to this relationship, which would, I believe, have to focus on the role of work and alienation in Marx's account of human being. I hope to discuss this issue in more detail the future, however Harrison (2009a) contains some brief comments on the topic. Further, where non-representational theories may provide alternative approaches to those which prioritise collective representation (however this is theorised) in the determination of the social, they often do so by emphasising the practical basis of the social. While this may be an interesting and important development it can contain a number of assumptions about the nature of practice and the human. See Harrison (forthcoming) for a critique of the 'ontologisation' of practice. Recently, in geography, Carter-White (2009a; 2009b), has given a nuanced and insightful account of the

informants speak; I am addressed, and I listen and I read. I start to interpret, to analyse, to code, to transpose; to put these in-coming words in their context, in their place (in the social), and thereby find their implicit thesis or rationale; their 'true voice' and so their truth.

I have become particularly interested in moments or 'instants' in given pieces of testimony which confound, resist or simply withdraw from such engagement and which, I believe, do so on quite a profound level, to the extent of calling into question the founding gestures of much of our heritage.[2] The instants in testimony with which I am concerned would, it seems fair to say, normally be identified as at best moments of failure, elision or obscurity, or at worst of obfuscation and potential avatars of fraud or perjury. They are moments when the addressor tells you, and often does so with somewhat alarming frankness, that they cannot tell you what they are about to tell you, or, having just told you of their experiences, that the account you have just been given is deficient and that, in fact, they have not told you anything yet.

Instants when, for example, Wiesel writes that he 'knew he must bear witness', however 'while I had many things to say, I did not have the words to say them. Painfully aware of my limitations, I watched as language became an obstacle', such that 'Deep down, the witness knew then, as he does now, that his testimony would not be received' (2006, vii-ix). Or, when Levi writes; 'our way of being cold requires a new word. We say "hunger", we say "tiredness", "fear", "pain", we say "winter" and they are different things' (1987, 129). Or, when reflecting on his experiences in Breendonk in 1943, Améry writes; 'Qualities of feeling are as incomparable as they are indescribable. They mark the limit of the capacity of language to communicate' (1980, 33). Or, when Delbo writes on the first page of her trilogy *Auschwitz and After*; '*Today, I am not sure that what I wrote is true. I am certain that it is truthful*' (1995, 1 original emphasis).[3] Marginal in the texts from which they are drawn and easily passed over as rhetorical or habitual gestures, these instants seem to me now to be of great importance – indeed I shall argue below that they are constitutive of testimony as such. Where the vast majority of analyses of these and other comparable texts focus on their intent, contexts, content and meaning, and rightly so, *these* instants threaten precisely both the object and practices of such analysis. They open gaps in the text, within the words of the text itself, between the referent and reality, between representation and experience and, perhaps first and foremost, a gap between the addressor and the addressee. For an

---

space and logic witnessing, unfortunately, a more extensive engagement with his work has been impossible due to its being published during the revision phase of this chapter.

2   On the 'instant' in the context of testimony and witness see Derrida (2000a), see also Baer (2005). The cleaving between two 'times', that of 'the instant' and that of the 'on-going', will become important as the discussion develops in section three.

3   As this selection of quotes demonstrates, the testimonies I have been working with are, for the most part, Holocaust 'survivor' testimonies, a term and a category which itself has a complex history, see Waxman (2006) and Wieviorka (2006).

instant these passages threaten to turn the words heard or pages read into a tessaract or hieroglyph, a sealed crypt impenetrable to the analytic eye and interpretive ear.[4] My questions on testimony are concerned with how such instants forestall and resist the dialectical and hermeneutic gestures of social analysis, and with what might happen to our modes of analysis if we accepted the confession of a loss of propriety over language, their somewhat dizzying profession of an alienation, deconcatonation, symbolic collapse or aphasia? What, in *this* instance, would be the responsible response?

Working via the negative, I think such an investigation has the potential to tell us much about our inherited frameworks, economies and structures for understanding the social and, through this immanent critique, approach what may be happening in these instants. While there should be no doubt that the texts in question are clearly heterogeneous in terms of their respective contexts, content, form and authorial intent (and there is much that could be said about these differences), what has surprised me, and what has pushed me along this path of investigation, is how in each of these accounts, and in one way or another, the author remarks on *the impossibility of their testimony*. As described above, each, if only for a moment – and often it is only for a moment – places into question the very 'transaction' and 'warrentability' of meaning which is both already underway and which guarantees the remainder of the text. Each marks and remarks upon the ultimate 'failure' of the testimony to convey. We shall consider the nature of this 'failure' below; suffice to say that what counts as 'failure' is always determined by the definition of 'success'. If these are moments of 'failure', of awkward or quasi-poetic 'performative contradiction', it is because, knowingly or not, our modes of analysis have already defined the conditions of 'success' to which failure emerges as a dialectical corollary. Our modes of analysis are already prompting a disavowal of another communication. As such these moments demand that we – as addressees – listen and read again, not only to try and understand better, more accurately and exactly, but also to try to hear and read our *inability* to hear and read, and so begin to outline our systems of thought and trace their hidden sources.

The chapter is divided into two substantive parts plus a conclusion. While attempting to demonstrate the contemporary currency of the claim, section one looks to classical sources to account for how the 'problematic' nature of testimony and witness became framed as a matter of content and so cast as an issue of 'failed' or 'flawed' speech. From this framing unfold a number of assumptions or

---

4   On crypts and processes of encryption see Abraham and Torok (1994; 1986). However, following Derrida (1986; 2005) and, in certain way, Laplanche (1999), I understand processes of 'encryption' as not being about the burying of a secret or original trauma to be brought to light and restored to the continuity of reason and sense, but rather as marking and remarking an instant, anasemically and quasi-poetically. This shift marks a shift away from hermeneutic or dialectical approaches to discourse, towards one orientated to, perhaps, the *giving* of messages. See Davis (2007) for a lucid account of the relation between Abraham and Torok and Derrida, see also footnote 12 below.

'protocols' in the analysis of testimony which, taken together, constitute the *telos* of analysis as such (and, in the same gesture, the disavowal noted above). As this chapter is primarily concerned with setting out the issues around language *qua* testimony in relation to social analysis, I shall discuss below only a short, though crucial, section from one 'testimony': the opening pages of Delbo's *Days and Memory* (1990). So, section two steps back from the foregoing account to give a reading of the first section of *Days and Memories* and reflect on how Delbo describes the process of 'Explaining the inexplicable' (1990, 1), as the opening words of her book have it. The chapter concludes with a comment on the *non-* of non-representational theory and reflects on some of the issues which the discussion has raised.

The Problem of Testimony

Discussion of the status of testimony has a long, a *very* long history. It is not, nor could it be, my aim here to review this history, instead in this section I want to outline the 'problem of testimony' in more formal and abstract terms; describe how this problem has been framed and outline the 'protocols' which follow from this framing. Importantly this framing of testimony reiterates, in a classical setting, the account given of social science above. We can therefore trace, albeit in shorthand, a genealogy of analytic gestures concerned with the relationship between the individual and the social which take testimony as their object of concern.[5]

While it has many incarnations, the 'problem of testimony' has perhaps been most forcefully restated in recent years by Lyotard in *The Differend* (1988). A key element of Lyotard's account is the apparently simple observation that testimony poses a problem. A problem insofar as the phrases strung together by the witness (or by the one 'wronged') require 'new significations and new referents in order for the wrong to find an expression' (ibid. no. 22), insofar as each wrong, each suffering or loss, is, as it were, non-transferable. Lyotard's term 'différend' describes the disquieting moment prior to the institution and recognition of the witness's address in nominalisation; 'the unstable state and instant of language wherein something which must be able to be put into phrases cannot yet be'; 'In the différend, something "asks" to be put into phrases, and suffers from the wrong of not being able to put into phrases' (ibid. no. 23). The witness and their testimony are caught between, on the one hand, the singularity and particularity of the event to which witness is to be borne and, on the other hand, the historicity, generality and substitutability intrinsic to all systems of signification. And this is why, in

---

5   While, in the context of this chapter, this is very much a *suggested* genealogy, it is one which follows Levinas's account of the thinking of the 'social relation' and the 'common' on the basis of *fusion* in the Western tradition, which, across various texts, he traces from Plato's *Republic*, through Hegel and others into more contemporary thinkers such as Durkheim, Marx, Bergson and Buber (Levinas 1991; see also Abensour 2002).

analytic terms, testimony poses a problem; its sending opens a semiotic circuit which is as yet unresolved but which demands resolution (see Ophir 2005, 133 *passim*).[6] As Carter-White suggests, to testify is not to 'prove' but to '*introduce a problematic that will not be closed down, to present something that is incomplete, yet to be properly articulated*, thereby indicating the *limits* of the external language game in which it is read' (2009b, 166 original emphasis). Simply put, testimony is disquieting. A claim, an obligation or a demand of some kind is announced, is in-coming, however the nature and status of this claim is by no means evident; it has yet to be assessed or accredited. With this formula in mind we may understand how testimony constitutes or poses a specific series of problems for any calculative or interpretive exercise, the premise and primary aim and task of which is to recognise, repair and resolve any such absence or abeyance in the continuity of meaning and sense, to provide a reason and give account.

Drawing on the work of Rancière (1999), Palladino and Moreria (2006) summarise concisely the classical and arguably still conventional context for understanding the 'demand' or 'problem' which apparently 'unresolved' speech poses to calculative or interpretive exercises (at least within the Greco-Christian tradition).[7] They note how rational 'argument has long been constructed as a public dialogue aiming to summon into existence the most equitable political community' (ibid. para 1). From this setting the primary moral and political task of rational thought and argument is to discern and decide upon 'the most agreeable distribution of rights and responsibilities' (ibid. para 1) amongst the members of the community and, therefore, the primary principle of rational and theoretical reflection is '*mathesis*'. That is to say, it is the task and responsibility of such thought to draw out, weigh and come to open, just and accountable determinations of the relative significance and worth of competing claims to public goods and apportion each their due. This is the 'count of reason', reason as giving an account and as accounting, as the dissolution of disagreement and discrepancy through the application of principle.

However, should we proceed too quickly to the point of calculation and judgement (be it via the respective epistemo-political-juridical metrics of, for

---

6 By way of analogy here we may think here of how Spivak (1988; 1999) asks the question of 'can the subaltern speak?', a question which is, she notes, primarily a semiotic problem; one which asks, academics (both postcolonial and otherwise) and institutions (i.e. the state, the family) *how* they are to hear, translate, represent, apprehend an 'odd', 'strange', 'illegitimate', 'silenced' or 'absent' speech.

7 These brackets contain a whole problematic of their own which I can address only implicitly here. The account of the 'count of reason', the framing of testimony in terms of content and the specific enactment and institutionalisation of public space and the social relation, given in this section are clearly located within a certain inheritance of Greco-Christian thought. This account seeks to problematise and so work to delimit this inheritance. The question of other apprehensions of testimony, (of other modes of accounting, of arrangements of the space of testifying, and so of the status of the first person singular), therefore remains open.

example, liberalism, utilitarianism or communism), Palladino and Moreria remind us that it is 'language, insofar as it alone would seem to enable such calculation, [which] founds the political community' (ibid. para 1). The apparent 'transparency' and 'straightforwardness' of any such calculative or interpretive exercise is both effected and affected by the specific instituted forms and protocols of speech; the putative 'rationality of dialogue' determined by the identification and institutionalisation of a particular relationship between speakers; in this case one of (idealised) equality and symmetry. The pragmatic effect here is to render the 'content', (the meaning or the referent), of speech and the consistency thereof as the primary if not exclusive concern. We could say that dialogue becomes framed as dia-*logos* (through-*reason*) at the expense of *dia*-logos (*through*-reason), the cognitive content privileged at the expense of the fact that speech is given (see Carter-White 2009a; 2000b; Cixous 2009b; Dudiak 2001; Lyotard and Thébaud 1985). So, as Rancière observes, in the ideal institution of rational dialogue speakers address 'each other in the grammatical mode of the first and second persons in order to oppose each other's interests and value systems and to put the validity of these to the test' (1999, 44). To be able to speak 'well' in such a 'public arena' at least three aspects of one's speech must be in order; one's topic and oneself must be acceptable to that arena, (of sufficient seriousness and significance); one's speech and choices must be 'reasonably argued' according to accepted standards of veracity and sincerity; and one must speak openly in front of an audience 'who thus participate, even [if] passively, in the operation of justification' (Callon and Rabeharisoa 2004, 6-7). Hence, and this is the key point, a speaker in such a public arena must, 'for the success of their own performance', submit this performance 'to conditions of validity that come from mutual understanding' (Rancière 1999, 45). Simply put, if one is to speak well in a public arena one is bound by an 'imperative to justify' (see Boltanski and Thévenot 2006): speakers must *give an account* of their utterance or claim *beyond* the simple fact of uttering it or holding to it, or else undermine the force of their utterance and the legitimacy of their claim and risk performative contradiction. This is a public or democratic space based on universally accessible knowledge, translation and transparency, within which what or whosoever 'resists the coercion of and to identity, to consistent self-presentation, necessarily assumes the character of a contradiction' (Adorno 2008, 8; see also Gasché 2009, 150 *passim*).

The inheritance here is classical. Plato (1993) refers to the requirement to 'give an account' as *logon didonai* (λογον διδοναι), a literal translation of which is, according to Held, '"to give a *logos*," whereby "to give" is intended as "lay out or show before other humans"' (2002, 83). Goldhill explains further:

> By this [*logon didonai*], Plato means not just having a true belief about something which you wish to communicate; nor indeed being able to declare that such and such a thing is true (when it is). Rather, he means being able to give a systematic account, which is explanatory and which is open to testing and which after testing can be demonstrated to be the case. The outcome of this

Plato calls *episteme*, which can be translated as 'knowledge' or 'science'. Its root meaning is 'knowing (how)'; but in Plato's hands, it implies disciplinary or systematic knowledge, a knowledge which is privileged as true knowledge. Sometimes he calls this process of the production of knowledge *logismos aitias*, 'the calculation of cause/reason' (2002, 96 original emphasis).

It is hard to underestimate the importance of this principle. As Held observes, the proper observance of *logon didonai* was crucial to the formation and functioning of the 'free-space', the 'open' or 'public space', that 'emerges among the Greeks in the *agora* and that transcends the survival-space of the [private] "house"' (2002, 95). Further, just as it marks a highly differentiated social and strongly gendered topography, this move from 'private' to 'public' enacts Aristotle's division between 'voice' and 'speech'; the division between the 'mere' animal and 'man' [sic] as the 'political animal' or 'the living being whose nature is to live in the *polis*' (see Elden 2006; see also Agamben 1999; Heidegger 1997; Rancière 1999):

> For nature, as we say, makes nothing in vain, and man is the only animal who possesses speech [*logos*]. The *voice* [*phônê*], to be sure, signifies pain and pleasure and therefore is found in other animals ... but *speech* is for expressing the useful and the harmful, and therefore also the just and the unjust. For this is the peculiar characteristic of man in contrast to the other animals, that he alone has perception of good and evil, and just and unjust and the other such qualities (Aristotle 1981, 1253a 9-19).

The giving of account, the supplementation of 'voice' (*phônê*) (or 'noise' as it is sometimes translated) by reasonable 'speech' (*logos*), allows for the consistent reckoning of the useful and the harmful, the true and the false, the good and the bad. To be consistent this speech (*logos*) must take place within an understanding of the law (*nomos*) bound nature of the shared world (qua *kosmos*) insofar as a speaker's 'awe' and 'respect' for the *nomos* protects the community of speakers from fragmentation and self-destruction (Held 2002, 94).

Where the aim of reasonable and responsible speech is understood to be clear, full and systematic manifestation, silence becomes the absence or inhibition of manifestation and, as such, is cast as either insignificant (absence as absence) or as a contingent blockage which prevents the 'proper' realisation and assessment of a claim (absence as defect). Silence becomes that which arrests all that might flow from rational argument, including the justice of the public sphere. The same may be said of 'failed' or 'broken' speech, of testimony in which the speaker is unable, for whatever reason, to 'fully' and 'successfully' give voice to their claim in a form which is acceptable or sufficient to the lodging of that claim. In such instances the witness becomes the victim of a wrong not, or not only, in the sense that they may have undergone a specific loss or harm, but because they are in some way inhibited from expressing the loss or harm suffered (see Derrida 2000a; Lyotard 1988; Ophir 2005; Rancière 1999). Following Aristotle's division of voice (or noise) (*phônê*) and speech (*logos*), the problem here is that, as Rancière observes, 'logos is never simply speech, because it is always indissolubly the *account* that is made of this speech' (1999, 22-23 original emphasis). From here speech which does not give an account of itself is, in fact, not 'speech' at all but 'noise' which stands in need of supplemental rationalisation and dialectical clarification. And so from here testimony is apprehended as problematic speech *due to* its unaccredited status; problematic *only insofar as* its lack of account of itself stands *as a flaw in need of repair*. Framed thus, it follows that the basic methodological task which testimony poses to us, as social scientists, is to complete the semiotic circuit opened by the testimony in question *by providing the testimony's missing account of itself* and thereby recover the testimony's true but otherwise partial, embryonic, hidden or elided significance. The analyst is to overcome blockages and inhibitions and make manifest the true meaning of the testimony in question; to move the speech in question from 'the private world of noise, of darkness and inequality, on the one side, [to] the public world of the logos, of equality, and shared meaning on the other' (ibid., 116).

A number of 'protocols' – the word is awkward but the connotations are useful – follow from the framing of the problem of testimony as 'flawed speech'.[8] While presented abstractly here, I believe that one may find these 'protocols' deeply embedded in many of our interpretive practices, indeed I would suggest that in many respects they describe the *constitutive telos of interpretive analysis* as such. To explain; these 'protocols' outline the general actions through which the exteriority of testimony *qua* 'problematic speech' is apprehended as 'flawed speech'; as speech with a discernable but elided set of functions and meanings, functions and meanings which *may be* decoded and recovered *through* social analysis (see Spargo 2006, 259). To put this another way, the protocols describe the general moves through which the exteriority or the singularity of the testimony in question is negated; where the testimony's paradoxical or impossible 'being-in-itself' – as given in phrases which have the form of 'I cannot tell you what I just told you' – are appropriated, translated and shown to be effects of a wider pre-existing structure (see Adorno 2008, 16). How is this achieved? First, through a *revelatory* understanding of truth, wherein analysis is understood as the progressive and systematic removal of barriers to shared understanding and clarity of meaning; analysis as the undoing of knots or dissolving of complications – a task which has as its goal the recovery and restoration of the continuity of reason, meaning and sense.[9] Second, by a *reparative* method, where analytic and interpretative

---

8   Along with its 'non-specialised' meaning as 'procedure' or 'code of behaviour', in this context the term protocol has the added benefit of meaning (according to the OED) in the context of computing and telecommunications 'a (usually standardised) set of rules governing the exchange of data between given devices, or the transmission of data via a given communications channel'.

9   Analysis; ἀναλύ–ειν, to unloose, undo; analytic; ἀναλυτ–ος, dissolved, dissolvable (see Derrida 1998).

work consists in the disclosure of hidden but recoverable factors (ignorance of which, it is presumed, deflects us from appreciating the 'truth' of the testimony in question). Speech is understood as flawed or even 'failed' insofar as it does not convey or make manifest its meaning, thesis, claim, concept or idea. Seeking to overcome this 'failure' consists in reparative supplementation of 'voice' into 'full speech'. Laudably, and with many important and positive consequences, such an effort aims to set right a wrong by bringing into the illuminated 'public' sphere what had up until this point been consigned to the inchoate realm of the 'private', the 'marginal', the 'personal' or the 'subjective'. Third, what we might call a contextual or *topological* reduction, insofar as analysis is an attempt to repair and reveal the truth of the testimony in question it works to place the testimony back into the continuity of meaning and sense; to return it to and situate it in its 'proper' or 'appropriate' place in terms of, for example, its location in a psychic, symbolic, historic or economic system or in a sensible, practical or material world. To return the (now repaired) testimony to the site in which it is 'at home' with itself, identical to itself, and so the place from which its true meaning or thesis may be disclosed, its value assessed and its 'due' apportioned (see Rancière 1998; 2004).

In these ways the 'problematic speech' of testimony becomes apprehended as a *contingent* problem; as *flawed*; challenging certainly but ultimately resolvable. Hence the testimony in question, along with its torsions, elisions and intervals, is apprehended as essentially '*homogeneous to the order of the analysable*' (Derrida 1998, 4 original emphasis). Any secrets it contains are of the order of 'the hidden secret, the dissimulated meaning, the veiled truth: to be interpreted, analysed, made explicit, explained' (ibid., 10). The testimony is, as it were, pre-comprehended by the systems and systemisations of analysis; sense is anticipated and so any idiomatic phrases and impossible concatenations are available to perpetual negation, translation and re-contextualisation into the order of the 'analysable', the manifest and the representable (ibid., 16 *passim*). In this way we may understand how the revelatory act of recovering or 'saving' the meaning of the testimony in question is also, primarily and necessarily, *the act of saving the order of analysis* (see ibid., 1986, xxiii-iv). Hence the barb in Spivak's question 'what is at stake when we insist that the subaltern speak?' (1999, 309); for in *insisting* that the other speak, in insisting that they make manifest their truth, we are, *first and foremost*, insisting on the correctness and legitimacy of our systems of accounting. When we *insist* the subaltern speak, when we *insist* that our informants make sense, knowingly or not, we are insisting on the priority of our explanatory systems. This, therefore, is what is at stake; precisely the truth of the other as it becomes, through the passage to manifestation, just another truth.

'If, However, You Would Like Me To Talk About It ...'

Charlotte Delbo was born on 10 August 1913 in Vigneux-sur-Seine, Seine-et-Oise. In September 1941, while in Buenos Aires, Delbo learnt that her friend André

Woog had been guillotined in Paris. Woog was a communist, he had been arrested in April that year however the death penalty was not a legal option at that point; the sentence was handed down retroactively by a special court established in August. Against the advice of her companions Delbo felt compelled to return to France. She sailed, arriving in France in November. In Paris she lived with her husband Georges Dudach under assumed names and, with him, worked for the resistance producing anti-Nazi leaflets and printing copies of *Letters françaises*. On 2 March 1942 their apartment was raided; they were arrested and imprisoned at la Santé. Dudach was shot on 23 May at Mont-Varlérien, Delbo had been allowed to say goodbye to him that morning. On 24 August she was moved to Romainville and then on 24 January 1943, with 229 other female prisoners, mostly political, she was deported to Auschwitz. The convoy arrived at the camp three days later on Wednesday 27 January. Delbo spent the next 27 months in Auschwitz-Birkenau, Raisko, and Ravensbrük. At the end of June 1945 she was released to the Red Cross and taken to Sweden and then to France (Delbo 1995, 1997, Trezise 2002).

After the war Delbo wrote and, somewhat later, published a number of volumes of memoirs, collected in the trilogy *Auschwitz and After* (1995) (comprised of 'None of Us Will Return' [1965], 'Useless Knowledge' [1970], 'The Measure of Our Days' [1971]) and a historiographic account of the women on the transport of 24 January, *Convoy to Auschwitz* (1997, [1965]), as well as a number of plays, poems and essays. *Days and Memories* (1990, [1985]), a volume which is often regard as a companion to those collected in the trilogy, was her last completed work, published posthumously after her death from lung cancer on 1 March 1985.

Auschwitz and After and *Days and Memoires* are extraordinary texts. The method of composition across the volumes is comparable, each being made up of a series of discontinuous and discreet pieces of prose and poems, some written in the form of a monologue, some as a dialogue between the author and a friend or the author and a stranger, and some as fragments of uncertain provenance. While the trilogy and *Days and Memoires* follow a rough chronological sequence, from deportation through to life in the camps and after repatriation, the form of the texts works against a linear narrative. Indeed time, the time of the text, the time of events and the time of memory, are constantly at issue. As Trezise (2002) observes in his insightful reading of *Auschwitz and After*, from the very first words – the title of the first volume, 'None of Us Will Return' – a temporal fracture is opened between a then and a now, a there and here, both within the text and between Delbo as the addressor and us, her addressees. Despite being published some 20 years later, we may read the title *Days and Memory* in the same way. The title remarking the disjunction between the passing of days, the time of clocks and calendars, and the insistence of memory, a time of an event which punctuates and splinters. As we shall see, in this context 'memory' does not simply mean recall or recollection, as if remembering were like telling a story, but rather the return of an event in such way as to threaten the possibility of objectification, representation and narratvisation (see Caruth 1995a; 1995b; 1995c; 1996; Koffman 1998).[10]

In the opening section of *Days and Memory* Delbo explicitly addressees the tension between the day-to-day, on-going and integrative nature of existence and the persistence of an event which resists assimilation and remains inert in the face of the passing of time. The first paragraph of the book describes the process of 'returning', of the gradual reintegration of both herself as a coherent subject and herself into everyday life:

> Explaining the inexplicable. There comes to mind the image of a snake shedding its old skin, emerging from beneath it in a fresh glittering one. In Auschwitz I took leave of my skin – it had a bad smell, that skin – worn from all the blows it had received, and found myself another, beautiful and clean, although with me the molting was not as rapid as the snakes. Along with the old skin went the visible traces of Auschwitz: the leaden stare out of sunken eyes, the tottering gait, the frightened gestures. With the new skin return the gestures belonging to an earlier life: the using of a toothbrush, of toilet paper, of a handkerchief, of a knife and fork, eating food calmly, saying hello to people upon entering a room, closing the door, standing up straight, speaking, later smiling with my lips and, still later, smiling both at once with my lips and my eyes [...] It took a few years for the new skin to fully form, to consolidate (1990, 1).

A few pages later Delbo comments how 'fortunate' she feels. She has been able, for whatever reason – 'I continue to look for an answer, and find none' – to recompose a life, not simply to live-on but also to live; 'fortunate' to not be like 'those whose life came to a halt as they crossed the threshold of return, who since that time survive as ghosts' (ibid., 2-3). The imagery of a snake shedding its skin is commonly associated with rebirth and, along with that of the snake itself, with the effervescence of life, and this is clearly part of Delbo intention. Still, within the process of reintegration Delbo also describes a partitioning or splitting, a division without which, significantly, she 'would not have been able to revive' (ibid., 3). Her 'fortune' comes at a cost;

> To return was so improbable that it seems to me I was never there at all [...] I feel that the one who was in the camp is not me, is not the person who is here facing

---

10   Readers familiar with Freud will recognise in this formula the outline of his model of trauma; trauma as an event which, for whatever reason, exceeds the capacity of the subject to absorb and which the subject returns to in various ways (for example, in dreams or through 'acting-out') in the attempt to come to terms with it and so reintegrate the self (*qua* ego) (see Freud 1984). However, as indicated, my comments here are largely informed by Caruth's (1995a; 1995b; 1996) re-readings of Freud's account, though see also Abraham and Torok (1994; 1986), Derrida (1986; 1998) and Laplance (1999).

you. No, it is all too incredible. And everything that happened to that other, the Auschwitz one, now has no bearing upon me, does not concern me (ibid., 3).[11]

So, she writes; 'the skin of memory. It clings to me yet'; 'I live next to it, Auschwitz is there, unalterable, precise but enveloped in the skin of memory, an impermeable skin that isolates it from my present self' (ibid., 1-2). And while

> The skin enfolding the memory of Auschwitz is tough. Even so it gives way at times, revealing its contents [...] It takes days for everything to get back inside, for everything to get shoved back into the skin of memory, and for the skin to mend itself again. I become myself again, the person you know who can talk to you (ibid., 3).

Days *and* Memory; 'Explaining the inexplicable'; two times (at least). The time of explaining and of giving reasons; the time of co-presence, of the 'person you know who can talk to you', of speech which makes present and gives account; the worldly, synthetic and synchronic time of 'eating food calmly, saying hello to people upon entering a room, closing the door, standing up straight, speaking'. And the time of the inexplicable; the time of ghosts, an unworldly or *unworlding* time, a diachronic splintering which resists the integrative nature of representation. And so, drawing the first section of the book to a close, Delbo tells us that 'when I talk to you about Auschwitz, it is not from deep memory my words issue. They come from external memory, if I may put it that way, from intellectual memory, the memory connected with the thinking process' (1990, 3). As Davis suggests, the witness has 'good reason – indeed she has no option other than – to give false testimony. The truth of her experience is unavailable to her because it belongs to someone else' (2007, 95). Still, we can, with Delbo, trace the outline of this other time as it silently fissures the text, dividing and doubling each word. As when, for example, Delbo writes; 'Otherwise, someone who had been tortured for weeks on end by thirst could never again say "I'm thirsty. How about a cup of tea". This word has also to be split in two. *Thirst* has to be turned back into a word for commonplace use' (1990, 4 original emphasis). Here, then, we touch on the reason

---

11   Without wanting to assert this interpretation, we may note that the concept of splitting (*Splatung* in Freud) has a long history in psychoanalysis. As Laplanche and Pontalis (1988) describe, the process of the division of the ego were outlined early in Freud's work and relates to his accounts of neurotic repression and disavowal. The process is also central in the development of both Klein's and Winnicott's thought and plays an important role in Lacan's account of foreclosure. Therapeutic processes would normally aim at the reintegration of the ego through progressive objectification and working-through and what follows does not deny the tangible benefits of such work, however it does, following Carter-White (2000b), Caruth (1996), Langer (1991) and Spargo (2006) amongst others, raise questions concerning the overarching *telos* of such work and its hermeneutic metaphysics, in line with the comments of the previous section.

why the instants of testimony with which we have been concerned pose such a problem to and for analysis. 'This word has also to be split in two' – on the near side, in 'commonplace use' and 'connected with the thinking process', on the side of the 'explicable' and the reasonable; the word 'thirst'. And then, on the other, on the side of the 'inexplicable' and the 'incredible', anti-mimetic and anasemic, without support, without a world; the word 'thirst'.[12] Another word silently encrypted within the first. Sealed in such a way that both affirms and problematises any passage which the analyst or interpreter would seek to secure between the text in front of them and a world of common meaning, potentially irrecoverably. Split, angled, doubled – at once speakable and unspeakable, at once readable and unreadable. Here silence and speech, representation and non-representation, failure and success, are no longer opposed but inseparable. The performative speech act of testimony is a 'failed' performative in the sense that it does not convey or deliver, or, rather, what it delivers is unrecognisable and unknowable: 'Non-manifestation manifests itself (*perhaps*) *as* non-manifestation' (Derrida 2005, 90 original emphasis). Another communication, disavowed: The truth of the other as otherwise than truth.

The horizons of our common world have been reached; 'something asks to be put into phrases, and suffers from the wrong of not being able to put into phrases'. These instants in testimony return us to testimony as *problematic* speech as opposed to 'flawed' or 'failed' speech; not to speech in need of repair but to testimony as disquieting. To another communication mutely encrypted into existent syntax, terms and phrases; another, impossible communication lodged within our dialogue. Like a mote floating on the eye or dust caught in the throat. Like Delbo's ellipses dots, which shadow the migration of her words across the skin of memory and the skein of the text: 'If, however, you'd like me to talk to you about it ... This is why I say today that while I know perfectly well that it [the book, her account] corresponds to the facts, I no longer know if it is real' (1990, 4 ellipses in original).

Conclusion: Another Communication

Rather than conclude with a review of the discussion so far I want to use the space remaining to step back slightly and suggest three broader lines of thought which

---

12 The term anasemic is drawn from Abraham and Torok (1994, especially chapter three), wherein Abraham describes 'anasemic discourse' as discourse which works, insofar as it works, by *unworking* signification. Anasemic discourse is an antisemantics, a process of designification, but not one carried out for the sake of a series of alternative, better or more successful names or signifiers. As Derrida writes, 'Anasemia creates an angle. Within the word itself. While preserving the old word to submit it to its singular conversion, the anasemic operation does not result in growing explicitness, in a regression toward original meaning [rather] a change in direction abruptly interrupts the continuity of the process of becoming explicit and imposes an anasemic angulation' (1986, xxxiv).

lead off from the discussion so far, each of which follows on from the one before. First, and perhaps somewhat overdue given the wider volume, a brief comment on the 'non-' of non-representational theory. Second, returning to the characterisation of the social sciences as dialectical given in the introduction, a comment on the relationship between 'the individual' and 'the social'. Third, a general comment on testimony as problematic speech and the demands of enigmatic signifiers and anasemic phrases.

Firstly then; there are, as this collection demonstrates, multiple ways of thinking the 'non-' of non-representational theory. In this chapter the 'non-' does not refer to the pre-conscious process of the brain, the tacit habituations of the body, the inter-objective relations of non-human fields, the autonomy of affect, not even the performative 'push' or the promising excess of the world's potential over its actuality. No, here, in this chapter, the 'non-' is impassive, inert, it indicates that which remains aside from processes and economies of representation, indeed from any integration within a semiological system. To the light of representation, the glory of the word – *En arkhē ēn ho logos* – and of knowing, the 'non-' is a trace of the perpetually undisclosed, the immemorial, it indicates a cut or an interval, a falling out of phase or a dis-location.[13] The 'non-', therefore, indicates another beginning, one which never quite takes-place, as it is, insofar as it is, *atopic*, and which never quite happens, as it is, insofar as it is, *diachronic*. So, not quite another beginning but rather, perhaps, the beginning of the other. Of the other who interrupts and places into question the continuity and spontaneity of our systems of naming, identifying and knowing. Like Xenos, the foreigner or stranger in Plato's *Sophist*, who appears primarily as someone who 'doesn't speak like the rest, someone who speaks an odd sort of language' (Derrida 2000b, 5), but who, aware of the threat which he poses, asks not to be taken for a 'parricide'. Aware of the potential violence which may come to him, he avows that he is not attempting to do away with the *nomos* of ancient Hellas, all the while knowing that thanks to his odd words 'he is already put into question by the paternal and reasonable authority of the *logos*. The paternal authority of the *logos* gets ready to disarm him, to treat him as mad' (p. 11 original emphasis).[14] This is what, for me, the 'non-' indicates; that not all thought is knowledge and that 'this ab-sense is essential' (Derrida 2005, 76), this interruption of knowledge and manifestation,

---

13   '*En arkhē ēn ho logos*' – 'In the beginning was the word' (John 1:1), quoted in Derrida (2005, 69), from which this paragraph takes much of its bearing.

14   Without wanting to make too much of it, and recalling the 'reparative' and 'topological' protocols described in section two, a brief comment on the etymology of the word 'repair'. Repair, from the modern French *repairer* or *repérer*, and the earlier *repadrer*, meaning to return to one's country, *re-* + *patria* fatherland, hence *repatriate*. Hence the argument implicit throughout this chapter that to repair is in some sense to compensate and repatriate, to return a statement to itself, to its proper place under the patriarchal authority of an origin and identity, and hence the concern for 'failed' statements, for those which are without recognised authority, account, thesis or place.

this break in sense, is not an irrationalism but essential to thought (see Harrison 2007; 2009).

Secondly, and following on, in the introduction I suggested – after Adorno – that the founding move of social science was the assertion of the explanatory priority of the social (however defined) over the consciousness or will of the individual existent. To be clear, the discussion above should not be read as describing moments in which an individual somehow asserts themselves *over* the context, as if through strength of will or genius they somehow managed to exceed their conditions; my interest in testimony and commentary on social analysis are *not* carried out in the name of the individual. I have, in this chapter, tended to describe testimony in terms of an 'act', however it is also, inseparably and always already, a 'relation'; addressor to addressee. What I have been attempting to indicate are moments within the testimonies in question which return us to this situation. That is to say, moments which demonstrate how any testimony is, however marginally, irreducible; that it cannot 'be reduced to narration, that is to descriptive, informative relations, to knowledge or narrative' (Derrida 2000a, 38; Carter-White 2009a; 2000b). This is not an assertion of the sovereignty of the individual but rather of what Levinas (1991) calls the 'intrigue of inter-subjectivity'; of the exteriority and proximity of the other *qua* other, there, where 'There is no longer any world to support us, to serve as mediation, ground, earth, foundation, or alibi' (Derrida 2005, 158). As Cavarero writes:

> this ethic desires a *you* that is truly other, in her uniqueness and distinction. No matter how similar and consonant, says this ethic, your story is never my story. No matter how much the larger traits of our life-stories are similar, I still do not recognise myself *in* you and, even less, in the collective *we*. I do not dissolve both into a common identity, nor do I digest your tale in order to construct the meaning of mine. I recognize, on the contrary, that your uniqueness is exposed to my gaze and consists in an unrepeatable story whose tale you desire. This recognition, therefore, has no form that could be defined dialectically; that is, it does not *overcome* or *save* finitude through the circular movement of a higher synthesis. The necessary other is indeed here a finitude that remains irredeemably an other in all the fragile and unjudgeable singularity of her existing (2000, 92 original emphasis).

Thus before either the social or the individual, the non-relation: an anarchical pluralism. The term anarchical here is used after Levinas (1991), key to its use in this context is the reference it contains to a disruptive temporality; an-archy is 'before', 'before principle' and 'before beginning' (from the Latin *anarchia*: an – privative, arche – foundation, in Greek ἀυ – privative, ἀρχός – leader or chief). It indicates, for example, a time outside the time of the sovereignty and the state, in Benjamin's (1969) terms, a time outside the time of historicism, or in Adorno's (2008), a time outside that of identity and totalisation. As Levinas writes:

> The notion of anarchy we are introducing here has a meaning prior to the political (or anti-political) meaning popularly ascribed to it. It cannot, under pain of contradiction, be set up as a principle (in the sense that anarchists understand it). Anarchy, unlike *arché*, cannot be sovereign. It can only disturb, albeit in a radical way, the State, prompting isolated moments of negation *without any affirmation*. The state, then, cannot set itself up as a whole (Levinas 1991, 194 quoted in Critchley 2007, 122 original emphasis).

Thirdly, and to finish, a return to testimony as problematic speech. In the introduction to the chapter I suggested that, in analytic terms, testimony poses a problem insofar as its sending opens a semiotic circuit which is as yet unresolved but which demands resolution. The preceding comments and this demand should belie any suggestion that the arguments presented here simply concern the inability to fix the meaning of any given text. Following Caruth, I would suggest rather that the enigmatic or anasemic nature of the instants we have been considering bespeak the 'insistence of reference'; 'it may indeed be in those moments that are least assimilable to understanding that a referential dimension can be said to emerge [but] where it does not occur as knowledge' (1995b, 3). To refold and recuperate such moments to the field of meaning and representation is to impute a rational horizon, normally cultural or historical, within which the testimony takes place but *from* which, once we accept it as given, the testimony in question decisively departs. The paradoxical nature of the instances with which we have been concerned indicate this momentary absence of encircling horizon, an absence which gives their disquieting urgency. If the world is gone 'Perhaps there is nothing but the abyssal altitude of the sky [...] I am alone with you, alone to you alone' (Derrida 2005, 158). If, as Koffman (1998) has it, the 'ethical exigency' of testimony felt by the witness lies in how to 'phrase the unphrasable', then this is a demand which passes over, has already passed over, to the addressee: How are we to hear the other? How and what could it mean to talk in the absence of world? To what is reason owed?

> the impact of reference is felt, not in the search for an external referent, but in the necessity, and failure, of theory (Caruth 1995c, 103).

Acknowledgements

This chapter is a compacted version of work which has been ongoing for some years now. Over this time I have had the benefit of presenting to and experiencing the hospitality of many people. My thanks to the seminar organisers and participants at the Departments of Geography at the Open University, University of Bristol, University of Exeter and to the Philosophy Department and the Social / Spatial Theory group at the University of Durham. As for individuals, thanks especially to my co-editor Ben Anderson, and to Angharad Closs-Stephens, J-D Dewsbury, Charlotte Harrison, Adam Holden, Mitch Rose and John Wylie.

References

Abensour, M. (2002), 'An-archy between metapolitics and politics', *Parallax* 8(3), 5-18.

Abraham, N. and Torok, M. (1986), *The Wolf Man's Magic Word*. Trans. N.T. Rand (Minnesota: University of Minnesota Press).

Abraham, N. and Torok, M. (1994), *The Shell and the Kernel: Renewals of Psychoanalysis, Volume 1*. Trans. N.T. Rand (Chicago: Chicago University Press).

Adorno, T.W. (2008), *Lectures on Negative Dialectics. Fragments of a Lecture Course 1965/1966*. Trans. R. Livingstone (Oxford: Polity Press).

Agamben, G. (1999), *Remnants of Auschwitz. The Witness and the Archive*. Trans. D. Heller-Roazen (London: Zone Books).

Améry, J. (1980), *At the Mind's Limits*. Trans. S. Rosenfeld and S.P. Rosenfeld (London: Granta Books).

Aristotle (1981), *The Politics*. Trans. T.A Sinclair (London: Penguin).

Baer, U. (2005), *Spectral Evidence. The Photography of Trauma* (London: MIT Press).

Benjamin, W. (1969), *Illuminations*. Trans. H. Zohn (London: Fontana).

Boltanski, L. and Thévenot, L. (2006), *On Justification. Economies of Worth*. Trans. C. Porter (London: Princeton University Press).

Callon, M. and Rabeharisoa, V. (2004), 'Gino's lesson on humanity: genetics, mutual entanglements and the sociologist's role', *Economy and Society* 33(1), 1-27.

Carter-White, R. (2009a), 'Auschwitz, ethics, and testimony: exposure to the disaster', *Environment and Planning D: Society and Space* 27(4), 682-699.

Carter-White, R. (2009b), *Outside Auschwitz: History, Responsibility, Witnessing* unpublished PhD thesis (Exeter: Department of Geography, University of Exeter).

Caruth, C. (1995a), 'Introduction' in Caruth, C. (ed.), *Trauma: Explorations in Memory* (Baltimore: Johns Hopkins University Press), 3-12.

Caruth, C. (1995b), 'Introduction: The insistence of reference' in Caruth, C. and Esch, D. (eds), *Critical Encounters: Reference and Responsibility in Deconstructive Writing* (New Jersey: Rutgers University Press).

Caruth, C. (1995c), 'The Claims of Reference' in Caruth, C. and Esch, D. (eds), *Critical Encounters: Reference and Responsibility in Deconstructive Writing* (New Jersey, Rutgers University Press), 92-105.

Caruth, C. (1996), *Unclaimed Experience: Trauma, Narrative, and History* (Baltimore: Johns Hopkins University Press).

Cavavero, A. (2000), *Relating Narrative. Storytelling and Selfhood*. Trans. P.A. Kottman (London: Routledge).

Cixous, H. (2009), 'Jacques Derrida: Co-responding voix you' in Cheah, P. and Guerlac, S. (eds), *Derrida and the Time of the Political*. Trans. P. Kamuf (London: Duke University Press), 41-53.

Critchley, S. (2007), *Infinitely Demanding: Ethics of Commitment, Politics of Resistance* (London: Verso).

Davis, C. (2007), *Haunted Subjects: Deconstruction, Psychoanalysis and the Return of the Dead* (Basingstoke: Palgrave Macmillan).

Delbo, C. (1990), *Days and Memory*. Trans. R. Lamont (Evanston: Northwestern University Press).

Delbo, C. (1995), *Auschwitz and After*. Trans. R.C. Lamont (New Haven: Yale University Press).

Delbo, C. (1997), *Convoy to Auschwitz: Women of the French Resistance*. Trans. C. Cosman (Boston: Northeastern University Press).

Derrida, J. (1986), 'Fors: The Anglish words of Nicolas Abraham and Maria Torok' in Abraham, N. and Torok, M., *The Wolf Man's Magic Word*. Trans. B. Johnson (Minnesota: University of Minnesota Press), xi-xlviii.

Derrida, J. (1998), *Resistances: Of Psychoanalysis*. Trans. P. Kamuf, P-A. Brault and M. Naas (Stanford: Stanford University Press).

Derrida, J. (2000a), *Demeure: Fiction and Testimony*. Trans. E. Rottenberg (Stanford: Stanford University Press).

Derrida, J. (2000b), *Of Hospitality*. Trans. R. Bowlby (Stanford: Stanford University Press).

Derrida, J. (2005), *Sovereignties in Question: The Poetics of Paul Celan* (Stanford: Stanford University Press).

Dudiak, J. (2001), *The Intrigue of Ethics: A Reading of the Idea of Discourse in the Thought of Emmanuel Levinas* (New York: Fordham University Press).

Elden, S. (2006), *Speaking Against Number: Heidegger, Language and the Politics of Calculation* (Edinburgh: Edinburgh University Press).

Freud, S. (1984), *On Metapsychology*. Trans. J. Strachey (London: Penguin).

Gasché, R. (2009), 'European memories: Jan Patočka and Jacques Derrida on responsibility' in Cheah, P. and Guerlac, S. (eds), *Derrida and the Time of the Political*. Trans. P. Kamuf (London: Duke University Press), 135-157.

Goldhill, S. (2002), *The Invention of Prose* (Oxford, Oxford University Press).

Harrison, P. (2007), '"How shall I say it ...?" Relating the nonrelational', *Environment and Planning A* 39, 590-608.

Harrison, P. (2009), 'Remaining Still', *M/C Journal* 12.

Harrison, P. (forthcoming), 'In the absence of practice', *Environment and Planning D: Society and Space*.

Heidegger, M. (1997), *Plato's Sophist*. Trans. R. Rojcewicz and A. Schuwer (Bloomington: Indiana University Press).

Held, K. (2002), 'The origin of Europe with the Greek discovery of the world', *Epoché* 7(1), 181-105.

Koffman, S. (1998), *Smothered Words*. Trans. M. Dobie (Evanston: Northwestern University Press).

Langer, L.L. (1991), *Holocaust Testimonies: The Ruins of Memory* (New Haven: Yale University Press).

Laplanche, J. (1999), *Essays on Otherness* (London: Routledge).

Laplanche, J. and Pontalis, J-B. (1988), *The Language of Psychoanalysis*. Trans. D. Nicholson-Smith (London: Karnac Books).

Levi, P. (1987), *If This is a Man* and *The Truce*. Trans. S. Woolf (London: Abacus).

Levinas, E. (1991), *Otherwise than Being or Beyond Essence*. Trans. A. Lingis (Dordrecht: Kluwer).

Lyotard, J-F. (1988), *The Differend*. Trans. G. Van Den Abbeele (Manchester: Manchester University Press).

Lyotard, J-F. and Thébaud, J-L. (1985), *Just Gaming*. Trans. W. Godzich (Manchester: Manchester University Press).

Ophir, A. (2005), *The Order of Evils*. Trans. R. Mazali and H. Carel (London: MIT Press).

Palladino, P. and Moreria, T. (2006), 'On silence and the constitution of the political community', *Theory and Event* 9(2), no pagination.

Plato (1993), *Republic*. Trans. R. Waterfield (Oxford: Oxford University Press).

Rancière, J. (1994), *The Philosopher and His Poor*. Trans. J. Drury, C. Oster and A. Parker (London: Duke University Press).

Rancière, J. (1999), *Disagreement: Politics and Philosophy*. Trans. J. Rose (Minnesota: University of Minnesota Press).

Spargo, R.F. (2006), *Vigilant Memory: Emmanuel Levinas, the Holocaust and the Unjust Death* (Baltimore: Johns Hopkins University Press).

Spivak, G.C. (1988), 'Can the subaltern speak?' in Nelson, C. and Grossberg L. (eds), *Marxism and the Interpretation of Culture* (Illinois: University of Illinois Press), 271-313.

Spivak, G.C. (1999), *A Critique of Postcolonial Reason. Towards a History of the Vanishing Present* (London: Harvard University Press).

Trezise, T. (2002), 'The question of community in Charlotte Delbo's *Auschwitz and After*', *MLN* 117, 858-886.

Waxman, Z.V. (2006) *Writing the Holocaust: Identity, Testimony Representation* (Oxford: Oxford University Press).

Wiesel, E. (2006) *Night*. Trans. M. Wiesel (London: Penguin).

Wieviorka, A. (2006) *The Era of the Witness*. Trans. J. Stark (Ithaca: Cornell University Press).

Interlude

Chapter 10

'The 27th Letter': An Interview with Nigel Thrift

Nigel Thrift, Paul Harrison and Ben Anderson

Note on the Text

The interview on which the following text is based took place on the afternoon of 8 April 2009 in the Vice-Chancellor's Office, University of Warwick, UK. We had two broad aims when planning the interview. First, to ask Professor Thrift about his intellectual biography in relation to non-representational theory; his initial influences and how his concerns and interests have shifted up to the present day. Second, to take the opportunity to ask about a number of issues within Professor Thrift's work, such as the fate of the human subject, approaches to social difference and the relationship between his thought and contemporary politics and capitalism. More than an academic paper or book, an interview inevitably bears the trace of its context and moment. The editing process mitigates this to an extent and, in preparing the text, our aim has been both to preserve a sense of the improvised nature of the discussion while, at the same time, organising the material into distinct thematic sections. However the 'momentary' nature of interviews is also their strength. More than a paper or a book, they preserve a sense of being caught between past and future; of being on the 'cusp' where, for Professor Thrift, much of human life takes place. Hence, what emerged from our pre-planned structure was, in actuality, a wide-ranging discussion on the nature and place(s) of the human, one which demonstrates a profound scepticism about the centrality of the conscious human subject to the way the world is, or, rather, to the way that 'natures' go-on. However, this scepticism is not disabling or pessimistic; rather it is dedicated to engaging the diversity and complexity of life and, through this, opening up multiple new opportunities, spaces and registers for thought, action and intervention.[1]

Paul Harrison and Ben Anderson

---

1   The editors would like to thank Nigel Thrift for his time and hospitality, and Jennifer Laws for preparing the initial interview transcript. All footnotes have been added by the editors.

On Process, Practices and Representation

You have published lots of genealogies and itineraries of your thought in regard to non-representational theory and we are not going to ask you to do this again! However looking back now, from this moment, what do you think are the key influences on your thought and on non-representational theory?

The starting point was probably Pierre Bourdieu's *Outline of a Theory of Practice.*[2] The book became a kind of junction point for my thinking about theory; it immediately links to all kinds of other things, which I subsequently read. So, firstly, you have particular theoretical motifs, including Heidegger and others. Secondly, an interest in an ethnographic approach to empirical enquiry. And thirdly, an interest in time, and in process generally; there's a lot in that book, for example, about strategic use of calendars and time more generally. So, in a way, you have to see that book as a kind of start point, mixed with, strangely enough, some of Edward Thompson's work, and for some of the same reasons, partly because of Thompson's interest in time, and partly because of his reaction against Althusser, who in many ways is still an extraordinarily interesting figure, I think, but, in my generation, had caused immense difficulties because of the sheer dogmatism of a lot of his followers, who demonstrated a particular way you could do theory, which was through instilling fear of looking stupid.[3]

And that's something that I've continued to react against. Thompson had this sturdy historian's view of theory, which can be caricatured as some kind of throwback to yeomanry, though I don't think that's fair. But, at the same time, he also had a view that he wasn't going to be shoved around by some fancy French theorist who issued instructions on how to practice the world on the basis of having read a lot of philosophy. And I could sympathise with him after being in a few Althusser reading groups, and politically I had that same reaction too. I'm not that keen on enormous theoretical programmes and I'm not that keen on the macro-diagnostics that come from them, which tell you the way the world *is* – as if you could ever know this with certainty – and gleefully legislate against error. That explains a lot of the reactions I've had subsequently, politically and theoretically. So it's those two things that probably started the whole thing off and the way in which, for me at least, they knocked corners off so many of the apparent certainties of the world.

Going on from that, through the eighties and into the nineties, I read a lot of what would now be regarded at 'practice theory', although I'm not sure it was called that at the time. By the time you get to *Spatial Formations*, I'm in the thick of that style of work; *Spatial Formations* is a practice theory book in many

---

2   Bourdieu, P. (1977), *Outline of a Theory of Practice.* Trans. R. Nice (Cambridge: Cambridge University Press).

3   See for example; Thompson, E.P. (1978), *The Poverty of Theory and Other Essays* (London: Merlin Press).

ways.[4] It's one in which I tried to derive a means of doing theoretical work but one which would have some empirical character to it. I genuinely do believe in the combination; I'm not keen on *really* abstract abstraction!

I'm very unkeen, for example, on the idea that you can dictate political programmes from abstract theory. In fact, I'm pretty sure you can't. Indeed part of the whole issue of politics it that actually that there is an enormous degree of uncertainty involved in is and that's a good thing, as well as a bad thing. So, for example, I am both very attracted to French styles of theory and always have been, and at the same time, I am also extremely suspicious of them – which is probably a bit like Thompson. There's also an issue for me in terms of theory when it gives an account of the world that just assumes that this account is 'how it is': continental philosophy is egregiously guilty of this sin when it comes to pronouncements about capitalism. Which is why I still regard myself as a social scientist; we all know that once you get to social science, you spend an enormous amount of time trying to show that something is – you cannot just assume it. So, that takes me up to *Spatial Formations*.

A critique of or departure from certain ways of thinking about representation within the social sciences was clearly central to the formulation of non-representational theory; was there a particular way of doing social and cultural analysis that you were writing against at the time Spatial Formations *was published?*

Quite simply I wanted to put process into social life in a real way – as opposed to, simply saying, 'we've got some categories, people challenge them and they might change', to actually say that this is a continual process which goes on and on and on without any kind of end. The impetus here was Bourdieu's point that social action takes place in time, and it is tactical. Therefore the idea that somehow or another you can have categories which are immutable doesn't make any sense. So, that was the primary problem at the time; the extremely static notion of culture and in particular, the inability to take into account the on-going or improvisatory element to social life which I became more and more interested in. For me, that improvisatory element is simply a given about how people do things, that is the way that you get by and the way things change – it's the classic ethnomethodological point that a lot of the time, people have to really improvise their way out of a situation. Just as you need to do in an interview, in a way! I wanted to bring that improvisatory moment back.

I tried to achieve that in various ways, but undoubtedly the best way was dance, basically because people reacted so strongly against the idea of dance as being able to say anything serious about social life at all.[5] I still remember presenting

---

4   Thrift, N. (1996), *Spatial Formations* (London: Sage).

5   See for example; Thrift, N. (1997), 'The still point: resistance, embodiment and dance', In Pile, S. and Keith, M. (eds), *Geographies of Resistance* (London: Routledge) pp. 124-151.

a paper on dance at one conference and I thought 'oh well, this will cause a bit of a reaction'. Well, you're not kidding! It was partly a kind of offence that this was just not sufficiently serious; we can't do this because it will somehow or another threaten our political pedigree, and in some cases our masculinity, and it might even cause us to have to take risks. Then there's the second issue with dance; the need to work in every register of the body. One of the problems with representation is that it tends to be – tends to be – a visual register. Of course the visual is important, but it is only one of the registers through which people sense things and in some cases it clearly is not the most important. And all that was being deleted.

The third issue with dance was that if you really were to take some of these things that were being deleted as important, you might end up in a different political place, or at least a more interesting political place. And you could see people actually doing politics naturally grasping this. Just think of a demonstration! There are banners but also there's drums and noise, there's physical force, there's all sorts of things going on. Why we have to reduce it to representation, I don't know.

And then I suppose the last issue with dance was that at that point focusing on dance didn't half annoy a lot of people and that's usually a good sign! When people start to get offended and when they start to react very strongly, you are likely to be on fertile ground. Not always, by any means, but it's more likely to be the case than not. Now this is not a license for saying you should go out and serially offend people or anything of the kind. But it made people uncomfortable and I'm still not entirely sure why it did, but it did. I think part of it is because non-representational theory has built into it a sense of uncertainty and a sense of openness, which people say they want, but I'm really not convinced that they do a lot of the time.

Might it also annoy people because of the manner or style which non-representational theory has offered theoretical propositions? It has tended to be didactic, for example, even if unintentionally ...

I think that's quite fair, actually. I think there is a reason for that, (or at least there *was* a reason for it, I'm not sure there is any more); when it started out a lot of the kinds of literatures that were being drawn on were not familiar in Geography at the time. They are now, without a doubt. So there was a didactic element, there's no doubt about that but I think that has changed now. Certainly I have tried to move away from being didactic; I'm trying to move towards actual concrete moments; the *Halos* paper, for example, is trying to talk about specifics in quite a strong way.[6]

---

6   Thrift, N. (forthcoming), 'Halos: finding space in the world for new political forms', in Braun, B. and Whatmore, S. (eds), *The Politics of Stuff* (Minneapolis: University of Minnesota Press) – available for download from http://nigelthrift.org/downloads/.

On Phenomenology and Life

With the focus on practices there is a strong phenomenological influence in Spatial
Formations, *however since then your focus seems to have shifted to 'life'. What
are the reasons for this change of focus?*

There *is* a strong phenomenological strand in terms of what I do, but I think since
Spatial Formations I've been trying to work out how you can still do that kind
of work in a world which isn't really like that anymore, if I can put it that way.
This is why recently I've been attracted to work such as that of Peter Sloterdijk,
in that what he and others like him are trying to do is say is that well we *can* still
do phenomenology, but only if we change practically everything about it![7] And I
think that's right: there's something I still like about phenomenology but you have
to reinvent practically all of it for it to fit the now! I'll give you two reasons.

First, phenomenology is a profoundly nineteenth century moment in all sorts
of ways, even down to the kind of terminology it uses, based on an encompassing
'environment'. There is, for example, the idea of different kinds of worlds
which dates from von Uexküll and others which comes from around that time;
the idea that there are all-encompassing kinds of worlds, of which the idea of
phenomenology in many ways is just a part.[8] The problem, now, is to try and think
about these worlds again, not as hermetically sealed worlds in which flows can't
take place from one world to another but as moving processual worlds, worlds
whose natures are never static but are always moving and changing and mutating
and communicating.

Second, phenomenology was so centred on the human subject. Even though
we could argue about what kind of subject *is* at the centre, I think that such a focus
is increasingly untenable in the kind of world we are now in. Thus, I think there
is increasing scepticism about quite *what* inhabits the world. 'Life' could be seen
as a kind of place holder in a way, for all these other non-correlationist things that
there are around, which aren't us, and which don't necessarily interact with us
exclusively – and which indeed may sense us as profoundly unimportant bits of
dark fibre, e.g. the world generally, objects, some – *some* – human technologies,
what Latour calls 'plasma' which consists of all the unformatted stuff that still has
force, and so on.[9] All sorts of things in which humans only figure as Rosencrantz

---

7   See for example; Thrift, N. (2009), 'Different atmospheres: of Sloterdijk, China,
and site', *Environment and Planning D: Society and Space* 27(1), 119-138.

8   See for example; von Uexküll, J. (1957), 'A stroll through the worlds of animals
and men. A picture book of invisible worlds', in Schiller, C. (ed.), *Instinctive Behaviour:
The Development of a Modern Concept* (London: Methuen), pp. 5-80; Heidegger, M.
(1995), *The Fundamental Concepts of Metaphysics*. Trans. W. McNeill and N. Walker
(Bloomington: Indiana University Press).

9   Latour, B. (2005), *Reassembling the Social: An Introduction to Actor-Network-
Theory* (Oxford: Oxford University Press).

and Guildenstern.[10] In many ways we have to found a kind of inhuman humanism. One of the things that comes from that stance is that, from 'Afterwords' on, I've been trying to think through what the trace of individual subjectivity would look like under those kinds of circumstances, and I increasingly see it as fading, fading, fading [voice fades out].[11]

It follows that one of the things, consequently, I've become very interested in is what exactly is the use of consciousness? There is a strand of cognitive science, including the work of Metzinger and others, which argues that consciousness has no use at all and that in many ways it's probably a waste of time, evolutionarily speaking, and that, in actual fact, you might be better adapted as a species if you didn't have consciousness at all.[12] Everyone imagines that somehow or another consciousness is a fantastic attribute which actually makes the difference between us and animals whereas, actually, it might be the reverse. For example, there's a science fiction novel called *Blindsight* by Peter Watts; it's a wonderful novel precisely on what happens when humanity actually comes into contact with aliens who are not conscious, but are extremely successful.[13] They don't need it! What do we need it for?!

To come back to go forward: think of Foucault and Derrida's critique of phenomenology, which is partly based around the fact that in certain senses phenomenology doesn't take in life, and to some degree that's surely correct and so the problem becomes how you can actually correct that blind spot. Now, Foucault and Derrida do that in ways which I think are interesting but, in the end, I'm not sure they quite grip it because I think they are both using the tools of a previous era. So, I've been reading quite a lot of biological literature because it gives you ideas about other ways you might actually think about these issues, especially when read through the lens of writers like Muhlmann.[14] I have been parsing this work and trying to link it through in particular to spaces of various kinds, because I'm quite sure that one of the things that's clear about different species, for example, and the move we find difficult to make because of the subject discourse, is precisely the one where we don't see beings just as beings but we see them as beings-and-environments. But then how do you start taking that proposition on? Now, Sloterdijk comes up with one particular way of doing that. For him human beings

---

10   A reference to Tom Stoppard's 1967 absurdist play *Rosencrantz and Guildenstern are Dead* (London: Faber).

11   Thrift, N. (2000), 'Afterwords', *Environment and Planning D. Society and Space* 18(2), 213-255.

12   See for example; Metzinger, T. (2004), *Being No One: The Self-Model Theory of Subjectivity* (London: MIT Press).

13   Watts, P. (2006), *Blindsight* (London: Tor Books).

14   Mühlman, H. (1996), *The Nature of Cultures: A Blueprint for a Theory of Culture Genetics* trans. R. Payne (Wien: Springer), Muhlmann, H. (2005), *Maximal Stress Cooperation. The Driving Force of Cultures* (Wien: Springer/Verlag).

are surrounding-building animals.[15] And what they do is basically furnish a space so that they can turn it into an atmosphere in which they can breathe in one way or another. And there can obviously be all kinds of different ways of breathing. I'm very attracted to that idea and yet at the same time, it also seems to me quite conservative in spatial terms and I'd just like to take that on. I'm still thinking about this, but it seems to me key to think about the media that you're working with as part of the life that you're in.

So, as far as *life* is concerned, I would want to talk about speciation because you could talk about various forms of deep human history where some degree of speciation actually is probably taking place. But you can't see that just as genetic, you need to see it as this enormous cobweb of all sorts of different kinds of things that are happening all at once, including objects, habits, media, et cetera, et cetera, et cetera. It's about trying to think speciation not just biologically – we've reached a point where it almost makes no sense to think of it in that kind of way – but it also doesn't make any sense to think of it without the biological.

What is the relationship between your approach to the theme of 'life' and contemporary biopolitics as, for example, in the work of Giorgio Agamben or Michael Hardt and Antonio Negri?[16]

I did use some of Agamben's terminology however in retrospect I think that this was probably a mistake.[17] The reason for that is because his is an almost absurdly black and white account. That's the problem I had with Agamben's account; when you actually start trying to use that kind of work in any empirical sense it very rapidly breaks down, it is so abstract that it doesn't work very well. So, the issue is trying to find ways of proceeding that don't crumble in that way.

Equally, I am very nervous about Negri's thought on 'Empire', class, and contemporary capitalism, because it elides so much in order to produce a gripping account. The problem is that it makes a compelling story, and you can see why people are attracted to it, because in a way it's so all-encompassing it gives you the answers you need so that you don't need to think too much more, or your function is simply to elaborate what's already there, applying the same theoretical template

---

15   See for example; Sloterdijk, P. (1998), *Sphären I – Blasen, Mikrosphärologie* (Frankfurt am Main: Suhrkamp); Sloterdijk, P. (1999), *Sphären II – Globen, Makrosphärologie* (Frankfurt am Main: Suhrkamp); Sloterdijk, P. (2004), *Sphären III – Schäume, Plurale Sphärologie* (Frankfurt am Main: Suhrkamp). See also the recent special issue of *Environment and Planning D: Society and Space* 'The Worlds of Peter Sloterdijk', 2009, 27(1) (eds). Elden, S., Mendieta, E., Thrift, N.

16   Agamben, G. (1999), *Homo Sacer: Sovereign Power and Bare Life* trans. D. Heller-Roazen (Stanford: Stanford University Press); Hardt, M. and Negri, A. (2000), *Empire* (London: Harvard University Press).

17   See Thrift, N. (2000), 'Still life in nearly present time: the object of nature', *Body and Society* 6(3-4), 34-57.

over and over again. And, as I've already mentioned, I'm not keen on that kind of thing; I don't agree with the kind of theory that can act as kind of 'cover set' which does not produce challenges to its own ways of thinking. We could think about the postcolonial moment in this; for example, Hardt and Negri's ideas become *extremely* problematic when you think of the different kinds of cultures that exist in the world and the different kinds of ways they actually think about things: think about things that we take for granted in ways which are so *wildly* different that the idea that they can all be encompassed under a notion like 'Empire', even at a political level, strikes me as bizarre. Some of the political struggles going on in the world at the moment are very difficult for us to understand in the terms that *we* press on to politics and what it is.

On Life After the Subject

Mirroring the shift in focus from 'practices' to 'life', in Spatial Formations *and a number of other places,* The Still Point *for example, there is a refrain around 'practical self-fashioning' as potential ethico-political strategy, however by the end of* Non-Representational Theory, *and certainly in the most recent work, this theme seems to have disappeared. What's left over in the 'fading out' of the subject? Where does, for example, the political act come from without this?*[18]

I take this as a very serious issue in all sorts of ways, partly because there's the issue of what is it that we do leave behind, and I don't think that it's nothing, and it's also about working out what it is that's giving the 'push' to the world, especially when it comes to politics, if you're not willing to make the easy appeal to transcendence or immanence. And it's clearly the case that there is a something that is being fashioned, but I find it increasingly difficult to know what that might be. I have a feeling that we need something like a 'collective subjectivity' to be in there in some way, however framed, otherwise you do have immense problems. Part of that collective subjectivity is clearly bodies interacting with one another and passing certain things on through those alliances and the way that, at certain points, human bodies can have a privileged influence. However, at the same time, I think we need other kinds of ways of thinking about the way that the world is actually incarnated, other than just through the lens of subjectivity.

So, to take Sloterdijk again, in the end he's very interested in soul fashioning as being the way to go, ethically and politically. In his latest work he talks a lot about how one might become disciplined in good ways, which sounds to me, although

---

18   Thrift, N. (1997), 'The still point: resistance, embodiment and dance', in Pile, S. and Keith, M. (eds), *Geographies of Resistance* (London: Routledge), pp. 124-151; Thrift, N. (2008), *Non-Representational Theory: Space, Politics, Affect* (London: Routledge).

I'm sure he'd be somewhat horrified to hear this, as quite Foucauldian.[19] In many ways he's quite an admirer of a Wittgensteinian modernist asceticism. However the problem is, as he points out himself, that this tradition of asceticism all but disappeared after Wittgenstein and reinventing it will not be easy – notwithstanding Sloterdijk's attempts. In the end, I think that the notion of soul fashioning is interesting but it strikes me as going too far towards a kind of voluntarism based on the individual – though I am sure that this is not the intent. This is where I would agree with Marxists; I just think the idea that somehow or the other you're going to solve the world through fashioning yourself strikes me as too close to reinventing a kind of humanism, whatever the caveats.

In many respects I think I would prefer to replace the subject with other terms. At the same time you can see the pain this causes people when you try to do it, even though there are many philosophical forebears now. One of the classic critiques of what I do, and there is no doubt about this because I've had this argument in various bars with various people, is that by switching off the subject, you've switched off all the contributions that people have made to producing a particular kind of political movement and all the work of soul-fashioning that's been done to get to this particular point, which is undoubtedly better than before. Many feminists feel this particularly deeply – although, equally, others have clearly taken up the challenge of working the issues through. Whatever the case, you clearly have to be very careful. I am not trying to denigrate distinctive political subjectivities like feminism, but at the same time, I want to find a space for a different kind of inhuman vocabulary for what would have been called their agency.

What is then the relationship between acknowledgement of difference, of social or cultural difference, and your interest in speciation? How does your interest in speciation take difference into account?

Yes, well that's where I would come onto some recent work in anthropology, in particular that of Viveiros de Castro on multi-naturalism, as well as the rediscovery in philosophy of James's work on the pluriverse and Souriau's work on different modes of existence.[20] You might think of a kind of multi-naturalism in which you have a lot of different kinds of natures, if I can put it in that kind of way, they are *almost* mini-universes, which have very different terms of engagement arising out of different modes of existence. That's not to say that they're not all interacting as a cultural whole, but it is to say that, at the same time, they have fundamentally

19   Sloterdijk, P. (2009), *Du Must Dein Leben Andern. Uber Religion, Artistik und Anthropteknik* (Berlin: Marz),

20   See for example; Viveiros de Castro, E. (1992), *From the Enemy's Point of View. Humanity and Divinity in Amazonian Society.* Trans. C.V. Howard (Chicago: Chicago University Press); Souriau, E. (1943), *Les differents modes d'existence* (Paris: Presses Universitaires de France).

different natures/bodies: we are not just adding differences into some putative whole.

The more I've thought about it, the more keen I am on understanding the world as a series of naturalisms, and those naturalisms would be partly biological; they'd be all sorts of things at once. And the speciation that's involved with them would take in the environment that people are actually living in, the kinds of spaces they actually produce in order to cope with that environment, the kinds of descriptions they therefore produce, in terms of, even cardinals like up down, and sideways, for example, and the kinds of prospective commonalties that might be possible with other natures. So that's where I'm trying to get to, but I don't know where that means I'll end up at the moment. And added into all that is the technologies by which people themselves describe these kinds of worlds, which is why I have got so interested in classical Chinese culture because there you can find a nature, if I can put it that way, which really has very different means of making objects, very different means of understanding people, even very different means sometimes of counting and writing, which include in them all sorts of ways of perception and sensing that I don't think we would necessarily include as counting or writing at all.[21] And, of course, very different aesthetic hierarchies, some of which would strike us now as utterly bizarre. This is a nature as far as I'm concerned, a different mode of existence

It's inevitably the case that there will be tensions at this point because, of course, extant social categorical differences are themselves cultural artefacts involved in worlding in a quite strong way. So, you've got to work with those categories because those are the categories you're brought up with and through which, in some senses, you can make sense of the world. But you have to understand that the rest of the world isn't *necessarily* interested in those differences in the same way. The problem is that too often we get stuck at the level of people perceiving one world differently and I think that the deal is bigger than that. I think there are many worlds and they really are different! What's the 27th letter of the alphabet?

Isn't there a danger that by radically downplaying the subject you smuggle back in a normative subject?

Yes, I understand the objection, but at the same time I never thought I wasn't being normative! And I think it would be extremely foolish for anyone to think they weren't being normative. I'm not sure, to be honest, how that's possible. We either have unconscious or semi-conscious norms or conscious norms, but we certainly have them and I don't see any other way we could live. We can argue about what we *mean* by that and the degree of freedom we give it and, indeed there is a growing literature in political theory that considers precisely that issue. However, whether one needs to talk in terms of a *subject* in order to talk about these things is another

---

21   See Thrift, N. (2009), 'Different atmospheres: of Sloterdijk, China, and site', *Environment and Planning D: Society and Space* 27(1), 119-138.

matter altogether. Think of someone like Deleuze: no one could accuse Deleuze of being other than normative in all sorts of ways, he undoubtedly was, but he wasn't keen on 'subject talk'. In a way, I want to think in the same manner.

So I would want to put it in a different way, which would be much more about the old debates around habit. So, in much Christian thought there is the view that habit is something which is *bad*, you get stuck in it and it is habit which just simply rolls things over. A second view, the view of Aquinas and others, is that habit is a disposition which gives you freedom, within particular bounds, to be virtuous, and it's a habit you've chosen and which you choose as a virtue to stick to – a view which actually comes close to Foucault at some points.[22] So, in that sense I'm quite happy to believe that there's a normativism about what I'm doing but it doesn't deeply concern me – and I don't think I've ever tried to present it as anything else that is somehow being smuggled in.

On Capitalism and Politics

Paralleling the development of non-representational theory, has been a diagnosis of the contemporary of Euro-American societies in terms of a soft capitalism, including the engineering of invention, new forms of commodity and address, and the production of worlds not things. And yet this diagnosis has often been in the background. By way of an introduction, could you discuss the relation between non-representational theory and this diagnosis of the contemporary? What is it that makes non-representational theory particularly well placed to diagnose and engage the present moment? Also, more pointedly might we see, as some have claimed, non-representational theory more as a symptom of the present instead of a diagnostic?

The soft capitalism work, like my other strands of work on international finance or the history of time, has been taken mainly as separate by commentators, I think.[23] Only a few people have made the link. But I do mean there to be a link, let me say straight away. In a way, I'm enough of a Marxist to feel that there ought to be some kind of economic base to some of the things one's trying to understand. I mean I do think there are laws of supply and demand, for example! But what I've been trying to do is look at new currents in capitalism of one form or another – in Western capitalism on the whole, to be specific. New currents which have actually taken up in practical form exactly the same kind of motifs that inspire me and used it as a force to enliven the way in which business is done. Now, there are a number of different takes on this. You could direct it through the whole idea

---

22   Davies, B. (2002), *Aquinas* (London: Continuum); Foucault, M. (2000) *Ethics: Subjectivity and Truth: Essential Works of Michel Foucault 1954-1984, Volume 1* (London: Penguin).

23   See for example; Thrift, N. (2005), *Knowing Capitalism* (London: Sage).

of immaterial labour, you could take it through Virno and Lazzarato and authors like that and the way in which they argue that capitalism increasingly tries to call on the powers of the whole body to produce productive momentum.[24] You could say it's because gradually capitalism has developed a whole series of techniques which actually allow forces like semiconscious imitation to be explicitated. Once you can actually see them, you can operate on them. You can meter them. Then you can make them into a resource. I think capitalism is trying to stage the world literally, in different kinds of ways, so as to allow these things to be measured and captured – a bit like a series of honey traps. You can actually pull them in because you change the spaces, you change the environments, you change the spheres in which things actually work. And, in doing that, you've produced a kind of spiders' web, which allows these kinds of things to be picked up and amplified.

So, I tend to concentrate on new tendencies that I see appearing and ask what one might do about them. Of course that has downsides to it. The obvious criticism of it is that the constant searching after what one thinks of as novel in something skews the political diagnosis. In some ways, though, I think, 'well, fine!' There are plenty of people working on these other things and I don't think that we all have to do one thing. But there is another reason why I'm unapologetic. I think that most of what we count as human life exists right at the cusp between present and future. The problem being, of course, so is a large amount of capitalism. Look at what has happened with international finance. You could argue that there that cusp was just given free rei(g)n. In a way, in international finance what happened was that the cusp was instituted as the only thing that counted. But the other thing I'm interested in is the really mundane technologies which actually do frame these new tendencies. That's why I'm fascinated by what I've called the cultural forms of breathing; timekeeping, modes of locating address, repair and maintenance – and other processes and practices we can't get by without. The human version of Darwin's worms, if you like.

One thing following on from this stance that I think we need to understand – and this is the other thing which I have continually said – is that theory is now complicit in actually building this kind of world. You can't just walk away from it and say, 'this is all horrible and awful'. It's just too easy and too lazy a move. This produces a question, which I constantly want to ask, which is about the purity of critique. I don't think there's an 'outside'; little islands we can all sit on surrounded by the ocean of capitalism saying, 'God! That's bloody awful!'. Then you have to start asking, 'what is it you would be trying to do under these kinds

24   See for example; Virno, P. (2004), *A Grammar of the Multitude. For an Analysis of Contemporary Forms of Life.* Trans. I. Bertoletti, J. Cascaito and A. Casson (New York: Semiotext(e)); Virno, P. (2008), *Multitude: Between Innovation and Negation.* Trans. I. Bertoletti, J. Cascaito and A. Casson (New York: Semiotext(e)); Lazzarato, M. (1996), 'Immaterial Labour' in Hardt, M. and Virno, P (eds), *Radical Thought in Italy: A Potential Politics* (Minnesota: University of Minnesota Press) pp. 133-147; Lazzarato, M 1997 *Lavoro Immateriale: Forme di Vita e Produzione di Soggettività* (Verona: Ombre Corte).

of circumstances?' I think that there are several different answers to that. With Negri, for example, it seems to be a state of almost permanent networked activism leading to the recapture of the commons, achieved how? I have never been quite sure. With Sloterdijk, I think it's that one would simply try to be more disciplined but discipline has to be redefined as an aesthetic of abstinence, a reinvention of decorum. With Zizek, it is a grim kind of irony turned into a weapon that is able to foster the messianism of lost causes. With Badiou, it is a neo-Platonic certainty about the underlying revolutionary mission. So, there are all kinds of ways you could think about this, but what you can't do is step out of it. I am actually keen on trying to produce moments that might change things in some way or the other. That is, above all, about producing a tactical sense of propensity which is more than just saying, 'oh we'll give this a go', and less than a plan. If we are now in a constant war of modulation over the propensity of events, then how can we actually deal with it? What potentials does it have? What new tactics might have become available? The obvious criticism is you're being taken in by the machine by doing this. The obvious rejoinder is what would you *do* otherwise?

How does this diagnosis relate to the political dimensions of non-representational theory that you've drawn out over the last decade. These have often been phrased around a concern for democracy (i.e. 'What's Left? Just the future', 'Life, but not as we know it', 'Halos'*). In particular, concerns for 'more' democracy, new modalities of participation and innovative forms and fora which aim to 'address the deficit of felt powerlessness' and 'sidestep' existent 'behavioural codes'?*[25]

The only coherent thing I can probably say about non-representational theory in terms of its politics is that it's an attempt to try to increase the ways that we can act into the world but with that sense of uncertainty and openness that I mentioned earlier. Which is not an attempt to delete all the other ways of doing politics; I don't know why some people think it is, but some unfortunates clearly do. It is not. Not even slightly. Rather, it is saying there may be other political genres that we could invent, especially if we concentrate on building new kinds of space. Now, I don't know what they are precisely, but that doesn't seem to me to be any kind of reason to not set out. There are ways in which it is possible to be open to the world which can be positive and can do things – and I think non-representational theories tend to add to that kind of armoury of achievement. That's what I think it is about. It's not trying to do more than that. Other people might want to do more and that's fine, I've got no problems with that, but it's not what I'm trying to do.

---

25   Amin, A. and Thrift, N. (2005), 'What's left? just the future' *Antipode: A Radical Journal of Geography* 37: 220-238; Thrift, N. (2007), 'Life, but not as we know it' in Thrift, N., *Non-Representational Theory: Space, Politics, Affect* (London: Routledge) pp. 1-26; Thrift, N. (forthcoming), 'Halos: finding space in the world for new political forms' in Braun, B. and Whatmore, S.J. (eds), *The Politics of Stuff* (Minnesota: University of Minnesota Press – available for download from http://nigelthrift.org/downloads/.

Thinking about politics a bit, what I try to do is ponder on the ways in which you can actually concentrate things in such a way as to produce moments in which people do reflect. But what I mean by reflection is not necessarily the Greek philosopher sitting down on a stone somewhere. I'm talking about ways in which people might just think about the world a bit differently. I'm not anti-rationality. I'm quite pro-rationality, but I want to change what it is that we count as rationality. To do that, we have to get away from the notion of the demos in which everyone sits down and rationally reflects all the time on what's going on. I'm not saying that is not a good moment, but I suspect that it's always going to be a rare moment, given what Kant called humanity's 'unsocial social' character as a species. So I go the other way from the critical theory tradition: it almost formalises it too much for me, in that by exactly specifying in many cases what one means by rationality and how one does it one actually is in danger of doing the opposite. So if I've underspecified, I think critical theory over-specifies. We have to think about other ways in which we can do these kinds of things and the production of situations which allow that. That's why I'm so interested in the performing arts because I think if you look at the best performing arts they actually do jolt people, quite severely sometimes, into places they never thought they'd get. And they may well recoil from those places but at least they've been there. That's a great and wonderful moment in some cases. In some cases, it produces moments of epiphany; in some it still produces apathy. But at least you've had a go. And it's that ability to try to concentrate energies in various ways in order to produce those situations that I'm interested in. Now that's not to say there's no other things one should do. There are. But it's that that I'm interested in.

Going on from that point, why it is so important to do that is because it seems to me that democratic politics is often being subverted nowadays. I'm actually quite in agreement with Sheldon Wolin's work on totalitarian democracy in the sense that one of the problems I think we face is that democracy really can be manipulated in fairly predictable ways now, as well as the ways that people think about politics more generally.[26] Take the use of what we could call technologies of non-hesitation, technologies of making sure that people don't hesitate or think about things. We surely ought to be worried about that: in *democratic* politics at least, particular pre-treated conclusions can be produced through the concatenation of media and all sorts of other elements of life. Whereas the main condition of democracy so far as I'm concerned is anxiety. If there's one thing I think one should always be in a democracy it's anxious: about what might be going on and how it is going on. Now, I don't think that's negative, I think that's positive. You should have a kind of readiness/awareness which is an anxiety about what the future may bring. And it strikes me that certain technologies actually subvert that: they say, just act, get over it – get over the hesitation. Just do it. In those situations it seems to me that we have to produce other political genres as well, as

---

26   See for example; Wolin, S. (2009), *Democracy Incorporated: Managed Democracy and the Spectre of Inverted Totalitarianism* (Princeton: Princeton University Press).

well as fighting that tendency. In that sense I have a pessimistic view of western democracy, mixed with a view that we can actually think our way out of it by producing reflective situations.

I'll give you an example from the United States, which I suppose represents both the best and the worst of it. So the worst of it undoubtedly was George W. Bush's tenure of the presidency and the way that was won. It was won through fairly explicit media techniques of one form or another working on a relatively small group of voters who were crucial. We know that in democracies most people make their minds up about how they're going to vote really quite quickly and often on the basis of moments which are not really about explicit programmatic content as such, but more about feel. And so, you know, those kinds of technologies allowed Bush to gain the presidency with all of the consequences we saw afterwards. At the same time you can also see the best of it in Obama's campaign. I'm not saying that Obama will turn out to be a saviour of the American left or anything like that; it would be remarkable if he was. But at the same time, you can see the way in which a whole politics of hope actually did gather around Obama in ways which I actually thought were mainly and generally positive. And a lot of that was actually about structuring events in new ways so that people could participate in them and were allowed then to think differently. In particular I think what they did was to structure volunteering in ways in which it had not been structured before. It gave people a sense that this was something important, that this was something they could and should do. What struck me about the Obama campaign was the number of academics who actually became involved in it went out on the stump and worked for the campaign. Often academics are extremely sceptical as a constituency and they *did* become involved. So, there's the worst aspect of modern democracies of this kind but also the way in which actually you can use the media in ways which are positive about things as well. It takes people far more clever than I am to do this. For me it's the shuffle backwards and forwards which is important. For every awful tendency you can often find an equal and opposite one.

What becomes of judgement? How do different experiments in diagnosing the present effect our definitions of criticism and critique? How, in particular, can we to think about different modes of evaluation or judgment with the aim to help create (if only in minor ways) the possibility of different and better worlds?

Well, I think if you look at a lot of the skills that are now appearing, they are means of trying to better describe the cusp between present and future and how to inhabit it. You have to, in a sense, make a space for this activity, a tiny space in time in which judgement can thrive. Emancipation but with attachment. And I think that people are increasingly concentrating on constructing the kinds of practical skills that enable such emancipatory spaces to be built which can act both as interventions and as cultural probes. You can inhabit those new spaces in different kinds of ways which provoke different kinds of thoughts. And it's about trying to work out the different ways that we can inhabit these temporary worlds, not all of which have to be about being a good consumer. Which is why I'm especially interested in the new practical forms of geometry that are appearing, that are about precisely trying to *build* these spaces by simultaneously measuring them out differently and by producing new and unexpected alliances out of that work of measurement. If the practitioners of these new arts/sciences can get it right, we might be able to learn to breathe differently by discovering a lot more about the slight surprise of action found in every encounter. That's what I hope anyway.

PART III
Ethics

Chapter 11

Thinking in Transition: The Affirmative Refrain of Experience/Experiment

Derek P. McCormack

You do not run purposively through the world because you believe in it. The world, surprisingly, already runs through you. And that, really felt, is your belief in it. Virtual participation, really, brinking on truly, precedes actual cognition (Massumi 2000: 167).

Introduction

Experience is one of the most problematic of philosophical terms. It has been dismissed as a mere veil over the underlying truth of nature, and as the refuge for a kind of brute philosophical materialism. Its unproblematic affirmation has also been criticized on the basis that it assumes a shared set of values anchored in a universal and distinctively human subject. The championing of experience as a useful philosophical category is also often taken to assume the possibility of an authentic relation between self and world. Indeed, it is precisely the impossibility of this co-incidence that a range of post-structural thinkers have sought to place at the centre of philosophical and, indeed, ethico-political thinking. Moreover, as Giorgio Agamben (2007) has observed, various social, cultural and technological developments have eroded the stability, translatability, and authority of experience as a reference point for everyday life. For Agamben then, 'the question of experience can be approached nowadays only with an acknowledgement that it is no longer accessible to us' (2007, 15). Experience, at best, is something that can only be approached asymptotically.

In this chapter I want to consider how non-representational theories approach the question of experience. As Nigel Thrift reminds us, the category of experience is central to the practice and promise of non-representational theories. Indeed, the primary object of these theories is that which is 'present in experience' (2007 3). But what is present in experience for non-representational theories? Perhaps it is easier to begin by saying what is *not*. What is not present in experience for these theories is a representational picture of the world: indeed, these theories are relentlessly critical of any sense of experience as something that takes place through an act of cognitive representationalism. Experience is not our way of producing a synthetic facsimile of raw sense data. Nor is experience an after the event event: it is not (only) something we make sense of retrospectively through reflective contemplation. So non-representational theories move against a model of experience that would install a division between a perceiving subject and perceived object, and precisely because, in the process, experience is reduced to the status of something upon which a thinking subject reflects through an act of separation and transcendence.

So much for the nots of non-representational theory: experience is also something affirmed by non-representational theories, albeit in qualified terms. Such theories affirm experience as a kind of distributed, immanent field of sensible processuality within which creative variations give rise to modifications and movements of thinking. Experience, for non-representational theories, is sensible processuality without transcendent reflection. On one level, the aim of non-representational theories is the production of a diagram of this field, and to offer a critique of the contemporary techniques and technologies – from driving to dancing – organized to transform aspects of this field through generating new forms of experiential sensoria. But these theories also encourage, and indeed insist, on the necessity, where appropriate, of experimenting with such techniques and technologies in the hope that they may allow us to cultivate more ways of attending to the world. And they do so in anticipation that such experimentation can contribute to the elaboration of an expanded space of ethical and political potential, articulated through affectively layered 'dispositions' (Connolly 2002), 'sensibilities' (Bennett 2001a), and 'stances' (Gibson-Graham 2006) that might modify and work against the problematic tendencies of contemporary cultural and political life.

Non-representational theories can therefore be understood as simultaneously offering an affirmative critique of experience and an affirmation of an expanded sense of the experiential as a register of ethico-political experiment. This is not to say that the relation between critique and affirmation in non-representational theories is always equal. Given their differentiated and sometimes divergent tendencies, it is not surprising that the emphasis in this relation often varies quite significantly. My focus in this chapter is on strands of non-representational theories that can be said to have strong affirmative tendencies with respect to the question of experience: theories that share a commitment to finding ways of thinking – conceptually and otherwise – through the processuality of a world that, whether we like it or not, is always affirming its own becoming through the refrain of something which can be sensed in experience while always exceeding the actuality of this sensing. The strands falling under this heading include, amongst others: the speculative philosophy of Spinoza; the process philosophy of Alfred North Whitehead; the pragmatism of John Dewey and William James; and the transcendental empiricism of Gilles Deleuze and Félix Guattari. Clearly the work of these figures is diverse and divergent. Yet they share a number of common elements. They move away from a subject-centred account of affirmation, and away from versions of social constructionism in which world is conjured into being through the performative effect of discourse. Instead, they affirm world as a distributed and immanent field expressed through the becoming of difference. They also propose and enact a philosophy that is world and life affirming, while always recognizing that life is not limited to the figure of the human. And they affirm the fugitive temporality of world as a source of change and of novelty. In doing so these affirmative philosophies also lean heavily on the concept of the virtual, the necessary more than actual real excess of the world, without which, in the words of Alfred North Whitehead, we would be stuck on the 'narrow ledge of definite instantaneousness' (2004, 73) in a world with no more to come. As such, these thinkers are not so much interested in the question of what kind of world we should believe in before we can participate in that same world (and therefore make a difference). The question these theories pose instead runs something like this: how might we come to terms with the ongoing process of experience *expressing* us before we ever have time to reflect upon it through representational modes of thought?

To think about this question I want to begin by turning to the work of William James and John Dewey as exemplars of a tradition of thinking that seeks to reclaim experience as a valuable philosophical category, before considering how this emphasis on experience encourages an affirmative ethos of experiment. In the remainder of the chapter I exemplify how such an ethos of experience/experiment takes place through the generative constraint provided by the transition between relation-specific thinking-spaces, techniques for thinking with images, and encounters with concepts-in-movement.

Affirming Experience

North American pragmatism is one of the most affirmative of philosophical traditions. This is particularly so in the strands developed by William James and John Dewey. Their respective work is affirmative in a number of respects: in the faith it places in the promise of a world of change and becoming; through its vision of an ethics of immersive involvement within this world as the basis of a renewed philosophical orientation; and through the tone or style through which it is written. Yet what James and Dewey affirm, perhaps more than anything else, is the value of experience as a philosophical category: a concerted attempt at the renewal of this category is one of the defining features of their respective writings. Pragmatism, in deed, is nothing without experience.

While different, a number of overlapping themes characterize the respective visions of experience outlined in the writings of James and Dewey. Most obviously perhaps, for both figures experience is *of this world*: it is not a secondary reflection of the world apprehended from a distance. Experience, in other words, is part of the sensible materiality of nature. Dewey is especially emphatic in this regard: 'it is not experience which is experienced, but nature – stones, plants, animals, diseases, health, temperature, electricity, and so on. Things interacting in certain ways are experience; they are what is experienced. Linked in certain other ways with another natural object – the human organism – they are how things are experienced as well. Experience thus reaches down into nature; it has depth' (1958, 4). Similarly, for James, experience is part of the ontogenetic materiality of nature: it is the 'stuff of which everything is composed' (1996, 4). At the same time, this primal stuff is by no means homogeneous but is infinitely differentiated: 'there are as many stuffs [sic] as there are 'natures' in the things experienced' (ibid., 26).

If experience is in some sense co-extensive with the sensible yet differentiated materiality of the world then it becomes difficult to think of the process of experiencing as involving the activity of a mind representing (internally) to itself the details of an external environment. To affirm a pragmatist conception experience is, in other words, to take seriously the claim that experience can never be reduced to processes of representation. As James puts it;

> As 'subjective' we say that experience represents; as 'objective' that it is represented. What represents and what is represented is here numerically the same; but we must remember that no dualism of being represented and representing resides in the experience *per se* (1996, 23).

Similarly, Dewey argues that the work of experience is not to produce a copy of an environment external to itself. If this were the case, the 'experience' of an organism would actually be of a different environment to the one in which an organism lives and moves – a situation that would make its life infinitely more difficult. Dewey does not, however, deny the existence of cognitive experience: his claim is instead that 'cognitive experience must arise from that of a non-cognitive sort' (1958, 23). Cognitive *knowing*, as one mode of experiencing, emerges from a background of non-representational sense-making.

The conception of experience emerging through the pragmatism of Dewey and James is anything but an internal, subjective state: it is 'no slipping along in a path fixed by inner consciousness' (Dewey 1981, 63). Experience is, instead, *connective*: it is the ongoing product of a multiplicity of 'dynamic connections', involving all kinds of 'specific affinities, repulsions, and relative indifferencies [sic]' between things in the world (ibid., 65). Where Dewey speaks of connections James (1996) famously writes of relations. For James, 'pure experience' is fundamentally relational and the relations themselves are as real as anything else. Crucially, these relations are not just extensive or distributed – a relational conception of experience is absolutely not just a matter of drawing lines between different actors. The radical empiricism of James, and to a degree Dewey, hinges upon the claim that the relations of which experience are composed are also temporal. On one level this means that experience is never a static state of being – it is an active process of becoming in transition. But it also means that a pragmatist conception of experience has a particular orientation to futurity. For Dewey, this is expressed through the claim that anticipation is more primary than recollection, and projection more than the 'summoning of the past' (1981, 64). A similar orientation towards futurity is found in James' vision of radical empiricism. He puts it thus:

> We live, as it were, upon the front edge of an advancing wave-crest, and our sense of a determinate direction in falling forward is all we cover of the future of our path. It is as if a differential quotient should be conscious and treat itself as an adequate substitute for a traced-out curve (1996, 69).

Experience, in other words, is never all there, but is always leaning into a 'chromatic fringe' where the actual and virtual mix in a 'more that continuously develops, and that continuously supersedes them as life proceeds' (1996, 71).

Pragmatist experience is therefore conceived as the relational stuff of the world, a processual *thisness* that is only ever grasped in the course of its transitional immediacy, an immediacy that is always as virtual as much as actual. This is not to say that experience cannot be 'known' in the sense that we might 'recall' an experience. It is just that this process is not a matter of reflection or representation – it involves the addition of more relations, more transitions. Consequently, for James and Dewey, the actuality of experience can be understood as a process that is always becoming more than itself.

Experience/Experiment

To become radically empirical in a pragmatic sense is not to transcend experience through representation: it is not a matter of how mind can know, or represent, experience. Rather, the crucial question for pragmatism is how one thinks the moving event of transition, or how one finds 'different ways of being in and of the movement of things' (Dewey 1981, 91). Answering this question is a matter of experiment. The pragmatist conception of experience is linked closely to a renewed faith in the possibilities and promise of experiment as a way of going on, or moving in the relational midst of the world. As John Rajchman observes, 'pragmatism is a philosophy that, for certainty and invariable method, substitutes experimentation and belief in the world' (Rajchman 2001, 11). At various points in their respective writings both Dewey and James speak of experience as something with which one can experiment – James (1996), for instance, suggests that we can usefully experiment upon our ideas of experience before undertaking to change these experience through activity. Yet in a strong sense experience for Dewey and James is a kind of experimentalism in and of itself. Dewey's claims in this regard are based upon a kind of ethology of experience: for him the life of the organism is an ongoing process of testing and interacting within a shifting field of connections. Furthermore, this vision of experience as experiment is linked closely with they pragmatist orientation towards futurity in which anticipation has primacy over recollection. As Dewey puts it, experience in 'its vital form is experimental, an effort to change the given. It is characterised by projection, by reaching forward into the unknown; connection with a future is its salient trait' (Dewey 1981, 91).

What Dewey and James do is to affirm an ethos of experience and experiment in ways that reveal the mutual imbrication of both terms as part of work of thinking relations in transition. To think relations in transition is to experiment, and to experiment is provide possibilities for making more of experience – for adding more relations to the processuality of the 'tissue of experience' (James, 1996). Indeed, both thinkers would suggest that it is impossible to extract one term from the other without falling back upon a representational model of thinking from which experience precipitates as either subject or object. In this respect the writing of both thinkers resonates with the work of Alfred North Whitehead, whose process-philosophy is also an attempt to come to terms with the duplicitous nature of experience. For Whitehead, as for James and Dewey, experience is not something on which one acts. This is precisely the point made by Isabelle Stengers in a discussion of Whitehead's work:

> The verb 'to experiment' is here used in a sense akin 'to experience', that is, without 'on' or 'with', which would induce the idea of a separation between the experimenter and what she is experimenting on or with. It is thus a (French-inspired) neologism mean to signal a practice of active, open, demanding attention paid to the experience as we experience it. For instance, a cook would be said to experiment the taste of a new dish. In French, there is no clear distinction between the terms 'experience' and 'experiment' as there is in English. The neologism, when used throughout this article, signals Whitehead's particular empiricist stance that philosophy exhibits experience as experiment and vice versa (Stengers 2008, 109).

Affirming the potential for thinking to move in the midst of experience through a radical or transcendental empiricism is all well and good. But how might it take place? To read Dewey and James, it would seem that such experimentation can take place anywhere: at any given point something might happen to force us to think, to create new lines of thought, to allow us to sense the genesis in affirmation of the moving midst of things. And in a real sense this is precisely what their respective philosophies suggest: the sense of the world is as virtual as it is actual in any given occasion, however mundane. But sometimes the prospect of this potential is a little overwhelming. And affirming it can often seem like asking someone who has never played the piano to simply sit down and improvise. They may well be struck by a sudden rush of inspiration – who knows? – but it might be easier if that person has a rudimentary sense of the notes and scales, around which it then becomes possible to improvise. In other words, sometimes we need a little help, some way of foregrounding the experimental qualities of experience if only to allow experience to 'snowball', in James' terms.

Thinking-Spaces

One way of thinking about this is in terms of a generative constraint: that is, in terms of the establishment of limits that allow one to go on where otherwise this would be very difficult. In the remainder of the chapter I present a sense of how such constraint can be provided in three different ways. The first source of generative constraint are thinking-spaces designed to hold together while holding open the potential inhering in relations of experience/experiment. The lab is the archetypal model of this kind of space. Of course many labs are designed and organized precisely to exclude experience from the process of experiment for fear that the former might become a subjective contaminant in the in as far as possible objective workings of the latter. In these terms the construction of the lab becomes akin to the production of a kind of clean-room in which micro-particles risk upsetting the purity of the experiment taking place therein. Yet while the lab may often be understood as a space from which much of the world is excluded in order that certain kinds of experiments may be performed, it can and has also been understood as a space designed to construct and make worlds proliferate (Latour and Woolgar 1979). Furthermore, the lab can also become a site in which experience is not just isolated as an object for experiment but a space in which experience conceived in relational terms becomes the very milieu in which experiment takes place.

The lab was, of course, an important space for both James and Dewey. James' interest in questions such as effort and emotion can be situated in the context of the emergence of the lab as a space for experimental psychology, particularly at Harvard. While he acknowledged the importance of the experiments undertaken in the lab, he certainly did not seek to reduce the act of experimentation to a mechanical process. For James, the practice of psychological experimentation could not be divorced from the activity of philosophical speculation (see Bordogna 2008). Nor did James think that experiment be confined to such labs. Experiment, as a constituent part of radical empiricism, was for James something that could take place in multiple locations.

Dewey was also fascinated by the potential that certain spaces might have for facilitating a kind of experiential experimentalism. Pedagogical spaces were of particular importance in this respect. His diagram of an ideal classroom is exemplary here, depicting as it does a space designed to facilitate a rhythmic modulation of student attention through their movement within and between its different zones (see Kosnoski 2005). Such ideals were also given material form through Dewey's participation in the University of Chicago Laboratory School. Even if the design of this school did not replicate exactly Dewey's earlier plans, it nevertheless provided an opportunity for him to enact a set of ideas that affirmed pedagogy as one of the key practices through which experience is understood as a process of ongoing experiment.

One way of thinking about labs in the terms of pragmatism is as site-specific thinking-spaces whose organization is designed to facilitate experience/

experiment. Clearly, the physical organization of these spaces is important – their material architecture has the potential to play an important role in the enactment of pragmatism's radical empiricism (see also Kraftl and Adey, 2008). Yet these thinking-spaces are never reducible to the objects and materials from which they are constructed. They are also spaces in which concepts, affects, and percepts participate in unpredictable ways, adding to the relational matrices of which such spaces consist. As such, the kind of thinking that these spaces facilitate is as much a question of relation-specificity as it is of site-specificity (Massumi, 2003). Crucially, if we take seriously the argument of Dewey and James, such relation-specificity is not just defined in extensive terms. Radically empirical thinking-spaces consist also consist of relations in transition that can only be sensed in passing and in potential. As Massumi puts it:

> The relation always arrives, coming to us through a leading perceptual edge – usually visual – in advance of its next sequential unfolding. In other words, its arrival is a promised event that has yet to occur: an appointment with a known but not yet actually afforded outcome. To afford oneself of the outcome is to eventuate the relationship, to perform it: to follow through with its actual step-by-step unfolding (2003, 7-8).

How might you follow through this unfolding? Consider the following scenario. You are forwarded a call for participation in an event to be held in Montreal, from someone who thinks it might appeal to you.[1] The name of this event is *Dancing the Virtual*, and it is described thus:

> We would like to challenge the dichotomy between creation and thought/ research by establishing a working environment in which the emphasis will be placed on the ways in which research-creation reinvents collaboration and on the new modes of thought and action this makes possible. ... To engage actively in research-creation is not only to create movements of thought, it is also to instantiate new platforms of experimentation. This project proposes to create such a platform of experimentation.[2]

The themes of the event are interesting, and they seem to resonate with your own work: movement, abstraction, the virtual, research-creation. The key thing is that this is a call for participation, not for papers. It may not seem like much, but this change disturbs, however subtly, a well-engrained habit of thinking – the production of an abstract for a paper presentation. And you wonder – what can we do (as academics) at such gatherings if we don't present? And so a response to a call for participation becomes an opportunity to reflect upon the ethos through which

---

1 My thanks to Jane Bennett for this.

2 From the call for participation to *Dancing the Virtual/Danser le Virtual* organized by Senselab and the Workshop for Radical Empiricism, Montreal, 2006.

one tends to act into gatherings. And it becomes an opportunity to think upon how you might act into a situation without anticipating too rigidly the outcomes: how you might provide some orientation without refiguring the relation to come. And so you respond, outlining not so much a plan for participation: instead you frame your potential participation by posing three questions that you think might provide some constraint while opening into the concerns of the event:

- First, what conceptual devices allow us to enact a thinking-space in terms of forces of movement and affect?
- Second, how do we enact and construct a relational and experimental movement-space?
- Third, what techniques best facilitate the refiguring of research as a movement of thought?

In preparation for the event you read the suggested reading for participants: Bergson (2007), Deleuze (1994), Gil (1998), James (1996), Langer (1941), Whitehead (1927, 1967). And you take notes, identifying passages that seem to move and resonate with the concerns of the event to come. The event itself takes place in a large, bare studio space populated by a range of materials designed to facilitate forms of relational interaction, where again, these relations involve concepts, affects, and percepts: things in transition and things in the making. And, as things turn out, a great deal is afforded by this thinking-space (see Murphie, 2008): relational movement (Manning 2007), conceptual speed-dating, dancing with José Gil (1998). Indeed, so much happens that you go back, the following year, for a similar kind of event, albeit one with a different name – *Housing the Body*. And in preparation for the second event, you read some more: Deleuze (1993), Cache (1995), Lynn (1999), Arakawa and Gins (2002). And even if the second time round this kind of event is no longer as novel, it still has the capacity to generate surprises, and to do some work.

There is of course no guarantee that such thinking-spaces will produce anything. That, after all, is the nature of the experience/experiment matrix. Things fail and frustrate, deflate and disappear before going anywhere. Nor are such events necessarily naïve visions of happy togetherness: those organizing them, and those participating, need to be aware and careful of the potential violence of what might happen, to the fact that a surprising event can disturb and agitate as much as anything else. And one of the difficulties encountered after participation in such an event might be in telling the story of what happened, a difficulty borne of trying to make too many things cohere and add up even when they don't. And perhaps rather than doing so, it may become easier to think about how the after-affects of participation in events of research-creation become part of the minor variations in sensibility through which one thinks and acts.

OK, Go!

So let's say that at the very least the process of participation in these events generates a mood through which you become a little more responsive to the need to balance critique and affirmation with respect to a range of cultural practices. But if this mood exists as a background, how then might you foreground its participation in the process through which thinking takes place? Political theorist William Connolly (2002; 2005) provides some orientation here. For Connolly, techniques of thinking offer opportunities through which to experiment with the different 'layers' of experience with the view to foregrounding attachments and dispositions that already subsist in the visceral and affective fields from which this experience emerges. Techniques of thinking facilitate the enactment of a kind of 'immanent naturalism', a variation on radical and transcendental empiricism in which thinking emerges and can also be actively cultivated within an immanent field of sense and experience. Connolly's vision of immanent naturalism draws heavily upon the work of figures including Spinoza, Nietzsche, and Deleuze. Importantly Connolly affirms a pragmatist sense of experience, particularly the version developed by James, in which sensory experience always arrives in connectedness: a sense of the experience that is 'more comparable to the relation we have to our desks in the middle of a project than to the desk after the project has been completed' (2005, 73).

Connolly identifies a range of techniques through which to experience/ experiment. But he is especially fond of film because it affords ways of attending to the 'ubiquitous role of bodily affect in perception and judgement' (2002: 25) as well as affording opportunities for appreciating how images work upon registers of duration, memory and perception operating below thresholds of conscious, representational thinking. Furthermore, attending to film also provides one way of thinking about how affective energies travel across and within cultures, participating in processes of contagion and resonance. While Connolly concentrates on cinematic feature film of course it is also possible that encounters with shorter films might work as minor techniques for thinking.

How?

Consider how the scenario outlined above might 'snowball'. After participating in the second of those research-creation events you have a spare day in Montreal. In the morning you gaze awhile at the St. Laurence. And following that you seize the opportunity to visit the Montreal Museum of Modern Art. From the outside what catches the eye is an exhibition of work by the artist Bruce Nauman, perhaps most famous for his creation of politically charged neon sculptures and installations. Yet while such work is interesting, it is not the neon, clown-like figures that grab your attention most. What does is a short piece of video art in which Nauman sidesteps repetitively around a small square in a metronomic fashion. You sit and watch Nauman doing this for a number of cycles, marking time, making space. Perhaps watching this video amplifies your already heightened mood of responsiveness, because before leaving the museum, you decide to visit a rather unheralded exhibition in a room just off the bookstore, an exhibition you had noticed on the way in but had ignored. It is described thus:

> As part of the Projection Series, we are offering a new program of music videos this summer. Amazingly inventive in their concepts and images, music video artists are constantly coming up with new ways of doing things, redefining cinematic creation and enriching the visual arts as a whole.[3]

On the way in to the museum your response to such claims has been a degree of cynicism. Music videos? But you notice that 'Losing My Religion' by *REM* is on the list, and, because you have a special fondness for the minor scandal it caused in early 1990s Ireland (all that wound-poking with fingers), you decide to sit in on the exhibition and let the wave of nostalgia wash over you. But you come in at the wrong time in the sequence of videos to see 'Losing My Religion'. And so you sit and wait in a darkened projection room, watching a series of videos, many of which feature highly choreographed moving bodies. And you continue to sit there, in foot-tapping fascination until at some point 'Losing My Religion' does appear. And you find yourself a little embarrassed at its overwrought, mock anguish. But one video stands out. It features the four young men of a band called *OK Go* performing a highly choreographed dance routine on exercise treadmills. You've never imagined fitness machines might be used in this way (McCormack 1999).

On your return home you follow *OK Go*, dancing on the treadmills, through *iTunes* and *You Tube*. You discover that before making this video, *OK Go* had made another: just as highly choreographed, but without the treadmills. Released without knowledge of the band's record company, the video became one of the most downloaded clips on *You Tube*. And you also come to learn that the success of the treadmill video has generated a strange alliance between *OK Go*, *iTunes*, and *Nike*, in which the former supply a 30-minute soundtrack for a treadmill work out, which begins like this:

> This is Damian Kulash from *OK GO*, and I'll be your coach for this *Nike* treadmill speed workout. Let's cut right to the chase – I am a treadmill God and you, well you are not. Luckily however I can help you with this. I can't promise you'll be ice-skating on that thing nor vaulting over it but at least we can work on your speed, your stamina, and your general fitness. The workout will be an alternating thing – you'll speed up, you'll slow down and I'll tell you when. So, to begin ...[4]

Some months later, you see a variation on this video used as part of the advertising campaign for a health supplement called *Berocca*, marketed on the basis that it provides a 'hit of vitamins of minerals to help you have the best day possible. That

---

3   Montreal Museum of Modern Art, 2007.
4   From *Master the Treadmill with OK Go*.

sort of day when you catch that early train, make the lights, snatch the last parking space, get the phone number, nab the last one on sale, and score from your own half' (berocca.co.uk). In the ad four treadmills are located at a city crossroads. Four passers-by see the treadmills and begin to use them in a way that reprises the choreographed routine of *OK Go*. Their performance draws a crowd.

Running on a treadmill, listening to music, and consuming nutritional supplements are all, of course, simple techniques for modifying experience: they are designed not only to produce particular kinds of bodies, but also to generate positive affects. And they each participate in a wider assemblage of techniques and technologies designed to produce distinctive kinetic-affective experiences in contemporary life. How, then to respond to this alignment between *OK GO* and *Nike*, to say nothing of *Berocca*? You might be rather disappointed, seeing it as evidence of a certain selling out, of a deliberate complicity with a range of contemporary regimes designed to engineer the fit and healthy self. You may think that a moment of inventiveness has been reincorporated into a system of corporeal and affective value generation. But this critical disappointment might be interrupted by the exuberant affirmative energies of the movement of the video every time you watch it. These responses are not, of course, mutually exclusive. Indeed, as a number of thinkers have argued, such exuberant energies have the potential to participate in our response to the circulation of images in contemporary popular cultures. Jane Bennett's (2001b) enrolling of the moving bodies of a *GAP* commercial is particularly instructive in this regard because it foregrounds just what is at stake in this effort to make more of moving images of moving bodies. Bennett enrols such bodies as a way to think about the role that enchantment plays in cultivating an ethos of 'generosity towards others'. For Bennett, the moving bodies of the *GAP* ad tend to draw our attention to and exemplify a condition of enchantment as a 'mixed bodily state of joy and disturbance, a transitory sensuous condition dense and intense enough to stop you in your tracks and toss you onto new terrain, to move you from the actual world to its virtual possibilities' (ibid.,). As Bennett also argues, the moving bodies of GAP ads have this potential insofar as they can catalyze affirmative affective refrains that resonate through thinking in a way that makes the distinction between critique and affirmation difficult to sustain. Crucially, the point Bennett is making here is not that we have to suspend critical thinking when looking at such ads: rather, in order to really appreciate what is going on when bodies sing and sway, even as part of the business of selling, we need to moderate those more acidic, judgmental tendencies of thinking in which demystification is primary. And so, understood thus, you might work to balance your response to *OK Go* on the treadmills, seeing them as implicated in the production and reproduction of certain kinetic imperatives but also as indicative of a certain inventiveness with respect to the relation between bodies and technologies. You might hold on to the possibility that even the most manufactured of spaces – treadmills – can become sources for experimenting with the affective capacities of moving bodies, with their capacities to affect and be affected by other bodies.

Circulating Refrains

So you might modify your critical response to *OK Go*, dancing on the treadmills, becoming potentially more open to the affective energies of which they are generative. And you might become more response to the prospect of thinking with which moving images as a technique of experience and experiment. And how? The scenario snowballs again. Sometime later you receive an invitation to present a research seminar at a well-known geography department. You decide to speak about the affirmative ethos of which non-representational theories are generative. And you decide to do this through a presentation that explores how moving images of moving bodies might afford opportunities for experimenting experience. This focus on moving images of moving bodies might seem to run counter to the grain of non-representational theories. Such theories are not however about the acidic dismissal of images but about reconfiguring the terms of their use through the enactment of presentations as processes through which the world is made anew again and again and again. Difference differing (Dewsbury et al. 2002). And presenting a presentation is a way of experimenting experience as much as anything else: as valuable, and as potentially powerful a technique for thinking-with images as watching or making a film. And it has the advantage of being portable.

And so you decide to incorporate the *OK Go* video into a presentation about how moving images have an affective capacity – that is, a capacity to move other bodies, both off and on-screen. And you look for a way of giving conceptual coherence to the relation between the presentation as a technique of thinking-with images and the affective capacity of moving images with moving bodies. After some thinking you find this in the refrain, or ritornello, one of the key conceptual creations of Guattari's work with Deleuze (1988). When asked during a conversation if he and Guattari had created any concepts, Deleuze responded: 'How about the ritornello? We formulated a concept of the ritornello in Philosophy' (2006, 381). This formulation is articulated at great length in *A Thousand Plateaus* (1988). But it is also something about which Guattari writes in an essay called *Ritornellos and Existential Affects* (1996). It might seem odd to decouple Deleuze and Guattari's writing on the refrain in this way. Yet there is an earthy insistence on the pragmatics of experience/experiment in Guattari's writing that resonates particularly intensely with the writings of Dewey and James. If experience is the field of experiment for radical empiricism, it occupies a similar position in the transcendental philosophy articulated by Guattari.

The ritornello is a pragmatic concept for thinking through the relational processuality of experience, for thinking through transition. Guattari describes the ritornello as a kind of affective block of space-time, a duration that produces existential territories through the continual creation of heterogeneous durations of being' (ibid., 159). It is the rhythmic temporalization of affective-territorial complexes that move, an existential territory that always has the potential to resingularize itself. And while in *A Thousand Plateaus* (1988) Deleuze and Guattari suggest that the matter of expression of the ritornello is properly musical, Guattari makes it clear that the ritornello engages multiple registers: sensory, expressive, enuciative, problematic, and facial. Take religious icons, for instance. As Guattari suggests;

> the primary purposiveness [sic] of an Icon of the Orthodox Church is not to represent a Saint, but to open an enunciative territory for the faithful, allowing them to enter into direct communication with the Saint. The facial ritornello then derives its intensity from its intervening as a shifter – in the sense of a 'scene changer' – in the heart of the palimpsest superimposing the existential territories of the proper body (1996, 165).

Read through Guattari, the video of *OK Go* dancing on the treadmills can be understood as a kind of scene-shifting ritornello – it has the capacity to seize and shift the affective spaces of bodies, even if only for its approximately three and half minutes running time. Yet the *OK Go* video is also a different kind of scene-shifting ritornello to the religious icon. For one thing it is not nearly so fixed. And it circulates through a cultural economy that in many ways venerates kinaesthetic movement and mobility rather than quiet, solemn contemplation. Furthermore, as its alliance with *Nike* and *iTunes* suggests, it is very easy for such ritornello to be incorporated into efforts to generate and cultivate certain propensities to act-work-consume. In some ways then *OK Go* might be understood as an exemplar of one of the many ritornellos of and through which contemporary political and affective economies reproduce and resingularize themselves, generating value and interest in the process. And the experience of viewing it might well become part of the process through which existential territories become circumscribed by the affects of the screen.

But in so far as such ritornellos or refrains are scene-shifters then they have the potential to open an enunciative territory within which a kind of pragmatic process of experience/experiment can take place. As film theorist Amy Herzog has argued, 'looking even at the most hackneyed, clichéd films, the attentive, inventive thinker might see within their stutterings and pauses waves of affect that move against the prevailing current' (2001, 5). Used in the right way *OK GO* dancing on treadmills might just provide sources for moving in this way. Qualified, admittedly, but sources nonetheless. How then to modify the energies and experiences of one of these ritornellos, however modestly? Play what Guattari calls 'ritornello games'. Ritornello games are activities which 'fix the existential ordering of the sensory environment and which prop up the meta-modelizing scenes of the most abstract problematic affects' (1995, 128). What does this mean? Let's say that the 'abstract problematic affects' in this instance refer to the ethos of affirmative critique with respect to experience with which various strands of non-representational theory are concerned. And let's say the sensory environment refers to relation-specific thinking-space through which you experience/experiment. If you have to work with the materials at hand, then presentations provide as useful an opportunity as anything else to play these ritornello games. After all, we are all used to the deathly repetitions of many presentations. Intervening in them might provide ways of reworking the act of presentation as a process of experience/experiment. Put another way, it might provide 'a small act of repair' (Bottoms and Ghoulish 2007) in the presentational economies of academic knowledge insofar as it foregrounds the refrain of experience as something always in the making.

So in addition to *OK Go* you include other clips in the presentation. One of these is a famous scene from a Laurel and Hardy film called *Way out West*. The clip begins when Stan and Ollie arrive in a small western town after a long coach journey. They walk up to a saloon, on the steps of which a band is playing. As they watch the band they both begin to sway, ever so slightly, before beginning to dance in unison, a dance that gradually becomes more elaborate, ending with a final flourish as they enter the saloon. There is obviously a lot that can be read into this clip, and a lot that can be extracted from it through various tactics and techniques of critical interpretation. One could, for instance, situate it in relation to a genealogy of various kinds of odd couples and strange attractors across a range of cultural practices (see Sedgwick 1985; Livett 2001). Understood thus Laurel and Hardy might be conceived of as an affective-complex composed of multiple ritornellos that have crossed what Guattari calls a 'threshold of consistency': that playground jingle introduction; Stan running his fingers through his hair while puckering his face and whimpering; Ollie's face as he turns towards the camera with a look of knowing exasperation – can you believe this guy? And, of course, Ollie declaring: '*That's another fine mess you've got me into*'. These ritornellos cohere, they hold together, without necessarily precipitating a held, an object. And each of these elements, even if encountered in isolation, functions as a catalytic trigger for the affective complex that is Laurel and Hardy, in the same way that the swoosh functions as a trigger for the affective-complex that is *Nike*, a light sabre for *Star Wars*, and 'Feck' for *Father Ted*. Furthermore, this affective complex has been generative of other ritornellos that have come to circulate beyond their positioning within given narrative structures. The scene in *Way out West* in which Laurel and Hardy dance in front of the saloon is one such scene. Indeed it adds nothing to the narrative, and was not in the original script. So also is the scene in the same film, during which Stan and Ollie sing the *Trail of the Lonesome Pine*. It is precisely the gratuitous and gently interruptive quality of their affective duration that has allowed both scenes to circulate in a way that the films within which they were originally produced cannot. They have a mobility that is transitivist, or transversal to the narrative structures within which they sit. And because of this mobility, they are particularly interesting examples of ritornellos that move.

OK Go and Stan and Ollie dancing in *Way out West*. You want to use them in the same presentation, but how? In preparation you decide to juxtapose two blocks of space-time – the visual scene shifting refrain of *OK Go* on the treadmills and the soundtrack that accompanies Laurel and Hardy's dance scene. You play the *OK Go* video with the sound turned down and instead turn up Laurel and Hardy. And what you get is a modified affective refrain whose constructive interference links the pratfalls of the music hall with those of the contemporary music video.

With the help of Laurel and Hardy, OK GO slow down, no longer so expressive of a powered up kinetic capitalism. A gentle interruption modifies the affects of both, producing a minor variation in a minor refrain.

Why engage or indulge in such ritornello games? At the very least playing such ritornello games might encourage you to consider the possibility that experimenting with moving bodies in moving images does not necessarily work in opposition to thinking-with moving bodies off-screen. The affective ritornellos of one register of experience/experiment can fold into and inflect the other as part of an ecology of practices composed of multiple ritornellos: some of which work, some which don't; some of which cross a threshold of consistency, some which don't. This ecology is underpinned by what Guattari (1995) calls an ethico-aesthetic paradigm: an affirmative disposition towards the generative potential of experience/experiment. This disposition is aesthetic in the sense that it engages matters of expression across ecologies of practice (actual and virtual). Any number of obviously 'aesthetic' practices can provide opportunities for catalysing such ritornellos: Literature, Art, Cinema, Dance. Regardless of the source, the trick is not to be exclusive: aesthetic does not here name a threshold of judgement above or across which practices need to move. It refers, rather, to a certain kind of organization of expressivity, one that is never only a matter of individual self-creation. So an ethico-aesthetic sensibility refuses the tendency for judgement – or taste – to work as a kind of 'somatic marker': a memory imbued disposition to foreclose generosity towards other practices (Connolly 2002). Nor does it necessarily cringe or wince at country music; line dancing; Cajun dancing, (Stivale 1994); corn (Seigworth 2005); or treadmill running. And this sensibility is ethical in the sense not so much because it provides tools or guidelines for living but opportunities for the generation of new ritornellos through an ongoing process of experience/experiment. As much as the pragmatist empiricism of Dewey and James, this affirmative disposition is never, in the words of Guattari, 'given in and of itself' (1996, 166). It has to be worked on, for most of us at least. It is always a work in progress. And at most it has a certain loose consistency, something only 'attained through a perpetual flight in advance of inwardness, which conquers an existential territory in the very time that it loses it, and wherein, however, it strives to retain a stroboscopic memory' (ibid., 166-167).

Conclusion

For non-representational theories critique and affirmation are not mutually exclusive. As Deleuze puts it, the 'conditions of a true critique and a true creation are the same: the destruction of an image of thought which presupposes itself and the genesis of the act of thinking in thought itself' (1994, 139). To think experience beyond representation is to grasp the ongoing failure of a representational image of thought, particularly with respect to its capacity to grasp the fugitive participation of movement, sensation, and affectivity in thinking. At the same time it involves affirming experience as a process through which the world affirms itself within and through us. The question such theories pose is not so much about how we should define experience before we can act upon it. Rather, they pose another question: how might we think through how the experiential tissue of the world participates in us before we ever have time to affirm its presence through representational modes of thought. In other words, how might we sense the process of difference '*differing*'? (Deleuze ibid., 56).

As Nigel Thrift suggests, if we take seriously the claim that experience is difference differing, that it is always in process, then an ethos of 'affirmative experimentation' is an important element of non-representational styles of thinking (2004, 438). This is the ethos towards which the radical or transcendental empiricism of figures such as James, Dewey, and Guattari tends: a way of doing philosophy and geography that takes seriously and experiments experience without positing experience as the stable ground from which the world is affirmed. This ethos does not presuppose the possibility of transcending experience. Nor does it fall back on a solipsistic or subject-centred version of experience as the basis for an affirmative critique. Instead, it encourages the affirmation of something that is simultaneously always beyond the actual immediacy of cognitive representationalism, while also participating in that experience. It is also an ethos that holds to and hold open a variation on the Spinozist claim that we do not know in what ways the world might express itself through us, how it might participate in our becoming otherwise. That is not to say that it is an ethos which assumes anything is always possible: rather, it is an ethos of experiment/ experiment for which, within the constraints of any given set of relations one never knows what might happen, what kinds of ways in which the world might surprise us and make more of itself. It is an ethos that consists in experimenting with techniques for experiencing the many ways in which the world affirms us through the creative variations that can be sensed as differing capacities to affect and be affected. It is an ethos that foregrounds experience/experiment as the milieu through new refrains might emerge.

As I have tried to demonstrate in this chapter, this ethos can take place through the process of thinking relations in transition between thinking-spaces, techniques of thinking, and concepts. In the case of this chapter, participation in an event feeds forward into a mood of responsiveness to moving images before encouraging a degree of minor inventiveness with respect to the possibilities of presentation as techniques of experience/experiment. The point of presenting these relations here is not to claim that they add up, but to argue that much of the work of non-representational theory consists of thinking through things in the making, where these things are relations. Other relations and other transition might of course be added to the mix. The crucial thing is that the fact that relations are always 'arriving', and in ways that remix the spatio-temporality of thinking. Presenting a sense of this process, as I have done here, is sometimes labelled as 'auto-ethnography'. This, of course, is a misnomer. The radical empiricism of nonrepresentational theories is not auto-ethnography. It is more akin to an ethology of relational experience through which one learns to affect and be affected by variations in the relational 'tissue of experience' (James 1996). Crucially, the relations with which this ethology are concerned do not just exist between bodies: they cannot simply be mapped by cartographies of extension. They are also relations of transitional intensity. As James puts it:

> Within each of our personal histories, subject, object, interest and purpose are continuous or may be continuous. Personal histories are processes of change in time, and the change itself is one of the things immediately experienced. 'Change' in this case means continuous as opposed to discontinuous transition. But continuous transition is one sort of a conjunctive relation; and to be a radical empiricist means to hold fast to this conjunctive relation of all others (1996: 48).

Dewey, James, and Guattari all know the importance of thinking through this continuous transition, as do a range of more recent thinkers across disciplines including geography. They know the importance of thinking the world as always, at every moment, becoming more than it actually is through the ongoing arrival of relations. Difference differs. And in the process, experience snowballs, continuously. This is the 'more' to which Dewey and James refer, the 'and' that always trails the end of their every sentence. And this is the 'more' of non-representational theory, not a more that signals an addition to representation, but one that signals an affirmative refrain about the processuality of the world that is always in excess of itself.

References

Agamben, G. (2007), *Infancy and History: On the Destruction of Experience.* Trans. L. Heron (London: Verso).

Arakawa and Gins, M. (2002), *Architectural Body* (Tuscaloosa: Alabama University Press).

Bennett, J. (2001a), *The Enchantment of Modern Life: Attachments, Crossings, Ethics* (Princeton: Princeton University Press).

Bennett, J. (2001b), 'Commodity Fetishism and Commodity Enchantment', *Theory and Event* 5(1) (no pagination).

Bergson, H. (2007), *The Creative Mind* (New York: Dover Publications).

Bergson, H. (1988), *Matter and Memory*. Trans. N.M. Paul and W.S. Palmer (London: Zone Books).

Bordogna, F. (2008), *William James at the Boundaries: Philosophy, Science, and the Geography of Knowledge* (Chicago: Chicago University Press).

Bottoms, S. and Ghoulish, M. (eds) (2007) *Small Acts of Repair* (London: Routledge).

Cache, B. (1995), *Earth Moves: The Furnishing of Territories*. Trans. A. Boyman (Cambridge, MA: MIT Press).

Connolly, W. (2002), *Neuropolitics: Thinking, Culture Speed* (Minneapolis, MN: University of Minnesota Press).

Connolly, W. (2005), *Pluralism* (Durham, NC: Duke University Press).

Deleuze, G. (1993), *The Fold: Leibniz and the Baroque*. Trans. T. Conley (London: Athlone).

Deleuze, G. (1994), *Difference and Repetition*. Trans. P. Patton (London: Continuum).

Deleuze, G. (2006), *Two Regimes of Madness: Texts and Interviews, 1975-1995* (New York: Semiotext(e)).

Dewey, J. (1934), *Art as Experience* (New York: Minton, Balch and Company).

Dewey, J. (1958), *Experience and Nature* (London: Dover).

Dewey, J. (1981), 'The need for a recovery of philosophy' in *The Philosophy of John Dewey*, ed. J. McDermott (Chicago: Chicago University Press), 59-89.

Dewsbury, J.D, Harrison, P., Rose, M. and Wylie, M. (2002), 'Enacting geographies', *Geoforum* 33(4), 437-440.

Gibson-Graham, J.K. (2006), *A Post-Capitalist Politics* (Minneapolis: University of Minnesota Press).

Gil, J. (1998), *Metamorphoses of the Body*. Trans. S. Muecke (Minneapolis: University of Minnesota Press).

Guattari, F. (1995), *Chaosmosis: An Ethico-Aesthetic Paradigm*. Trans. P. Bains and J. Pefanis (Sydney: Power Publications).

Guattari, F. (1996), *The Guattari Reader* ed. G. Genosko (Oxford: Blackwell).

Herzog, A. (2001), 'Affectivity, Becoming, and the Cinematic Event: Gilles Deleuze and the Futures of Feminist Film Theory', Available at http://media.utu.fi/affective/herzog.pdf. Last accessed 7 October 2007.

James, W. (1996), *Essays in Radical Empiricism* (Lincoln, NE and London: University of Nebraska Press).

Kosnoski, J. (2005), 'Artful discussion: John Dewey's classroom as a model of deliberative association', *Political Theory* 33(5), 654-677.

Kraftl, P. and Adey, P. (2008) 'Architecture/affect/inhabitation: Geographies of Being-in Buildings', *Annals of the Association of American Geographers* 98(1), 213-231.

Langer, S. (1941), *Philosophy in a New Key* (Cambridge, MA: Harvard University Press).

Latour, B. and Woolgar, S. (1979), *Laboratory Life* (Beverly Hills, CA: Sage).

Livett, J. (2001) 'Odd couples and double acts, or strange but not always queer: some male pairs and the modern/postmodern subject', *Australian Humanities Review* 21(2) (no pagination).

Lynn, G. (1999), *Animate Form* (Princeton, NJ:, Princeton University Press).

McCormack, D. (1999), 'Body-shopping: refiguring geographies of fitness', *Gender, Place and Culture* 6(2), 155-177.

Manning, E. (2007), *Politics of Touch: Sense, Movement, Sovereignty* (Minneapolis: University of Minnesota Press).

Massumi, B. (2000), 'The ether or your anger: towards a pragmatics of the useless', in *The Pragmatist Imagination: Thinking About Things in the Making*, ed. J. Ockman (Princeton: Princeton Architectural Press), 160-167.

Massumi, B. (2003), 'Urban appointment: a possible rendez-vous with the city'. In *The Art of Databases*, ed. J. Brouwer and A. Mulder (V2: Rotterdam), 28-55.

Murphie, A. (2008), 'Clone your technics: Research creation, radical Empiricism and the constraints of models', *Inflexions: A Journal for Research-creation* 1(1) (no pagination).

Rajchman, J. (2000), 'General introduction', in *The Pragmatist Imagination: Thinking About Things in the Making*, ed. J. Ockman (Princeton: Princeton Architectural Press), 6-15.

Sedgwick, E.K. (1985), *Between Men: English Literature and Male Homosocial Desire* (New York: Columbia University Press).

Seigworth, G. (Dec. 2005), 'The affect of corn', *M/C Journal* 8(6). Retrieved 23 Oct. 2007 from <http://journal.media-culture.org.au/0512/12-seigworth.php>.

Stengers, I. (2008), 'A constructivist reading of process and reality', *Theory, Culture and Society* 25(4), 91-110.

Stivale, C. (1994), 'Spaces of affect: versions and visions of Cajun cultural history', *South Central Review* 11(4), 15-25.

Thrift, N. (2004), 'Summoning life', in *Envisioning Human Geographies*, ed. P.C. Cloke, P. Crang and M. Goodwin (London: Arnold), 81-103.

Thrift, N. (2007), *Non-representational Theory: Space, Politics, Affect* (London: Routledge).

Whitehead, A.N. (2004), *The Concept of Nature* (New York: Prometheus Books).

Whitehead, A.N. (1927), *Symbolism: Its Meaning and Effect* (New York: Fordham University Press).

Whitehead, A.N. (1967), *Adventures of Ideas* (New York: First Free Press, Macmillan USA).

Chapter 12

Encountering O/other Bodies: Practice, Emotion and Ethics

Kirsten Simonsen

Introduction

Within Human Geography, otherness[1] has in the latest decades been approached through Said's (1978) seminal work on *Orientalism* and 'imaginative geographies', which from its basis in Colonialism has been translated into 'national' and 'local' scales. While this 'exterior' approach to the construction of otherness is extremely important, it is equally important to understand how it occurs in everyday life, how everyday experiences and bodily encounters at the same time respond to and produce otherness. In this chapter, I will argue that a turn to practice can inform such an understanding. I will explore and seek to illustrate how embodied encounters construct differentiated bodily and social spatialities and how these practices of differentiation include emotional as well as ethical aspects.

The theoretical standpoint from which I work is a phenomenological inspired theory of *practice* understanding the social in terms of embodied, materially interwoven practices organized around shared practical understandings. That means, as will be developed below, that I adopt an ontological stand that privilegew practices as constitutive features of social life. Certainly, practice theories have in many ways informed British non-representational theory (a.o. Thrift 1996), there are however differences in approach. They might be clarified by way of Schatzki's (2002) distinction between 'practice theories' and 'theories of arrangements'; a distinction of social ontology in which the latter denotes theories which take arrangement of entities (apparatus, assemblage, network) to be the principal compositional feature of social life. They prioritize a 'configurational order' of the social: 'the involuted lacing of human and other phenomena into extensive arrangements that determine as well as bind together their character and fates' (2002, XIII). In distinction, practice theories takes nexuses of practice – bodily doings and sayings – as the building blocks of social life. They constitute meanings and identities and they establish social orders.

---

1 When I talk in this chapter about otherness, the Other and other bodies, I am referring to human others. Some of the arguments might be extensible to other kinds of life, but that is beyond the scope of this chapter.

This also involves a certain 'residual' or 'agential' humanism, emphasizing socio-cultural experience and the unique social significance of human agency.

From that starting point, the chapter proceeds in three sections. The first one very shortly outlines the central elements of the practice theory put into work. The two following sections are extensions from that discussing 'emotional encounters' and 'ethical encounters' respectively. The purpose of that is twofold; to add to the theoretical account as regards an understanding of embodied emotions and ethics, and to explore how such understandings can be used to grasp encounters with O/other bodies. In order to put some 'flesh and blood' to the discussion, in these sections I try to insert some empirical examples to illustrate the argument. They all come from an analysis performed in Copenhagen under the title of 'Multiple faces of the city – construction of the city in practice and narrative' (Simonsen 2004, 2005b).

Practice, Embodiment and Spatiality – A Starting Point[2]

The starting point taken in this chapter is a social ontology of practice, that is, an account of social life maintaining that human lives hang together through a mesh of interlocked practices. That means that practices constitute our sense of the world, and that subjectivity and meaning are created in and through practice. This account of social life takes its initial inspiration from phenomenology as for instance formulated in Merleau-Ponty's 'slogan': 'Consciousness is in the first place not a matter of "I think that" but of "I can"' (1962, 137). One of the first philosophers to take such a view on social existence and subjectivity is Heidegger – in particular in *Being and Time* (1962). In this work, 'Dasein' or human 'being-in-the-world' is described as an existential 'facticity' – as a practical, directional, everyday involvement. Our concern with the environment, he argues, takes form by way of tools and articles for everyday use as well as useful products and projects – all together designated as 'equipment' (Zeug). 'Being-in-the-world' is the everyday skilful coping or engagement with an environment including things as well as other human beings. That means that our 'environment' does not arrange itself as something given in advance but as a totality of equipment dealt with in practice. Heidegger thus demonstrates that the only ground we have or need to have for the intelligibility of thought and action is in the everyday practices themselves, not in some hidden process of thinking or of history (Dreyfus and Hall 1992).

From a social point of view, however, it may be important just to touch on the way in which this conceptualization relates to social order. Following Schatzki (2002) social order can be seen as co-existence or 'arrangements of people, artefacts and things', where social practices constitute both the meaning of arranged entities and the actions that bring arrangements about. Seen in

---

2   For more developed versions of this section, see Simonsen (2003, 2007).

relation to my former distinction between practice theories and theories of arrangement, the point is that arrangements are not primordial. Social order is established within the sway of social practices in accordance with their practical intelligibility and their institution of meaning.

Of particular importance in the present connection is however the way in which lived, everyday practices are intrinsically corporeal. Following Merleau-Ponty (1962), I shall see the body as part of a pre-discursive social realm based on perception, practice and bodily motility (Merleau-Ponty 1962). Lived experience, then, is located in the 'mid-point' between mind and body, or between subject and object – an intersubjective space of perception and the body. In that, perception is based on practice; on looking, listening and touching etc. as acquired, cultural, habit-based forms of conduct. It is an active process relating to our ongoing projects and practices, and it concerns the whole sensing body. This means that the human body takes up a dual role as both the vehicle of perception and the object perceived, as a body-in-the-world – *a lived body* – which 'knows' itself by virtue of its active relation to this world. This 'relation' might be taken even further, Merleau-Ponty (1968) suggests, by meshing the body (subject-object) into the perceptible world. His term for this extended conception of the body is *flesh*; a generative body of being and becoming that touches, sees, hears, smells and tastes both itself and 'other' flesh and becomes aware of itself in the process. The flesh of the body belongs to the flesh of the world, where the flesh of the world refers to the perceptibility that characterizes all worldly reality. In this way Merleau-Ponty is blurring the body-world boundary.[3] However, since all of the perceptible flesh of the world (which includes our own) is not self-sentient in the way percipient (or human) flesh is, he at the same time manages to preserve that percipience and maintain some 'thickness' of human flesh constituting a distance or difference to other forms of worldly flesh. I have found it suitable to recognize that distinction and talk about an ongoing intertwining between the flesh of the body, the flesh of others and the flesh of the world. Probably the most important character of the flesh is its *reversibility* – its dual orientation inward and outward – by which bodies are folded into an interworld or 'intermundane space'. Body-subjects-objects are visible-seers, tangible-touchers, audible-listeners etc. For a theory of practice, the most significant consequence of this 'mediation through reversal' is the way in which it grounds a principle of exchangeability or *intercorporeality*. It is however important to appreciate that it is not only the sheer sensibility of the body that institutes intercorporeality. It is also the meaning involved in the bodily practices of the other. You do not just perceive another body in its materiality; you are affected by the meaning of its appearance. The other body is animated and its animation communicates and calls for response. You do not contemplate the communications of the other, they affect you and you reply to them.

---

3 These insights from Merleau-Ponty have (in a different way) been drawn upon within geography by John Wylie (e.g. 2005, this volume).

However, the corporeality of social practices concerns not only the sensuous and generative nature of lived experiences, but also how these embodied experiences themselves form a basis for social action. Bourdieu (1977, 1990) can add to Merleau-Ponty as regards the cultural forming of the body, its gestures and its actions. He does so by introducing the concept of 'habitus' as embodied history, which is internalized as a second nature. As a result of this, social structures and cultural schemes are *incorporated* in the agents and thus function as generative dispositions behind their schemes of action:

> Adapting a phrase of Proust's, one might say that arms and legs are full of numb imperatives. One could endlessly enumerate the values given body, *made* body, by the hidden persuasion of an implicit pedagogy which can instil a whole cosmology, through injunctions as insignificant as 'sit up straight' or 'don't hold your knife in your left hand', and inscribe the most fundamental principles of arbitrary content of a culture in seemingly innocuous details of bearing or physical and verbal manners, so putting them beyond the reach of consciousness and explicit statement (Bourdieu 1990, 69).

The notion of incorporation, then, opens up to an understanding of the cultural differences between bodies. It involves how different cultural schemes and norms dispose for specific bodily practices but also how bodies are marked by the incorporation of assumptions made about their gender, race, ethnicity, class and 'natural' abilities. Some of these issues I shall return to later on in this chapter.

Another important quality of practice and of the body, which is indispensible in the present connection, is their intrinsic spatiality. Basically, it is about the situatedness of the body in space-time and about the way in which the body itself is spatial. To explore that we can once more start from Merleau-Ponty. Initially, he states that the spatiality of the body is not a spatiality of *position*, but one of *situation*. This goes for temporality as well, and it means that we should avoid thinking of our bodies as being *in* space or *in* time – they *inhabit* space and time. Or with a more active phrasing drawn from Lefebvre (1991), we could say that each living body both *is* and *has* its space; it produces itself in space at the same time as it produces that space. The presupposition for this production of space is a *spatial body* that is

> a practical and fleshy body conceived of as totality complete with spatial qualities (symmetries, asymmetries) and energetic properties (discharges, economies, waste) (Lefebvre 1991, 61).

This means that active bodies, using their acquired schemes and habits as well as their gestural systems, position their world around themselves and constitute that world as 'ready-to-hand', to use a Heideggerian expression. These are moving bodies 'measuring' space and time in their active construction of a meaningful world. They are also material bodies constituting what Lefebvre calls a 'practico-sensory realm' in which space is perceived through sight, smell, tastes, touch and hearing.

In this short discussion of the spatiality of the body, I have tried to integrate ideas from Lefebvre into the ones of Merleau-Ponty. The reason is that while the two of them shared views of the role of the body in lived experience (and lived space), Lefebvre can add to Merleau-Ponty when it comes to social order and power relations.[4] Like Bourdieu could socialize the body, Lefebvre can socialize the ideas of bodily spatiality. In his now widely discussed conceptual triad of social space – composed of spatial practice, representations of space and spaces of representation – social practice, social order and symbolic-material creativity merges into a constructive power-play in the production of space. This means that the issues to come, such as the marking of bodies through the marking of space or the construction of borders and boundaries in the encounters between different bodies, can be grasped by the theory of space.

What I have tried to establish so far, then, is an indispensable relationship between practice, body and time-space. In the following I shall use that as a basis for an understanding of emotional and ethical dimensions of bodily encounters.

Emotional Encounters

Seeking an understanding of emotion connected to practice and the encounters between different bodies once more leads me to phenomenology.[5] This account, as a start, suggests that we are never 'un-touched' by the world around us; our relations with others (and with objects) are always 'mooded' (Merleau-Ponty 1962, Heidegger 1962). Moods are basic human attributes, but they are not inner physical and psychic states. We should rather see them as an attunement – a contextual significance of the world, associated with practices, lifemode and social situation. Situatedness and the collapse of the distinction between 'inner' and 'outer', then, are crucial dimensions of emotion which we, as a first approximation, can summarize in a notion of *situated corporeal attitudes* (Crossley 1996). Emotions are inseparable from other aspects of subjectivity, such as perception, speech/talk, gestures, practices and interpretations of the surrounding world, and they primordially function at the pre-reflexive level.[6]

---

4   For a more thorough discussion of Lefebvre's treatment of the body in relation to the production of space see Simonsen (2005a).

5   Within geography issues of emotion and affect have gained increasing attention during the later years – see e.g. *Social and Cultural Geography* (2004), Thrift (2004), Davidson, Bondi and Smith (2005). With a few exceptions (see e.g. Davidson 2003), however, phenomenology is not used as a theoretical ground.

6   Taking this stand means that I depart from the recent tendency to distinguish between affect and emotion, most explicitly stated by (or inspired from) Massumi

They are neither 'purely' mental nor 'purely' physical phenomena, but ways of relating and interacting with the surrounding world. This is a broad account that might be specified through the elaboration of two dimensions or sides of emotional spatiality. This double conception describes different modes of emotional percipience, but it is of course an analytical distinction where the two sides are never supposed to exist separately from each other.

One side is an *expressive space* of the body's movements, which might be seen as a *performative* element of emotion. Here, emotions are connected to the expressive and communicative body. Our body, Merleau-Ponty argues, is comparable to an expressive work of art, but expressing emotions in the form of *living meaning*. These meanings are communicated and 'blindly' apprehended through corporeal intentions and gestures that reciprocally link one body to another:

> Faced with an angry or threatening gesture, I have no need, in order to understand it, to recall the feelings which I myself experienced when I used these gestures on my own account ... I do not see anger or a threatening attitude as a psychic fact hidden behind the gesture, I read anger in it. The gesture *does not make me think* of anger, it is anger itself ... The sense of gestures is not given, but understood, that is, recaptured by an act on the spectator's part. The whole difficulty is to conceive this act clearly without confusing it with a cognitive operation. The communication or comprehension of gestures comes about through the reciprocity of my intentions and the gestures of others, of my gestures and intentions discernible in the conduct of other people. It is as if the person's intention inhabited my body and mine his ... One can see what there is in common between the gesture and its meaning, for example in the case of emotional expression and the emotions themselves: the smile, the relaxed face, gaiety of gesture really have in them the rhythm of action, the mode of being in the world which is joy itself (1962, 184-186).

Even if feeling is an integral part of emotional experience, Merleau-Ponty obviously does not identify it with feelings in the form of 'inner states' (e.g. pleasures and pains) explicable only in terms of bodily systems. It is something

---

(2002). This distinction between affect as 'pre- and postcontextual, pre- and postpersonal, an excess of continuity invested only in the ongoing of its own' (2002, 217) and emotions as 'subjective content, the sociolinguistic fixing of the quality of an experience which from that time is defined as personal' (2002, 28) from my point of view bears some problems. It on the one hand tends to miss Bourdieu's point that what is not personally recognized might still be mediated by past experiences or bodily memories, on the other hand it risks to 'cognitivize' emotions and then cut them off from the living meaning of being and having a body (for a similar argument see Ahmed 2004). Therefore in this chapter I refrain from using affect as a 'thing' in itself (as a noun) but of course not from exploring how human flesh is affected by the surrounding world.

'in-between' – situated in the perceptibility of its gestures. Emotion as living meanings in this way relates to meaning in the above-mentioned sense of being created through practice and the experience of mobile bodies, but also to meaning as it is found in artwork, expressed in poetic or musical meaning. Emotional meanings are 'secreted' in bodily gestures in the same way that musical/poetic meaning is 'secreted' in a phrase of a sonata or a poem. This also endows them with a cultural situatedness, nearly laconically expressed by Merleau-Ponty as 'feelings and passionate conduct are invented like words' (1962, 189), thereby emphasizing that both speech and gestures are significant uses we make of our bodies.

The other side of emotional spatiality is *affective space*, which is the space in which we are emotionally in touch – open to the world and to the different ways it is affecting us. Affective space is intermingled with situational space, that is, it is where emotional feelings are palpably *felt* and where we are in touch with the sense of our situational surroundings (Cataldi 1993). This means that emotions are not just actions, something that our bodies express or articulate. Another side of them are about the way that we – as for instance when experiencing or appreciating a work of art or a landscape – are possessed by them or swept into their grasp. It is the felt sense of having been moved emotionally; the more passive side of emotional experience. Cataldi describes it as follows:

> It coincides with the senses in which we might say that we have fallen into love, are gripped with fear or seized by terror, burdened with remorse, overcome by shame, filled with joy, cast into despair, and so forth. It coincides, that is, with a view of emotions as 'passions' (1993, 106).

These notions suggest an active-passive duality (or rather circularity) as a complementary relation. Emotions are neither 'actions' (something we do) nor 'passions' (forces beyond our control that simply happen to us) – they are both at once. The active-passive circularity relates back to the idea of reversibility. It suggests that like the two sides of perception there are two sides to emotional experiences and that neither of those sides is intelligible apart from the other. They overlap and cross over into each other, but they never completely become the same. For example, 'actively' fearing some danger and 'passively' being endangered by it is such a two-sided and reversible emotional experience.

The phenomenological account of emotion forwarded here, then, first of all emphasizes the embodied nature of emotional experience; involving situated and moving bodies, the expressive character of the body, active-passive circularity and the reversibility of 'feeling'. Furthermore, it essentially sees emotions as public and *relational*. They are formed in the intertwining of our 'own' bodily flesh with the flesh of the world and the intercorporeal flesh of humanity. But what happens when we pursue the emotions in the meeting with O/other bodies? Merleau-Ponty suggests that to be embodied essentially involves an affective opening out of bodies to other bodies. This mutuality should of course not be mistaken for harmony; the emotions involved can take form of a range of different feelings as for instance love, desire, hate or fear. Having said that, however, the phenomenological approach remains limited insofar as it does not appreciate differences among bodies and power relations involved in intercorporeal meetings. One step in the direction of overcoming this deficiency has already been taken through Bourdieu's notion of 'incorporation' and Lefebvre's one of social space, but they do not address the question of encountering the O/other. Help to do that can most expediently be sought in feminist and postcolonial theory (e.g. Said 1978, Irigaray 1984, Young 1990a, 1990b, McClintock 1995, Ahmed 2000, 2004, 2006). They challenge the generality of Merleau-Ponty's idea of the social body as a body opening up into the fleshy world of other bodies by maintaining that this world is not a general world of humanity, but a differentiated world.[7] In such a world, Ahmed argues, what sociality is about is more often than not 'precisely the effect of being with some others over other others' (2000, 49). I would argue that notwithstanding the universalist bias in the phenomenological approach, it does open to such an extension – in two ways. The first one is the exchangeability or intercorporeality constituted in particular in the thesis of reversibility. The self-sentient human flesh holds the embryo of otherness; by having an 'outside' experience of itself, it also anticipates the way in which it will experience and deal with the other. Secondly, I would add to that the general phenomenological principle of *orientation*. By arguing that our attention is always directed towards concrete (objects or) other bodies, a phenomenological approach opens to questions of both the particularity of these bodies and the emotional spaces created in the meeting.

More than anything else, the contribution from postcolonial analysis is the demonstration of the strength and inertia of binary us/them distinctions in imaginations of Other people and Other parts of the world.[8] From a practice point of view these 'imaginary bodies' are not only institutional constructions; they are as well created and reinvented in relations between bodies.[9] Encounters with other bodies involve practices and techniques of differentiation – as for instance in relations with bodies 'recognized' as familiar and/or strange. In such encounters familiar bodies can be incorporated through a sense of community, being with each other as like bodies, while strange bodies more likely are expelled from bodily space and moved apart as different bodies. In this way 'like' bodies and 'different' bodies do not just precede the bodily encounters of incorporation or expulsion, likeness and difference are directly produced through these encounters. This is what Sara Ahmed calls S*trange Encounters*

---

7   This 'generality' is rather (as in much philosophical writing) an implicit generalization of the white, male experiences.

8   The affinity to Lefebvre's representation of space is obvious.

9   It might be argued, then, that the combination that is pursued here, in the same move searches to include a social differentiation into phenomenology and a practical embodiment into postcolonialism (see also Haldrup, Koefoed and Simonsen 2006).

(2000) and describes as a visual and tactile process: just as some bodies are seen and recognized as stranger than others, so too are some skins touched as stranger than other skins. It of course involves 'situated corporeal attitudes' or emotions. Various familial relations involve particular forms of emotion and ways of touch, while the recognition of some-body as a stranger – a body that is 'out-of-place' because it has come too close – might involve disgust or fear of touching. This is part of more general practices of otherness where marginalized groups are culturally imprisoned in their bodies. They are defined in terms of bodily characteristics and constructed as ugly, dirty, defiled, impure, contaminated, or sick (McClintock 1995). Young (1990b) uses Kristeva's notion of 'abject' and one of an ambiguous 'border anxiety' to account for the double emotional experience in such exclusive situations; on the one side group-based fear or loathing and on the other one a group-connected experience of being regarded by others with aversion. The notion of abject is figured in a manner similar to the way Merleau-Ponty figured flesh; that is, as prior to the emergence of an opposed subject and object and as making possible that distinction. The abject takes an ambiguous position between subject and object; 'the abject' is distinct from the object and other than the subject. It is just at the other side of the border, next to the subject and too close for comfort (1990b, 144). Kristeva discussed three broad categories of abjection – abjection toward food and thus toward bodily incorporation; abjection toward bodily waste; and abjection toward signs of sexual difference (Grosz 1994). When the notion here is translated from Kristeva's psychological and subjective register into a socio-spatial one,[10] it concerns the presence or proximity of groups from which the majority emotionally strives to distance themselves. It suggests, that 'the others' are not as different from us as objects; they are like us but are affectively marked as different. The point is that in the aversive reactions to 'others' – even those whom we hate, loathe, and strive to distance ourselves from – we do not experience them as entirely separated from ourselves. In Merleau-Ponty's terms, we are part of a common flesh. The 'border anxiety' in the construction of the Other involves intercorporeality, the permeability of bodily space, and a spatialization of emotion in the form of an ambiguous relationship between distance and proximity.

As suggested in the introduction, the examples to come in this and the next section do not aspire to give a coherent analysis. They are supposed to give 'flesh and blood' to the argument and to introduce some conceptions developed between the above and the empirical analysis.

---

10   Or translated back, it could be argued, since Kristeva developed the notion with inspiration from Mary Douglas's seminal anthropological text on *Purity and Danger.*

Example 1   Practical Orientalism

> When I walk in the street and meet those of another ethnic origin (it is awful that you don't know what to call them, isn't it?) but people that are different from us. But I don't know what I'm meeting. He can come from Iran, Iraq, he can be from Turkey, he can be from Morocco, he can be from Lebanon, Syria, it is not written on them ... What I am fed up with is that they keep on whacking me with their Koran and say 'our Islam'... Because then you can say; one is supposed to wear a scarf, another one is not supposed wear a scarf ... one is allowed to shower together with other people, another one is not. It is rather difficult. And then I'll not call it religion; I'll call it culture.
>
> I think that the greatest problem is that we in Denmark are afraid of contact. We are horribly afraid of being called racists. But it is not that we won't give them their rights. But we shall not just say, 'well, since you are allowed to come to live here, then we will of course also adjust our society to you'. That won't do – that's what I mean by fear of contact. It can't be right that we in our primary schools just say 'of course we rebuild the showers because there are two children who are not allowed to shower'. Then they damned well must refrain from showering ... And the same about their *halal* meat, it is ridiculous. What about the Danish children? Are they then supposed to eat *halal* meat? (Karin, police officer, 39)

This is a case of expressing *orientalist* feelings (cf Said 1978) – an imagination of difference building upon and reinventing binary dichotomies between 'us' and 'the others'. We meet it most distinctly in the phrase telling that we are dealing with 'people that are different from us'. The imagined geographies are translated from external to internal 'strangers' and construed in overlapping scales of proximity and distance. A significant point is how difference and 'strangeness' are created through emotional reactions on everyday, banal, bodily practices such as choice of dress ('one is supposed to wear a scarf, another one is not supposed to wear a scarf'), showering ('of course we rebuild the showers because there are two children who are not allowed to shower') and eating ('What about the Danish children? Are they then supposed to eat halal meat?'). That is why I, as a development of Said's institutionalist perspective, suggest the notion of *practical orientalism* (see also Haldrup, Koefoed and Simonsen 2006) as a means to grasp how everyday sensuous experiences such as the sight of an 'other' outfit, the hearing of 'other' languages/sounds and the taste of 'other' food give rise to feelings of anger and anxiety that come to permeate cultural encounters.

Emotional responses do however not only take the 'abjective' or dismissive form. Another side of orientalist feelings is the fascination with the 'exotic' habits of the other.

Example 2   Affective cosmopolitanism

And then I love the strangeness of Copenhagen in the way that you have so many different people. There are a whole lot of people from all sorts of countries. In the start I was very confused, when I saw a so-called guest worker: 'Oh man, I have never seen those people at home!' Yes, we had a Polish or two, but nobody from Turkey or Iran and everywhere else ... Here at work there are many people from different cultures. It is great. I soak up every day. And I love that people come from all over the world, because it is very exciting to hear about the places they come form, what they eat, how they live, what they have earned and family relations and ... I think about, there is so many people who are afraid of strangers, because they have some preconceived ideas about them. Instead of trying to give yourself the chance and the joy it might be to learn to know them (Birgit, cleaning inspector, 47).

Birgit functions in a 'multicultural' working community – a community which at the same time is practically given and something she appreciates and tries to strengthen. The working community creates a familiarity and an emotional obligingness between the members. Birgit talks about how she 'love that people come form all over the world' and about the 'joy in learning to know them'. She also tells how she has had the whole group on a visit at her home, at which occasion each of them brought a dish from their own 'kitchen'. As in many other cases, food and meals appear as a practical/sensuous symbol of difference, here working as an object for gathering and mutuality as well. To describe these emotions, I adopt Nava's (2002) notion of *affective cosmopolitanism*, which emphasizes the attraction of different and exotic products and body cultures.

The other side of the process of encountering 'other' bodies is the experience of being exposed to oppressing vision or emotion. It is about the process of 'internalizing otherness' or the development of a 'double consciousness' due to the enculturation of the body.[11] Fanon, in *Black Skin, White Masks* (1967), talks about how for black men and women 'consciousness of the body is solely a negating activity. It is a third-person consciousness. The body is surrounded by an atmosphere of certain uncertainty' (1967, 110-111). The additional 'task' faced by men and women of colour is one of reconciling their own 'tactile, vestibular, kinaesthetic, and visual' experiences with the operation of a 'historical-racial schema' providing 'racial parameters' within which their corporeal schema is supposed to fit. In this way, persons of colour are overdetermined both from within and without insofar as racist attitudes and actions penetrate the skin and are incorporated into both the white and the black's bodily images (Weiss 1999).

---

11   The affinity in these formulations to Bourdieu's notion of incorporation is obvious.

Example 3   Incorporation of otherness

> To be honest, I think that Christiania, that's a wonderful place. Not because of the cannabis, but because of the atmosphere ... And, do you know what I like? The people out there, they can actually communicate, because there is many nationalities coming in there and so. The people in there, who lives there and are in there, they – if I go in there, I don't feel that people frown at me or something like that.
>
> If you for example walk on Strøget, ... if you enter into a shop, then I feel – it's not that I will accuse people in there of something – I just feel that people keep an eye on me; do you know what I mean? I mean, you are allowed to go into a shop and have a look and leave again, if you don't see anything you want to buy. And there I feel the difference, where people think 'they are going to steal something' ... If you are in my situation, like I am, and enter, that's how I feel.
>
> Also the way people reply to you ... People often make the mistake, if you have another colour, and if somebody has done something wrong, then they place you in the same group as them (Neezan, cleaning assistant, 23).
>
> For Neezan, the experienced 'otherness' and the emotions it raises in him enter into his construction of different parts of the city. He distinguishes between places where people 'can actually communicate' and 'don't frown' at him, and those where they 'keep an eye on' him because they think he is 'going to steal something'. For him, it becomes a distinction between spaces having an accommodating and unaccommodating 'atmosphere' – in this way contributing to an affective mapping of the city. He obviously has met racist bodily expressions; 'people keep an eye on you', 'if you are in my situation' and 'if you have another colour'. He experiences how the construction of 'the other' works through a process of homogenization where the individual body-subject is met, not as an individual, but as a collective identity or an archetype characterized by the 'sign on the body'.

In this way, the recognition of differentiation in encounters with 'other' bodies renders visible the necessity of modification of the phenomenological approach to the expressive and affective spaces of emotions. Notions applied here such as abjection, border anxiety, practical orientalism, affective cosmopolitanism and incorporation of otherness are some of the ones that can be helpful in doing the job.

Ethical Encounters

The investigation of bodily encounters and encounters with 'other' bodies would in many ways be incomplete without also considering their ethical dimension. It has put emphasis on the ambiguity of the distance/proximity relationship in living with the 'other' – the way she/he at the same time is distant or uncommon in our lives and yet close enough to haunt us – in this way raising an ethical question on how these encounters are performed. In accordance with the above, this will render necessary an account of ethical/moral consideration that emphasizes the body, emotion, expressiveness and affectivity. The very act of connecting ethical questions to bodily meetings precludes some positions in the landscape of moral philosophy; it departs from the mind/body dualism inherent in accounts exclusively identifying morality with reason and the mind. Many traditions, and not least feminist ones, have been engaged with the overcoming of this rationalist bias. Many of them start from the idea of contextual, particularized 'ethic of care', which was introduced by Gilligan (1982) as a non-exclusive alternative to an 'ethic of justice' and developed within 'materialist feminism' (e.g. Ruddick 1989) as one that emphasizes the role of the body in our moral interactions and the significance of bodily demands, needs and desires. It takes off from the mother-child relationship or 'maternal interests' arguing that this foundational relationship in which one body seeks to realize interests for the other can serve as a model and be extended to our more distant moral relations with others.[12]

Ethic of care, then, provides a framework stressing contextuality, particularity, and pragmatism as a distinctive way of dealing with real life rather than theoretical situations. However, due to its reliance on the caring subject and its basis in derivations from the mother-child relationship, it also carries some inherent flaws. For one thing, it might turn out as a one-way (hierarchical) relationship in which the 'caring' or 'responsible' figure is in the position of power to define and meet the interests of the other. For another, relations of care still carry connotations of proximity, if not in space then in alikeness. For example, Scandinavian research has shown the degree to which relationships of care within the welfare state have been build on presuppositions of culturally homogeneous populations (see Koefoed and Simonsen 2007), and also transnational relations of care seem to rely on differential identifications. In order to remedy these flaws and move in the direction of a more 'mutual' or 'reversible' ethics of responsibility, in the following I look toward three contributions.

The first one comes from Levinas (1979, 1985) who in his ethics of alterity particularly locates ethics in encounters with the other. An important step in his analysis focuses on the face, which he describes as an ethical (non-physical) resistance to the destruction of life. Here it is the naked and defenceless face of the other that through its very expression condemns violence and calls for social

---

12   Also within geography ideas of care ethics have gained increasing attention, not in order to reject notions of justice, but to foreground an understanding of social relations as contextual, partial, attentive, and responsible (Lawson 2007). Part of this work, such as Brown's (2003) work on hospice and geographies of dying, works rather directly from care work, care relations and their gendered aspects while other parts seek to extend them or combine them with other issues. McDowell (2004) extends the discussion to the sphere of work and work/life balance. Smith (2005) combine feminist ethic of care with the ethic of mutual responsibility involved in the welfare state, even arguing for an extension of both to the (housing) market. And Massey (2004, 2005) argues for an extension of care to distant others. In her 'geographies of responsibility' she challenges the tendency to associate care with proximity and tries to situate it in transnational chains of ordinary actions and inequalities.

justice. From that, Levinas develops an ethic of responsibility that is beyond and before being. Responsibility for others is for him not only absolute and unconditional, it is the very condition of possibility for subjectivity. This means that subjectivity has to be described in ethical terms. Ethics/responsibility is not just a quality in subjectivity as if subjectivity already existed before the ethical relationship; it is in ethics that subjectivity is constituted. Subjectivity is being *for* an-other, and the other is close to me – not in the form of spatial proximity or familiarity – he/she is approaching me and rendering me responsible for him/her. By this move, Levinas relocates the ethical relationship from the caring subject and relations of familiarity to the multifarious ways of encountering others. He does so somewhat sliding between the descriptive and the prescriptive; on the one hand providing poetic descriptions of how the other is encountered in the forming of the self, and on the other hand indirectly suggesting that some ways of encountering are better than other ones. This does not take the form of rules or codes of behaviour, but is described through the preference of some sensuous practices over others (e.g. caress and erotics of touch (desire) over eating and digesting (need)). The point is that the former ones encounter the others in a way that neither reifies nor incorporates them, but rather allows them to retain their otherness. The responsibility involved in these encounters is *infinite*: it is a total responsibility, encompassing all others and everything in the other; and it is an imperative that can never be fulfilled, there will always be a call for a future response to an other who is yet to be approached.

My second source is Ahmed (2000), herself inspired by Levinas but whom she criticizes for abstracting the other from particular others. Through that abstraction, she says, the other becomes a fetish; it is assumed to contain otherness within the singularity of its form. Therefore, she defines her task as considering how it is possible to maintain Levinas's idea of an infinite responsibility, for everything and for everyone, and at the same time include the role of the particular and finite: 'We need to recognize the infinite nature of responsibility, *but the finite and particular circumstances in which I am called on to respond to others*' (2000, 147, *emphasis in original*). The solution Ahmed suggests is to introduce particularity not as a description of different others, but at the level of encounters. Particularity becomes a question of *modes of encounter* through which others are faced; encounters that at the same time flesh out and call for specific responses to the other. Even if it cannot fulfil our responsibility, our infinite responsibility in this sense begins with the particular demands we face encountering an other.

In order to underline the embodied character of such ethical encounters, as the third and last source I look to Weiss (1999) who (paraphrasing the Kantian categorical imperative) introduces the notion of *bodily imperatives*. To be embodied, she says, is to be capable of being affected by the bodies of others and, therefore, to be embodied is both a necessary and a sufficient condition for the generation of a bodily imperative. The bodily imperatives emerge out of intercorporeal exchanges and both demands and responses will therefore depend upon the specific contexts in which our encounters with others are situated. Ethics in this sense involves developing a moral agency that can only be experienced and enacted through bodily practices, practices that implicate and transform both the bodies of others and our own body images. Thus the path through these three contributions holds to the ideas of ethics of care and responsibility as regards the situatedness of ethical agency as an embodied and context-sensitive phenomenon. It does however, in accordance with the principle of reversibility, suggest a corrective in the form of a relocation of the ethical relationship from the caring subject to encounters and the way in which they are performed.

Example 4   Practical imperatives

It's always a challenge to meet people with another cultural background ... How shall I describe it? The hospitality is big, even if you are wearing a uniform. In my work, I sometimes have to visit parents, and the hospitality is always big. When you visit their homes, whether they like it or not when such a uniform arrives, their norms are that they *have to* be hospitable. This is definitely something you feel ... But it is also about respecting the way that they live, isn't it? For many of the ethnic families, for example, shoes are something you put outside the door. You don't just enter. Well, I do bring the shoes in; I don't dare to leave them on the staircase. But I put them just inside the door, and people are glad to see that you respect their habits ... In my work, I don't come out here and now to solve a violent police task. I have to talk with the young people and their parents, so I have time enough to pay regard to all those things. It is a fascinating work where you meet many different views of life. That's also what makes it challenging (Sten, police officer, 47).

Sten works with crime prevention in community police organization. For him the 'multicultural city' is a (fascinating) work place. He is dealing with young boys on the edge of a criminal career, working on the streets and involving them in different kinds of activities. An important dimension of that is communication with the parents. The task is, he says, to draw their attention to the 'life on the street', outside the 'safe walls' of the home – a public space with its own rules that you as a parent has to be aware of. It is primarily through this contact he describes his relation to 'the other'. For example, he is both glad and surprised to be met with such a great hospitality, even if he shows up (wearing uniform) in order to discuss problems with the children. It is also interesting to see how he represents cultural difference and ways to meet it in terms of banal, bodily practices such as taking off the shoes when entering people's apartments. It is his way of showing his respect and answering to the rules of their private space. What he describes, then, is a mode of encounter where banal practical imperatives are forming (and easing) the concrete, everyday meetings of this specific sensitive relationship.

The bodily imperatives are however also spatially and temporally mediated, insofar as the immediacy of the bodily and practical encounters is laden with broader social processes that also operate elsewhere and in other times.

Example 5   Value imperatives

> ... It becomes impiously in some sense. And it must be important to maintain some social values, because that's what holds society together. The big debate is when you consider why we are so much against immigrants, why we are so much against Islam, why we are so much against foreigners. It is because we are very different. Culturally, the others, for example Islam, will be offended by the pornofication that is part of our everyday life. The way we dress on a summer day, the way our advertising looks, the way we act on the beach and so on ... We don't find it offending, but you have to remember that culturally you end up in a situation where you despise each other, because we are so different. And I think you should think about it, because it is exactly opposite to a development of social values that we share. Instead, you end up like going in different directions, and you build up confrontations. That's what happened in the later years, and that is no good. I think that we need to find a platform from where we can find some common values that we all care about. In some sense it is also why some Danish people suddenly convert their faith to other things. It is because they miss communities, they miss social values, unfortunately. It is the spirit of the time we live in. We live in the age of materialism. Before we get to the point where also non-material things gain value, it will be difficult to find the social values. It involves a philosophical element ... I definitely think that there shall be room for all of us. We have to welcome the strangers. We just have to understand why they might despise us at some points (Jan, financial adviser, 43).
>
> Here, we have a person that meets the 'multicultural city' in a mediated form. His response to the requests of 'others' does not have the form of immediate bodily practices but he does try to understand how 'our' bodily practices will affect 'others' – here talking about the 'pornofication' of public space and different practices of dressing as examples. None of that offends him, he says, but he understands the emotions it can generate between the 'others'. He describes Danish society through a pessimistic figure of loss; 'they miss communities', 'they miss social values', 'we live in the age of materialism'. This condition he renders responsible for 'our' problems encountering 'the others'. It is 'our' loss of values, 'our' lack of communities, that renders us insecure and undermines our readiness to meet difference and the other's different, but stronger cultural communities. It creates confrontational emotions; we drift into a situation where we 'despise' each other. He appeals to mutual responsibility and pleads for the necessity of developing a social and cultural 'platform' of common values – a mode of encounter that can (re)create values and in this way open to difference and respond to the others' emotions and value imperatives.

The notion of ethical encounters then covers a relational and non-individualistic approach to ethics grounded in responsibility, encounters and bodily imperatives. As suggested above such an approach refers to the principles of reversibility and intercorporeality and sets a close connection between practice, sociality, emotion and ethics.

Concluding Comments

In this chapter, I have, theoretically and empirically, explored how otherness, emotion and ethics can be approached through a theory of practice. Central elements in this exercise have been embodiment and bodily encounters. The proposed account is a theory of practice taking inspiration from phenomenology, and as such it suggests an understanding of emotion and ethics that is grounded in lived meaning, intercorporeality and 'the reversibility of the flesh'. Due to acknowledged lacks, however, it further seeks to bring phenomenology into a differentiated world – a world imbued with difference and power relations – in the present connection primarily by means of feminist and postcolonial writings. This procedure has led me to an understanding of ethics as an embodied phenomenon where the starting point for our moral practices is sensitivity to the bodily imperatives that issue from different bodies. That is, an understanding that locates ethics on the level of practice and encounter (or sociality), emphasizing its mutuality or reversibility. Such encounters are however always mediated; the immediacy of bodily encounters is always imbued with broader social processes and experiences of which some are occurring in other times or in other spaces. Accordingly, ethical encounters might be characterized as the ones that in the specific situation recognize how the encounter itself is implicated in broader relations (in time-space) and are able to respond to (even 'strange') bodily imperatives of the other without reconstructing him/her as 'the Other'.

Such an ethic of encounter, to my mind, can overcome the problem of what Massey (2005) critically calls 'Russian-doll geography of ethics' – one that is territorial and emanates from the local. Encounters and bodily imperatives can be mediated and do not confine themselves to physical proximity. Furthermore, as argued by Ahmed (2000), the focus on particular modes of encounter (rather than particular others) involves temporalities and spatialities that open up the encounter. The temporality of the encounter concerns not only its historicity but also its potentialities; what futures it might open up. The spatiality consists of the movement of arrival to the particular place of encounter as well as the link of this arrival to other places, to an elsewhere that is not simply absent or present. These movements further a geography of ethics that does not fix the other but anticipates the possibility of facing something or somebody different – the not yet and the elsewhere.

References

Ahmed, S. (2000), *Strange Encounters. Embodied Others in Post-Coloniality* (London and New York: Routledge).

—— (2004), *The Cultural Politics of Emotion* (Edinburgh: Edinburgh University Press).

—— (2006), *Queer Phenomenology. Orientations, Objects, Others* (Durham and London: Duke University Press).

Bourdieu, P. (1977), *Outline of a Theory of Practice* (Cambridge: Cambridge University Press).

—— (1990), *The Logic of Practice* (Cambridge: Polity Press).

Brown, M. (2003), 'Hospice and the spatial paradoxes of terminal care', *Environment and Planning A* 35, 833-851.

Cataldi, S.L. (1993), *Emotion, Depth, and Flesh: A Study of Sensitive Space* (New York: State University of New York Press).

Crossley, N. (1996), *Intersubjectivity: The Fabric of Social Becoming* (London: Sage).

Davidson, J. (2003), *Phobic Geographies: The Phenomenology and Spatiality of Identity* (Aldershot: Ashgate).

Davidson, J., Bondi, L. and Smith, M. (eds) (2005), *Emotional Geographies* (Aldershot: Ashgate).

Dreyfus, H. and Hall, H. (eds) (1992), *Heidegger. A Critical Reader* (Oxford: Blackwell).

Fanon, F. (1967), *Black Skin, White Masks* (New York: Grove Press, Inc).

Gilligan, C. (1982), *In a Different Voice: Psychological Theory and Women's Development* (Cambridge: Harvard University Press).

Grosz, E. (1994), *Volatile Bodies: Toward a Corporeal Feminism* (Bloomington and Indianapolis: Indiana University Press).

Haldrup, M., Koefoed, L. and Simonsen, K. (2006), 'Practical Orientalism – Bodies, everyday life and the construction of otherness', *Geografiska Annaler* 88B(2), 173-85.

Heidegger, M. (1962), *Being and Time* (Oxford: Blackwell).

Irigaray, L. (1984), *Etique de la difference sexuelle* (Paris: Minuit).

Koefoed, L. and Simonsen, K. (2007), 'The price of goodness: Everyday nationalist narratives in Denmark', *Antipode* 39(2), 310-30.

Lawson, V. (2007), 'Geographies of care and responsibility', *Annals of the Association of American Geographers* 97(1), 1-11.

Lefebvre, H. (1991), *The Production of Space* (Oxford: Blackwell).

Levinas, E. (1979), *Totality and Infinity: An Essay of Exteriority* (The Hague: M. Nijhoff Publishers).

—— (1985), *Ethics and Infinity: Conversations with Philippe Wemoi* (Pittsburgh: Duquesne University Press).

Massey, D. (2004), *For Space* (London: Sage).

—— (2005), 'Geographies of responsibility', *Geografiska Annaler* 86B(1), 5-19.

Massumi, B. (2002), *Parables for the Virtual: Movement, Affect, Sensation* (Durham and London: Duke University Press).

Merleau-Ponty, M. (1962), *Phenomenology of Perception* (London: Routledge and Kegan Paul).

—— (1968), *The Visible and the Invisible* (Evanston: Northwestern University Press).

McClintock, A. (1995), *Imperial Leather: Race, Gender and Sexuality in the Colonial Context* (London: Routledge).

McDowell, L, (2004), 'Work, workfare, work/life balance and an ethic of care', *Progress in Human Geography* 28, 145-163.

Nava, M. (2002), 'Cosmopolitan modernity: Everyday imaginaries and the register of difference', *Theory, Culture and Society* 19(1-2), 911-928.

Ruddick, S. (1989), *Maternal Thinking: Towards a Politics of Peace* (New York: Ballantine Books).

Said, E. (1978), *Orientalism* (London: Routledge and Kegan Paul).

Schatzki, T. (2002), *The Site of the Social: A Philosophical Account of the Constitution of Social Life and Change* (University Park, Pennsylvania: The Pennsylvania State University Press).

Simonsen, K. (2003), 'The embodied city: From bodily practice to urban life', in Öhman, J. and Simonsen, K. (eds), *Voices from the North: New Trends in Nordic Human Geography* (Aldershot: Ashgate), 157-173.

—— (2004), 'Spatiality, Temporality and the Construction of the City', in Bærenholdt, J.O. and Simonsen, K. (eds), *Space Odysseys: Spatiality and the Social Relations in the 21st Century* (Aldershot: Ashgate), 43-63.

—— (2005a), 'Bodies, sensations, space and time: the contribution from Henri Lefebvre', *Geografiska Annaler* 87B(1), 1-15.

—— (2005b), *Byens mange ansigter – konstruktion af byen i praksis og fortælling* (Frederiksberg: Roskilde Universitetsforlag).

—— (2007), 'Practice, spatiality and embodied emotions: An outline of a geography of practice', *Human Affairs* 17(2), 168-182.

Smith, S.J. (2005), 'States, markets and an ethic of care', *Political Geography* 24(1), 1-20.

Social and Cultural Geography (2004), 'Themed section: Embodying emotion, sensing space' 5(4), 523-633.

Thrift, N. (1996), *Spatial Formations* (London: Sage).

—— (2004), 'Intensities of feeling: Towards a spatial politics of affect', *Geografiska Annaler* 86B(1), 57-78.

Weiss, G. (1999), *Body Images: Embodiment as Intercorporeality* (New York and London: Routledge).

Wylie, J. (2005), 'A single day's walking: narrating self and landscape on the South West Coast Path', *Transactions of the Institute of British Geographers* 30(2), 234-248.

Young, I.M. (1990a), *Throwing like a Girl and other Essays in Feminist Philosophy and Social Theory* (Bloomington: Indiana University Press).

Young, I.M. (1990b), *Justice and the Politics of Difference* (Princeton: Princeton University Press).

Chapter 13

'Just Being There …' : Ethics, Experimentation and the Cultivation of Care

Jonathan Darling

Ethics is closer to wisdom than to reason, closer to understanding what is good than to correctly adjudicating particular situations (Varela 1999, 3).

Responsibility is connected with the notion of answering to the other, responding to the other's moves and the other's sensibility … It is more to do with a kind of sensibility – the way I perceive things, the way I make sense of the whole environment about me (Lingis 2002, 35).

In a recent review of ethics within geography Jeffery Popke (2009, 81) highlighted the diverse body of work that constitutes non-representational theory as an area which might offer 'a different set of resources for considering matters of ethics and responsibility'. In particular, Popke (2009, 84) suggests that non-representational styles of thought speak to 'a different kind of ethics, one that takes the form of an ethos rather than a morality or a set of principles grounded in universal norms or juridical constructs. Such an ethos works toward encounters that open us to a generous sensibility, one that might be capable of re-enlivening our affective engagements with others and fostering a heightened sense for what might be possible'. In this manner, non-representational theory is seen to orientate an ethics of performative dispositions, of practice and embodied judgement, creating an 'ethics of enactment' (McCormack 2005, 142), wherein moments of generosity and responsibility arise from within the unfolding of events (Thrift 2004a; Dewsbury 2000). The ethical impulses of non-representational thought have been most clearly articulated by Derek McCormack (2003) and Nigel Thrift (2003a, 2004a), in suggesting how an account of ethics might arise from 'unreflective, lived, culturally specific, bodily reactions to events' (Thrift 2000, 274). Here the negotiation of the immediate present provides a space not for the application of pre-given moral tenants, but for the emergence and cultivation of ethical sensibilities which value moments of generosity and open engagements with difference. Popke (2009, 84) though raises a question of this work, warning that 'our ethical vision is likely to remain stunted if we limit ourselves to a consideration of the affective potentialities lurking within events and encounters, without also posing the broader question of how events and encounters become constituted as the locus of a shared sense of conviviality and solidarity'. In this chapter I want to take seriously this concern in order to develop an account of non-representational ethics which arises from such a space of 'conviviality and solidarity'.

Drawing on Popke's (2009) account of non-representational ethics, this chapter considers the responses, sensibilities and negotiations present in a UK drop-in centre for asylum seekers. Drop-in centres themselves have received attention from geographers interested in notions, and spaces, of care (see Conradson 2003a, 2003b, 2003c; Johnsen et al. 2005a; Parr 1998, 2000), with a particular focus upon the ways in which practices of care actively create environments, and how drop-in spaces might make particular forms of identity and subjectivity possible. Here, whilst considering the spatial accomplishment of environments of care for the homeless, Cloke et al. (2005, 2007) examine the 'ordinary' ethics of the individual which prompts volunteers to care for others. Cloke et al. (2007, 1092) suggest that '[o]rdinary responsibilities for others – neighbours, strangers or sojourners – are the platform for more specific acts of ethical practice', acts which come into being through spaces of care. However, they also point out that while everyday sensibilities of 'ordinary ethics' may be harnessed in motivating volunteers to care for others, these sensibilities are themselves often 'questioned and confronted in the didactic practice of serving homeless people' (ibid., 1095). It is this sense of questioning, and uncertainty, which I want to foreground, for, as Popke (2006, 507) argues 'care is more than simply a social relation with moral or ethical dimensions; it can also be the basis for an alternative ethical standpoint'. As such the practice of caring for others can have profound ethical implications, it can alter the way in which we not only view ourselves, but also the world around us. It is this sense of the ethical, of a responsive sensibility towards others which might be deemed caring or generous, which I want to argue is worked upon within drop-in space.

In doing so the chapter develops three interwoven points of argument. Firstly, the chapter attends to McCormack's (2003) desire to extend the field of the ethical in which geographers might move, by demonstrating how the ethical dispositions and sensibilities that non-representational theory promotes might be practiced in the daily life of a drop-in centre. Secondly, I suggest that the practicing of such a situational ethics produces an affective sense of belonging and collective accomplishment which allows asylum seekers to feel comfortable within such space. Thirdly, I argue for an account of responsive ethical engagements which is realistic about both the possibilities, and the limits of this mode of thought, suggesting that while the drop-in centre provided many positive moments of relating across difference, there was still much that these relations and sensibilities could not address. This chapter therefore represents an attempt to flesh out some of the ways in which non-representational modes of ethical thought might be actively practiced through this particular space of care. Before considering these ethical negotiations in more depth however, I shall first introduce The Talking Shop.

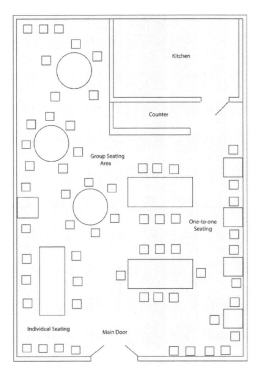

Figure 13.1   Plan of the Wednesday Talking Shop

The Talking Shop

The Talking Shop comprised two linked drop-in centres for asylum seekers and refugees in the centre of Sheffield. The first of these ran on a Wednesday for two hours and was housed in a church hall, while the second was on a Friday for three hours, also housed in a church hall. Both of these centres were run solely by volunteers and were partially funded through the regional charity The Northern Refugee Centre (NRC). At both centres 'service users' were welcome to come and go as they pleased, as were volunteers. Both sites provided a kitchen in which tea, coffee and biscuits were provided free of charge, spread out across a counter which connected the kitchen space to that of the halls themselves (see Figure 13.1 and 13.2). The halls were arranged around a series of small tables, normally with four to five chairs designed to facilitate small group discussions, conversations and meetings. The aim of The Talking Shop was to provide a space for asylum seekers and refugees to meet, to practice their English language skills and to gain some level of informal social contact. While there was not an explicit religious or political orientation at work it was notable that many volunteers drew upon a belief in Christian charity when describing The Talking Shop and that a series of

Figure 13.2   The Friday Talking Shop

Source: Author's photograph.

political activities, from arranging petitions and demonstrations to accompanying individuals to appeal hearings, took place through this shared space. In this sense The Talking Shop's official presentation, as a site of secular charity, was interwoven with a wide variety of different political, religious and ethical orientations, all of which interacted in the production of this shared space (Conradson 2003c; Cloke et al. 2005).

My own engagement with The Talking Shop came about through a project to study the varied ways in which Sheffield performed a sense of 'welcome' towards asylum seekers (Darling 2008). I attended this site as a volunteer for a 10-month period, with my presence as a researcher also being made known to those who attended. Throughout my months at the drop-in centre I kept a research diary which documented the many events, informal conversations and exchanges which helped to sustain this social space, along with the varied ways in which my relationships with a number of asylum seekers shifted over time. Alongside this situated diary, I also conducted interviews with a number of the drop-in centres volunteers, the centres organiser and with 12 of The Talking Shop's service users.

The Value of Presence

I want to begin by suggesting what The Talking Shop offered to those service users who made visiting it a part of their weekly routine. As I have said, there were no formal political processes behind The Talking Shop, this was not a space through which asylum seekers could defend their cases and fine-tune their appeals. The majority of volunteers here were not trained in the legal dimensions of the asylum system and as such the discussion of the merits of individual cases was not encouraged. Nor was it a space of social service provision, though volunteers were encouraged to signpost individuals to services that may be of use. What The Talking Shop offered was a space of social engagement. It provided an opportunity for asylum seekers to meet one another and to meet volunteers. In my experiences of attending The Talking Shop I found an array of different demands placed upon me, some individuals wanted to talk about themselves and their past, some wanted me to translate letters and newspaper articles, while others were happy to sit in silence over a cup of tea. Part of the openness of this space of care was in offering a sense of 'license' (Parr 2000) to all these different uses of the drop-in, while part of the skill of creating such a space was in attuning oneself to respond to these different performances, relations and bodily dispositions.

Over my time at The Talking Shop it was clear that relationships were built around the uncertainty of playing the asylum 'waiting game'. For the majority of those asylum seekers who attended The Talking Shop decisions on their asylum claims, and their status within the UK, were not yet made and as such their lives were held in a state of limbo, existing between the acceptance of refugee status or the rejection of detention and likely deportation. This position of waiting was further exacerbated by the UK government's decision to remove the right to work from asylum seekers leaving individuals 'warehoused' (Fekete 2005), often socially isolated and unable to develop any secure sense of connection to the city they had been dispersed to. It is this state of isolation and uncertainty over status and rights which not only denies asylum seekers the ability to consider a secure future, but also conditions much of what The Talking Shop came to mean to those asylum seekers I met. The drop-in centre was thus a crucial resource for tackling isolation and for feeling a part of something, as Adil describes in the following interview extract;

> Places like [The Talking Shop] give you more of a chance to get in contact with other people, so you can make friends and afterwards this is a place where you can go and they care about you, it's about just people being there really (Adil Interview, 2007).

The relations which came about here, of presence and contact, were therefore seen as central in developing the ability to cope with the daily struggles facing asylum seekers and with the sense of uncertainty that such a position imposed. Tinashe presented The Talking Shop as a site of release, a space through which to open oneself to new ideas;

> [The Talking Shop] is a big relief, it gives you something to do, to expend some energy and meet people once again. At these times you feel like life gets going again and by going there you make yourself useful and valuable again, you rediscover yourself again, and the person you find is often better for the experiences of having met some new people and opened up to many new and different perspectives (Tinashe Interview, 2007).

The Talking Shop was also recounted as a space to rebuild aspects of lives which had been forgotten, lost or hidden. Thus Ilya states that;

> Because of [The Talking Shop] I started playing the piano again and yeah, it was like getting onto new levels and you know, started living rather than just existing (Ilya Interview, 2007).

In both of these accounts, The Talking Shop represents a place through which new relations are made possible. Both with oneself, through re-approaching one's ideas and capacities, and with others, through encountering other perspectives and orientations. In this sense we might see much of the work of The Talking Shop as being in the situated and affective activation of lives that McCormack (2003) suggests arises through an account of positive affects, or relations which increase our capacity to 'go on'. Viewing The Talking Shop as such is to tap into that swirl of relations, emotions, affects and dispositions which Conradson (2003b) argues marks spaces of care as performative accomplishments. Here the drop-in centre takes on the role of a space where individuals begin to feel a sense of movement, of progression and change. This may be felt in the rekindling of past pleasures and interests, such as the playing of a piano, or an exposure to entirely new interests and ideas, but the key attribute to emerge here is a sense of development. The Talking Shop was about providing an environment through which asylum seekers could *feel* comfortable, could *feel* a part of something, as an affective response to the relations which took place. Within the context of an asylum system which confines individuals to a mode of warehoused existence however, this feeling of attachment created far more than simply a comforting space of care. That sense of positive attachments, of the will to 'go on' as McCormack (2003) terms it, also produced a will for life beyond the constraints of an uncertain existence. Through the attachments developed here individuals were able to begin to *feel* moments of hope and build plans for the future, as the relations of The Talking Shop offered a measure of how life did indeed progress from week to week as piano keys were learnt once again, English pronunciations improved and friendships developed. These small moments of sociality created a feeling that life did not have to be put on hold here and that the potential for discovering the new was never fully exhausted. I want to now consider one example of this practice of sociality.

One of the key tasks performed through the drop-in was that of translating letters for asylum seekers, and this act was central to the affective tenor of the drop-in as my research diary suggests;

> Over the past weeks I've noticed that it is common for asylum seekers to come to the centre with Home Office documents and letters for translation. Lynn [a co-founder of the drop-in centre and the volunteer co-ordinator] normally takes up this role, going to a table in the corner to do so. Lynn normally sits with an individual for between ten and twenty minutes, talking in a slow, measured tone and occasionally gesturing with her hands and using a range of facial expressions. Though these acts take place out of earshot they spill over into the rest of the room, a sense of stillness takes over other tables as hushed conversations wonder what is being said. Eyes dart across the room on the look-out for bodily reactions to the words on the page. A smile and an embrace imply a letter of acceptance. Shrinking into the chair, often crying, and a series of frenzied calls to solicitors, means a failed case and the forthcoming removal of welfare and home (Research Diary, 24 November 2006).

Such moments of translation illustrate two important aspects of this space of care. Firstly, they highlight the way in which embodied relations between individuals, texts, materials and wider discourses come to condition and alter the performative creation of drop-in space. The arrival of a Home Office letter, its opening, translation and impact, all illustrate how the asylum system and its discourses of belonging, rights and exclusions, permeates and alters this space. A Home Office letter acts as a 'spatial manager' (Hage 1998), for it dictates national limits of acceptance through its arrival, reading and interpretation. It defines and dictates the possibility of belonging. The act of translation has an affective force over the drop-in, either positively encouraging others to 'go on', to feel a sense of hope at the success of others, or negatively diminishing that sense of potential belonging. The act of translation also highlights the particular style of ethics which was being performed here. Alongside the act of reading came a series of embodied and responsive gestures which Lynn performed to place others at ease, she would maintain eye contact whilst speaking, she would gently place a hand on the arm of others to reassure them, she would hug those successful through joy, and those defeated through sorrow. These gestures formed the kind of habitual, embodied and responsive care that Lynn practised in The Talking Shop. These actions were not reliant upon an established series of moral codes and rationalities, rather they emerged precisely through the situation, the event, of translation. Moments such as this in The Talking Shop are indicative of its ethical style, one not confined to moral deliberation and regulation, but rather linked to ideas of practice and response and it is this style which I want to consider in more detail through recent accounts of non-representational ethics.

Practical 'Know-How'

A number of recent political and ethical claims have been made for an expansion of the arena in which we view 'politics' and 'ethics' as such (Connolly 2002; McCormack 2003; Thrift 2004b). Whilst these calls maintain a heterogeneous lineage, they all in some way draw upon recent reconsiderations of thought and practice within the field of cognitive science, most centrally work which has cast into doubt the centrality of rational deliberative judgement (Damasio 2000, 2004; Dreyfus and Dreyfus 1990; Varela et al. 1993). Here Varela (1999, 6) poses a central distinction between 'know-how' and 'know-what', between 'spontaneous coping and rational judgment', and argues that it is within this former experience, of learned, experiential and embodied 'know-how' that the vast majority of human practice is achieved. For Varela (1999, 17, original emphasis) we can only come to know the world through action, thus 'we can say that the world we know is not pre-given, it is, rather, *enacted*', any understanding of the world is therefore momentarily reached through our situated responses. Thus here actions 'do not spring from judgment and reasoning, but from an *immediate coping* with what is confronting us. We can only say we do such things because the situation brought forth the actions from us' (ibid., 5, original emphasis). Within this framing of the ethical, 'the practical activities of embodied human beings give priority to "know-how" over propositional knowledge' (Connolly 2002, 92). This is not an insignificant claim, for it implies that ethics must be considered as about more than purely rational judgements and knowledge, rather ethics becomes an embodied, lived stance towards the world in which ethical *sensibilities* are *enacted*.

Through this area of thought a number of geographers have become interested in how an appreciation of the momentary and embodied nature of judgements, responses and knowledge might impact our relationships to the spaces of everyday life (Anderson 2004, 2005; Harrison 2000; Latham 1999; Thrift 2005; Wylie 2002). In particular, Anderson (2005) has highlighted the ways in which practices of judgement act as the cornerstone of the prosaic accomplishment of everyday life. For Anderson (2005, 651) judgement is not a deliberative process of cognitive ordering, rather it is a practice of 'thought-imbued feelings', such that judgements are 'bodied-forth without deliberation'. Judgement is an embodied, situational and affective practice then, one which arises from an 'ethics of affection based on how bodies compose with other bodies' (Anderson 2005, 653). To provide an example of such thought, I want to draw on Brian Massumi's (2002a, 216, original emphasis) account of how we judge moments of anger;

> There's always an instantaneous calculation or judgment that takes place as to how you respond to an outburst of anger. But it's not a judgment in the sense that you've gone through all the possibilities and thought it through explicitly – you don't have time for that kind of thing. Instead you use a kind of judgment that takes place instantly and brings your entire body into the situation. The response to anger is usually as gestural as the outburst of anger itself ... An outburst

of anger brings a number of outcomes into direct presence to one another – there could be a peace-making or a move towards violence, there could be a breaking of relations; all the possibilities are present, packed into the present moment. It all happens, again, before there is time for much reflection, if any. So there's a kind of thought that is taking place *in the body*, through a kind of instantaneous assessment of affect, an assessment of potential directions and situational outcomes that isn't separate from our immediate, physical acting-out of our implication in the situation.

In the moment of judgement there are a myriad of possible responses, a vast array of potential to do otherwise, yet in the embodied nature of such thought and decision, occurring as it does before reflection, such potential often appears just out of reach. Because judgement occurs within a specific context, that 'instantaneous assessment of affect' which directs our response relies upon a vast archive of past judgements, bodily relations and affective outcomes, those 'potential directions' of the past, govern the response of the present. If we view judgement as such an embodied response to the world, then it becomes important to consider how we might effectively alter such resources of judgement and fine-tune our responses to situations both past, present and future. Thus, as Antonio Damasio (2004, 179, original emphasis) argues this is not 'a simple issue of trusting feelings as the necessary arbiter of good and evil. It is a matter of discovering the circumstances in which feelings can indeed be an arbiter, and using the reasoned coupling of *circumstances and feelings* as a guide to human behaviour'.

Such an emergent, situational ethics, would be concerned with 'working on the faculty of judgement as it is actually exercised – in the immediate present' (Thrift 2004a, 93), for it is in this immediate present that events and encounters have the capacity to surprise and to shock, to throw off guard previous ideals and open space for the new. This form of ethical thinking therefore calls for experimentation and a concern to focus upon the human capacity to flourish, and look to 'producing dispositions that are open to the moment' (Thrift 2004a, 97). Being open to the moment is about practising the ability to encounter others and thus to mould those pre-cognitive resources of 'immediate coping', for here 'ethics as sensibility or ethos demands an openness to the uncertain affective potentiality of the eventful encounter as that from which new ways of going on in the world might emerge' (McCormack 2003, 503). Ethics in this sense is a continual fine tuning of the sensibility of the individual towards others, so that this disposition might be usefully brought to bear on future encounters.

One way we might begin to conceptualise this form of ethics is through Romand Coles' (1997) account of generosity as a social, and inherently receptive, practice, wherein generosity might be 'recast as an embodied disposition that subsists in the practices and dispositions of attending and responding to others ... generosity not as a regulative ideal, but as a constitutive practice of sociality, community, and being together' (Barnett and Land 2007, 1073). Coles (1997, 3) suggests that any form of receptive giving must also be an act of 'receiving the other in agonistic dialogical engagements', it must therefore be an act of opening, an 'effort to erode *a priori* closures', which demands of the self a 'suppleness and interrogative comportment' (ibid., 22) that chimes with many of the experimental impulses of an ethics based upon sensate, and momentary, judgement. The transformative potential which Coles (1997) attaches to receptive generosity is therefore indicative of the uncertainty and openness which pattern a situational account of ethical experimentation. It is here that these divergent accounts of ethics, sensibility, response and generosity converge, and I want to now consider how far we might see such a convergence being enacted within The Talking Shop. I shall consider the ways in which this space was seen to foster an open orientation towards the world, before then examining how this might be achieved.

Opening and Questioning

During my interviews with those who attended The Talking Shop a sense of responsive generosity became narrated through casting the self as open to contestation and to question. As Omar comments;

> The British people as well I thought after two months, three months, she is changed or he is changed, because it is the way it works, it changes me as well as he or she, it is a way in which both is changed it is not just one way. I found you become more open in mind about some things (Omar Interview, 2006).

Thus for Omar;

> It is people coming together, and it is changing your view of things or becoming more soft not like this hard and rigid, in this moment unfortunately we don't have many of these places for people to come and talk to each other, and we can find we have much in common (Omar Interview, 2006).

In a similar fashion Ilya and I discussed these gestures towards opening in the following interview extract;

> *Jonathan*: What do you think binds the people who come here together?
> *Ilya*: I think there are two-way gains really, one way there are asylum seekers who needs the help and there are people who are willing to help them … just the interaction of cultures, interaction of nationalities it broadens them, because you can have views on certain countries and people, but when you meet them in person those stereotypes practically disappear and you give the other person the same feelings, the same impression, and you find a lot of these personal meetings can change a lot of things.
> *Jonathan*: And do you think that's a fundamental part of [The Talking Shop]?

Ilya: Of course, sometimes it's very hard to admit if you're defending something and you don't want to kind of abandon your position, but it's in how you defend something and allowing someone else to comment and accept that they have a fair point, maybe not that time and that conversation but you take it on board and you think about it and maybe next time when you meet other person or in other situation you won't defends those stereotypes and practically you are changed in that way (Ilya Interview, 2007).

For both Ilya and Omar the relationships which emerge through The Talking Shop act to alter individuals, they push and pull at the bounds of self-constitution. The work of slow alteration which Ilya refers to reflects what Connolly (2005) refers to as a disposition of 'critical responsiveness'. For Connolly (2005, 126, original emphasis) critical responsiveness takes 'the form of *careful listening and presumptive generosity* to constituencies struggling to move from an obscure or degraded subsistence below the field of recognition, justice, obligation, rights, or legitimacy to a place on one or more of those registers'. An outlook of critical responsiveness is therefore one through which individuals are opened to a relationship of 'agonistic respect' with those engaged in a struggle for political rights, recognition and justice. Clearly such a position requires much of that ethical practice and dispositional labour which Connolly and others argue opens us to others, thus as Connolly (2005, 126) suggests the 'cultivation of critical responsiveness ... is at once to work tactically on gut feelings already sedimented into you'.

The relations noted by Omar and Ilya are not simply captured through a sense of 'critical responsiveness' however. Rather, the encounters which define The Talking Shop not only produce gestures of generous engagement towards others, but bound into these gestures is an affective charge of being-with others, of being in common. For Ilya, The Talking Shop is a space in which you may 'give the other person the same feelings, the same impression', while Omar describes people 'coming together' to create The Talking Shop. In this manner the relations that sustained The Talking Shop do not simply promote a sense of generous engagement, they also create a feeling of commonality, of commitment to others who partly constitute what The Talking Shop is. The sense of sociality developed here, of The Talking Shop as a space attached to both particular relations and particular feelings, refers back to Popke's (2009, 84) challenge to such non-representational modes of ethical thought, that they might overlook the 'question of how events and encounters become constituted as the locus of a shared sense of conviviality and solidarity'. The affective relations of The Talking Shop discussed thus far offer a case in which such encounters with difference do indeed compel a sense of convivial accomplishment and collective attachment to both the place and the relations, of this drop-in centre. Here the relations of engagement which found this space do not simply promote moments of reflective cultivation as Connolly (2005) suggests, but they also actively 'produce a feeling of being in a situation together' (Thrift 2004b, 84), as this form of responsive ethical space 'takes an inclusive, non-judgemental approach to tending belonging-together in an intense, affectively engaged way' (Massumi 2002b, 255).

The Talking Shop thus provided a space through which that pattern of questioning attached to notions of responsive engagement, of actively working upon the relations which constitute the self, could be performed. Thus the 'active work' of attuning oneself to others which is represented through those moments of change that Ilya and Omar attest to, is accomplished through the drop-in centre as a site in which ethical questioning comes into being. Yet within such questioning of the self, The Talking Shop also created a feeling of commonality, of togetherness between individuals and of a shared commitment to the continued accomplishment of this space of care. The Talking Shop was a site which had to be actively created, practised and lived and as such that feeling of being with others, of being in something together, was central both to the comfort that some took from it, and to its continued creation.

Responding and Attuning

If we are to view The Talking Shop as a space of ethical production then what properties of such does it display? I think we can identify a number of pointers here. The first of these is felt through the unpredictable nature of this space, as The Talking Shop demands an ethics of the impromptu, akin to those demands of responsive and embodied judgement which Anderson (2005) considers. Lynn and I discussed this aspect of The Talking Shop in the following terms;

> *Lynn*: This is something that's just so organic, it's just grown and we never know quite what will happen, on any particular day for instance, it all depends on who comes in, what they want to talk about, it depends on how people respond...
> *Jonathan*: So is there a sense of spontaneity there?
> *Lynn*: Absolutely, absolutely.
> *Jonathan*: Do you think that makes it an enjoyable place to be?
> *Lynn*: Yes, I do, it's the reality of it, it's an opportunity to be oneself and to discover new things, hence there's a spontaneity to it, and the openness from everybody to be able to respond to each other (Lynn Interview, 2007).

The Talking Shop presents a constantly demanding environment, one in which one's very presence is structured around an expectation of response. The demand to listen, the demand to talk and to translate, all bombard you from an array of angles, yet it is in the moments of openness which responding offers that responsible ethical gestures come to be actualised. A generosity of response meant for many volunteers an alteration of their orientation towards both those strangers they encountered and the spaces in which these encounters took place. The Talking Shop allowed for improvised interactions to occur, for people to find a way of getting along with one another, as here 'the cultivation of "expertise" as judgement able to be fully attuned to each event rather than the application of set rules' (Thrift 2004a, 93) was worked upon. Such work presents a generosity of going into these encounters each time anew, of being open to the unexpected. Here Lynn's account of the drop-in centres openness and ability to respond fits well with Omar and Ilya's accounts of a perspective of critical responsiveness being generated. The Talking Shop was therefore a space of constant becoming, of shifting patterns, relations and responses, for as Parr (2000, 233, original emphasis) notes, 'the norms of the drop-in are both constant *and* changing. The interactions between members are dynamic, and therefore the atmosphere, tolerance and performances within the drop-in are always different from one day to the next'.

It was not however simply the unpredictable nature of drop-in space which was important but also, as Lynn suggests, precisely the lack of formal rules and procedures which brought this affective mood into play. Drop-in centres often train volunteers to show particular forms of care and empathy, providing systematic guidance on the role of volunteers and the codes of conduct which regulate drop-in spaces. Yet this form of training was largely absent from the relations of The Talking Shop, despite the establishment of a 'training day' for volunteers which centred upon both teaching volunteers about the complexities of the UK asylum system and running a series of 'role-playing' exercises in which the exchanges of the drop-in centre were simulated. I took part in one such series of exercises;

> During these exchanges we were encouraged to circulate the room and listen in on other conversations, witness the tactics, habits and responses of others. This was an uncomfortable process, one out of step with much of that which took place at The Talking Shop, for while the facilitator was keen to stress that there were no 'hard and fast rules', it seemed that this made little sense outside the relations brought together precisely through drop-in space. Attempting to translate some of these relations and practice them here, on a Saturday morning in a Sheffield seminar room, lost touch with precisely that affective charge which made this environment what it was (Research Diary, 27 January 2007).

Attending The Talking Shop training day highlighted that those ethical relations of concern which arise through this space ceased to make sense in any meaningful way once they were translated to another setting. Here they became modes of practice, yet there was a hollowness to their enaction, they lacked the affective spark which engaging in The Talking Shop brought into being. Naturally this was partly due to their nature as relations between volunteers, relations of acting a part. Yet there is more to be drawn from this. Rather, while Connolly (2002) and Thrift (2003a) suggest that cultivating a disposition of responsive generosity is undertaken through a combination of active engagements with others and critical reflection, this form of cultivation is only ever fully activated within those situations and contexts which force judgements upon us, those contexts in which improvisation and response is created through a lack of time for reflection and calculation. The uncertainty of The Talking Shop meant that any form of ethical training could

Figure 13.3   Responding across the table

Source: Author's photograph.

never be enough, rather at various unpredictable moments volunteers had to react beyond and before a series of trained judgement decisions took place. Such open responses were central to this space as Lynn highlighted earlier, and for her these responses were vital in accounting for the main outcome she witnesses through the drop-in centre, the ability to discover new things about others and oneself.

The final attribute of an affective ethical cultivation I want to highlight is the fact that this space, its performance, affective pulse and materiality matter to the kinds of ethical openings on offer here (see Latham and McCormack 2004). The rooms in which The Talking Shop took place were both filled with a series of small tables, each with four or five chairs designed to facilitate small discussions and conversations. Part of the notable geography of these often chaotic, noisy, rooms was the way in which each of these tables became, albeit briefly, a world in itself, an affective island connected and yet separated from the rest of the room. The table, and the immediacy of those around you, demanded your full attention. The table acted to draw people together. In the same way that the Home Office letter acted to define and delimit certain positions, relationships and intimacies, so this series of tables provided the material basis, the facilitation, for moments of ethical response.

Moving through the room, from table to table, became a matter of negotiating and traversing not only space, but also demands to respond;

> Today I moved around hectically as there were a number of people I wanted to talk to. As I moved from table to table, stopping at each for around twenty minutes, I became aware that as I moved, each time my position changed, I also had to change. I was responding to what I felt others wanted of me. Omar engaged me in a friendly chat about music, women and art, then I was sitting next to Tinashe as he forcefully made a case for why the British government had no sense of human rights, then Rubi was telling me how her children were getting on in a Sheffield school, then, finally, I was faced with Adil's account of his time in a detention centre (Research Diary, 3 February 2007).

At each of these points I became slightly different, I attempted to attune myself to the affective charge of those around me and adjust my responses accordingly. I would certainly not claim to have been successful in this and an array of awkward moments of misunderstanding, confusion and occasional annoyance came to the surface. However it is this sense of an environment which forces one to constantly adapt that I feel is important here. The acknowledgement that one is continually under demand to be different, to respond with a generosity which does not hold the self in place but rather opens it to others, not only presents that sense of continual ethical learning which Varela (1999) and Thrift (2004a) argue for, but also allows one to respond generously in the present to those one is faced with. I cannot claim success in this endeavour, yet some volunteers clearly engaged in such practices, they reacted to all demands with a generous and open disposition 'illustrative of what might be described as an expanded subjectivity; that is a way of being and relating to others that extended beyond [their] ... previous domain of being and affect' (Conradson 2003c, 516). We might think back here to the work of Lynn with which I opened this engagement with The Talking Shop for her mode of responding, of personal warmth, reflective engagement and prosaic openness suggests a sensibility born in, and sensitive to, the situational and shifting demands, relations and intimacies of drop-in space.

Conclusion

Throughout this chapter I have suggested that the varied relations, encounters and negotiations of The Talking Shop enact a different kind of ethics to that normally accounted for in environments of care. This is an ethics of dispositional cultivation, an ethics of active, practiced and embodied work, wherein ethical judgements are recognised as responsive, situational processes. Connolly (2002, 19-20, original emphasis) argues that such *relational techniques of the self*, represent 'choreographed mixtures of word, gesture, image, sound, rhythm, smell, and touch that help to define the sensibility in which your perception, thinking, identity, beliefs, and judgment are set'. Within The Talking Shop this myriad of connections, influences and mediators came together to not only create an ever-changing space of encounter between asylum seekers and volunteers, but also a space in which ethical dispositions could be attuned, worked upon and practiced. The Talking Shop organised attachments, it arranged people, objects and ideas into close proximity and allowed for the encounters which such proximity brought to be played out. Rekindling a sense of receptive generosity through this space is a reflection of one way this disposition might be considered, for here 'volunteers give freely in and of the moment without the expectation that service users should respond in specific ways' (Johnson et al. 2005b, 334).

This is not to suggest that the forms of ethical opening considered through this chapter represent solutions to the difficulties posed in encountering difference. Indeed, it is precisely the rejection of such all-encompassing, procedural and regulatory modes of moral certainty that I have sought to highlight in favour of a more open, situational and critical account of ethical judgement. However, I think we can begin to draw three central strands from this work in order to orientate future work on geographies of care, ethics and encountering difference. The first of these is in recognising the central role which these spaces of care play in the performance and actualisation of relations of ethics, generosity and concern. The rooms in which The Talking Shop took shape were not simply blank canvases or containers onto which narratives of care were imposed, rather, as I have suggested, their materiality, their unique combination of colour, texture, image, shape and furniture, all helped to create certain practices, to encourage distinct sensibilities and to energise an 'affective aura' particular to The Talking Shop. A situational ethics of response was not simply performed *in* this drop-in centre, but rather, this drop-in centre played an active part in *performing* such an ethics, in giving it license, potential and presence.

The second key concern to emerge from this work is in highlighting the importance of spaces such as The Talking Shop in the daily lives of asylum seekers in the UK. The Talking Shop provided a key space through which to face the pressures of the asylum process, the apparently simple provision of a space in which to meet others, socialise and exchange views, was viewed as a fundamental means to provide some measure of humanity, compassion and care to a social position so often demonised, derided and dehumanised. Whilst I would agree with Valentine's (2008) hesitation in ceaselessly affirming the positive potential of spaces of encounter, this chapter has suggested that such 'micro-publics' of engagement (Amin 2002) do indeed have a great deal to offer both asylum seekers and volunteers. Not least in providing a site through which attitudes towards difference might be worked upon, relationships forged and dispositions of critical responsiveness honed as earlier accounts suggest. Within a political system which continues to push asylum seekers to the margins of both the social and the political (Bloch and Schuster 2005; Darling 2009a; Sales 2002; Zetter and Pearl 1999), the importance of spaces such as The Talking Shop, spaces for 'just being there', should not be underestimated.

Finally, while it is important to valorise the role which such spaces of care might provide it is also crucial to remain focused on the challenges which simply 'being there' cannot meet. For the asylum seekers of The Talking Shop this meant the marginal social and political position they occupied beyond the gates of the church hall, their continued media vilification and their subjection to a repressive politics of detention and deportation (Tyler 2006). The dispositions of the drop-in centre could only do so much. Providing a respite from these pressures was undoubtedly a positive affective gesture, one which resonated in feelings of attachment, generosity and belonging, yet this should be viewed as representing only the beginning of the transformative potential which attaches to the relations felt in such spaces of care. To further such potential it is vital to consider ways in which such sensibilities are brought to bear upon the politics of asylum itself, whether that be through the wider promotion of micropolitical modes of affective engagement and political becoming (Connolly 1999, 2002), involvement in campaigns around asylum rights (Nyers 2003), or a more fundamental reconsideration of how asylum is cast as an issue of responsibility (Darling 2009b). The ethics that The Talking Shop brings forth is one built upon the possibilities that attach to the lived moment, for as Massumi (2002a, 218, original emphasis) argues, the 'ethical value of an action is what it brings out *in* the situation, *for* its transformation, how it breaks sociality open'. A non-representational account of ethical practice valorises precisely this moment of lived potential, of cultivating dispositions which are open to transformation, and what the relations of The Talking Shop suggest is a need for such openness to transformation to be actively and creatively inserted into the political contestations and injustices of the present. In considering the liveliness of non-representational theory as a mode of thinking the 'immediacy of the now', Thrift (2003b, 2020) argues that 'social scientists have taken less notice than they might of just how much of life is still lived on that cusp [of the now] and of the situational wisdoms it brings forth: of the body moving, of how to speak the right words at the right time, of how to arrange spaces so that they modify certainties, and so on'. The Talking Shop was a space in which such ethical gestures, of speaking the right words at the right time, founded a collective sense of affective attachment and belonging, it was a space for (mainly) positive transformations. The challenge that such a space poses is to enable those transformations to spread. The challenge of The Talking Shop is to extend the ways in which we respond with care and responsibility to others, and to extend a commitment to 'speak the right words at the right time'.

Acknowledgements

A version of this chapter was presented at the Doing Race Conference, Durham, January 2008. I am grateful to Daniel Swanton for the invitation to present my work there and to the participants for their comments and discussion. My thanks to all those at The Talking Shop, both past and present, who create such a unique space of care, and to Ash Amin, Mike Crang and Helen Wilson who all contributed to the development of this chapter. Finally, my thanks to Ben Anderson and Paul Harrison for their editorial guidance and to the Social and Spatial Theory Research Cluster at the Department of Geography in Durham for providing an environment in which to think through many of these ideas. All errors remain my own.

References

Amin, A. (2002), 'Ethnicity and the multicultural city: Living with diversity', *Environment and Planning A* 34, 959-980.

Anderson, B. (2004), 'Recorded music and practices of remembering', *Social and Cultural Geography* 5, 1-17.

—— (2005), 'Practices of judgement and domestic geographies of affect', *Social and Cultural Geography* 6, 645-660.

Barnett, C. and Land, D. (2007), 'Geographies of generosity: Beyond the "moral turn"', *Geoforum* 38, 1065-1075.

Bloch, A. and Schuster, L. (2005), 'At the extremes of exclusion: Deportation, detention and dispersal', *Ethnic and Racial Studies* 28(3), 491-512.

Cloke, P., Johnsen, S. and May, J. (2005), 'Exploring ethos? Discourses of "charity" in the provision of emergency services for homeless people', *Environment and Planning A* 37, 385-402.

—— (2007), 'Ethical citizenship? volunteers and the ethics of providing services for homeless people', *Geoforum* 38, 1089-1101.

Coles, R. (1997), *Rethinking Generosity: Critical Theory and the Politics of Caritas* (London: Cornell University Press).

Connolly, W. (1999), *Why I Am Not A Secularist* (Minneapolis: University of Minnesota Press).

—— (2002), *Neuropolitics: Thinking, Culture, Speed* (Minneapolis: University of Minnesota Press).

—— (2005), *Pluralism* (London: Duke University Press).

Conradson, D. (2003a), 'Geographies of care: Spaces, practices, experiences', *Social and Cultural Geography* 4(4), 451-454.

—— (2003b), 'Doing organisational space: Practices of voluntary welfare in the city', *Environment and Planning A* 35(11), 1975-1992.

—— (2003c), 'Spaces of care in the city: The place of a community drop-in centre', *Social and Cultural Geography* 4(4), 507-525.

Damasio, A. (2000), *The Feeling of What Happens: Body, Emotion and the Making of Consciousness* (London: Vintage).

—— (2004), *Looking for Spinoza: Joy, Sorrow and the Feeling Brain* (London: Vintage).

Darling, J. (2008), *Cities of Refuge: Asylum and the Politics of Hospitality*. Unpublished PhD Thesis, Durham University.

—— (2009a), 'Becoming bare life: Asylum, hospitality, and the politics of encampment', *Environment and Planning D: Society and Space* 27(4), 649-665.

—— (2009b), 'Thinking beyond place: The responsibilities of a relational spatial politics', *Geography Compass* 3(5), 1938-1954.

Dewsbury, J-D (2000), 'Performativity and the event: Enacting a philosophy of difference', *Environment and Planning D: Society and Space* 18(4), 473-496.

Dreyfus, H. and Dreyfus, S. (1990), 'What is morality? A phenomenological account of the development of ethical expertise'. In Rassmussen, D. (ed.) *Universalism Versus Communitarianism* (Cambridge: MIT Press), 237-264.

Fekete, L. (2005), 'The deportation machine: Europe, asylum and human rights', *Race and Class* 47(1), 64-91.

Hage, G. (1998), *White Nation: Fantasies of White Supremacy in a Multicultural Society* (Annandale: Pluto Press).

Harrison, P. (2000), 'Making sense: Embodiment and the sensibilities of the everyday', *Environment and Planning D: Society and Space* 18(4), 497-517.

Johnsen, S., Cloke, P. and May, J. (2005a), 'Day centres for homeless people: Spaces of care or fear?', *Social and Cultural Geography* 6(6), 785-811.

—— (2005b), 'Transitory spaces of care: Serving homeless people on the street', *Health and Place* 11, 323-336.

Latham, A. (1999), 'The power of distraction: Distraction, tactility, and habit in the work of Walter Benjamin', *Environment and Planning D: Society and Space* 17(4), 451-473.

Latham, A. and McCormack, D. (2004), 'Moving cities: Rethinking the materialities of urban geographies', *Progress in Human Geography* 28(6), 701-724.

Lingis, A. (2002), 'Murmurs of life'. In Zournazi, M. (ed.) *Hope: New Philosophies for Change* (London: Lawrence and Wishart), 22-41.

Massumi, B. (2002a), 'Navigating movements'. In Zournazi, M. (ed.) *Hope: New Philosophies for Change* (London: Lawrence and Wishart), 210-242.

—— (2002b), *Parables for the Virtual: Movement, Affect, Sensation* (London: Duke University Press).

McCormack, D. (2003), 'An event of geographical ethics in spaces of affect', *Transactions of the Institute of British Geographers* 28(4), 488-507.

—— (2005), 'Diagramming power in practice and performance', *Environment and Planning D: Society and Space* 23(1), 119-147.

Nyers, P. (2003), 'Abject cosmopolitanism: The politics of protection in the Anti-deportation movement', *Third World Quarterly* 24(6), 1069-1093.

Parr, H. (1998), 'Mental health, ethnography and the body', *Area* 30(1), 28-37.

—— (2000), 'Interpreting the "hidden social geographies" of mental health: Ethnographies of inclusion and exclusion in semi-institutional places', *Health and Place* 6, 225-237.

Popke, E.J. (2006), 'Geography and ethics: Everyday mediations through care and consumption', *Progress in Human Geography* 30(4), 504-512.

—— (2009), 'Geography and ethics: Non-Representational encounters, collective responsibility and economic difference', *Progress in Human Geography* 33(1), 81-90.

Sales, R. (2002), 'The deserving and the undeserving? Refugees, asylum seekers and welfare in Britain', *Critical Social Policy* 22(3), 456-478.

Thrift, N. (2000), 'Entanglements of power: Shadows?'. In Sharp, J., Routledge, P., Philo, C., and Paddison, R. (eds), *Entanglements of Power: Geographies of Domination/Resistance* (London: Routledge), 269-278.

—— (2003a), 'Practising ethics'. In Pryke, M. Rose, G. and Whatmore, S. (eds) *Using Social Theory: Thinking Through Research* (London: Sage), 105-121.

—— (2003b), 'Performance and...', *Environment and Planning A*, 35, 2019-2024.

—— (2004a), 'Summoning life'. In Cloke, P. Crang, P. and Goodwin, M. (eds), *Envisioning Human Geographies* (London: Arnold), 81-103.

—— (2004b), 'Intensities of feeling: Towards a spatial politics of affect', *Geografiska Annaler* 86(1), 57-78.

—— (2005), 'But malice aforethought: Cities and the natural history of hatred', *Transactions of the Institute of British Geographers* 30, 133-150.

Tyler, I. (2006), '"Welcome to Britain": The cultural politics of asylum', *European Journal of Cultural Studies* 9(2), 185-202.

Valentine, G. (2008), 'Living with difference: Reflections on geographies of encounter', *Progress in Human Geography* 32(2), 323-337.

Varela, F., Thompson, E. and Rosch, E. (1993), *The Embodied Mind: Cognitive Science and Human Experience* (Cambridge, MA: MIT Press).

Varela, F. (1999), *Ethical Know-How: Action, Wisdom, and Cognition* (Stanford: Stanford University Press).

Wylie, J. (2002), 'An essay on ascending Glastonbury Tor', *Geoforum* 32(4), 441-455.

Zetter, R. and Pearl, M. (1999), 'Sheltering on the margins: Social housing provision and the impact of restrictionism on asylum seekers and refugees in the UK', *Policy Studies* 20(4), 235-254.

Chapter 14

Ethics and the Non-Human: The Matterings of Animal Sentience in the Meat Industry

Emma Roe

Introduction to Non-Human Ethics

This chapter considers ethics and the non-human in the empirical context of animal production and meat processing. There exists a long-standing interest within non-representational theory in non-human agency (Thrift 1996), influenced at its outset by among others the work of sociologist of science and technology, Bruno Latour. As Thrift writes on Latour's work:

> [This] has meant a concern with a 'new classification of things' (Latour 1993) in which the bounds between the subject and the object become less easily drawn, … because the things we have conventionally depicted as objects, for example machines, are allowed into the realm of action and the actor (Thrift 1996, 2).

This category of 'things' not only refers to machines, but equally to animals like elephants (Whatmore and Thorne 1998), vegetables (Roe 2006) and prions (Hinchliffe 2001), all of which can be understood as hybrid entities – a co-production of the 'natural' and the 'social'. Each of these authors has explored how practices, whether for conserving animals, or eating food, or responding to metabolic risk in the food industry, have had to work with the specific capacities of non-human materialities. The empirical narrative at the centre of this chapter develops this work on non-human agency and ethical engagement by arguing we should be sensitive to different kinds of (non-human) processes of matter that generate the materialities we know and sense.

A relational ontology (Whatmore 1997) underpins the study of non-human agency; it has implications not only for how we conceive the non-human but also for how we conceive the human, since the non-human mounts a challenge towards human supremacy (Latour 1993, 2005) and the notion of the autonomous subject (Whatmore 1997). In this relational ontology, as Whatmore (ibid.) outlines, non-human actors in all guises attain a new ethical significance through the relations they form with humans, because whatever the non-human, and whatever the conditions, a fundamental shift has occurred about the role non-humans play in making up the world we know. Consequently, a different sort of ethical consideration has developed, called a 'relational ethic' (Whatmore 2002) or an 'affective ethic'

(Bennett 2001) that is imbricated into how we are 'making the world', a world increasingly populated by human artefacts, technologies and socio-materialities.

Take, for example, the fast food burger, fashioned out of matter from an animal's body, sold with a carefully cultivated branded aesthetic and multisensory taste to work *affectively* to resist the discourse of unhealthy eating despite personal self-reflection on what one should and should not eat. We can illustrate the relational ethic or affective ethic at work when considering the placing of fast-food adverts in the everyday environment to encourage desire and ultimately to meet the aim to motivate a purchase act. Or the affective ethic that continues to materially connect the burger to the birth, killing, cutting-up and processing of an animal's body. This relational or affective ethic matters because actions and ideas are constituted in part by the make-up of the environment (objects and artefacts within it included). Thus how the environment is made, what objects are within it and how they generate thoughts and actions ultimately constitutes world-making and with that society-forming activity. Thus what is thought and done is always relationally and affectively formed and imbricating the ethic. Returning to the burger, before it reaches an eating event, numerous practices have brought it into being, from artificial insemination, to meat processing techniques, and the skills of marketers; together they have contributed to fashioning a burger with a repeatable positive eating experience and brand-recognisable qualities. All these practices have worked with the biochemical and biophysical processes of matter to generate from the animal flesh a meat which when cooked constitutes through taste, touch, sight and smell a materiality we recognise as an edible burger, and perhaps can even be specified to a particular brand of burger. The three – the *processes of matter* or matterings (Barad 2007), which can be understood as the actancy of matter in itself, human *practices*, and *materialities* co-generate the burger. It is the energetic relationship between the processes of matter, specific practices in the livestock and meat processing industry, and existing meat carcass and burger materialities that co-generates new burger materialities and an associated set of ethical relations with the consumers, the animals, the marketers, meat processors etc. This example points to how non-humans of all scales, forms and kinds from burger to cow to energetic matterings are constitutive of the world and form relational ethical attachments with humans. This makes non-humans relevant actors in cultural and social geographies (Whatmore 2006), not only because they are shaped by particular (human) social and cultural norms and practices but also in the way they allow for, inform and 'act back' into such practices and norms.

Whatmore and other authors over the last 15 years have explored non-human agency and a 'relational ethics' (Whatmore 1997). Animals (Philo and Wilbert 2000; Whatmore and Thorne 2000), agri-food (Roe 2006; Whatmore and Thorne 2000; Stassart and Whatmore 2003), technologies (Bingham 1996; 2006), and props and gesture in dance movement (McCormack 2003) are active in some of the geographies produced to explore the implications of relational ethics. These examples in various ways map out the ethical relation between bodies affecting and affected, or the identification of a normative ethical concern as emergent within the network, or the term I prefer to use, the assemblage. For when a human responds to another their embodied practices indicate an affective proximity to the personal in a '*thoroughly*' (Anderson and Wylie 2009) materially or emotional way – this in itself is a relational ethic. Thus the ethic is identified as not orientated solely at the human(s) 'feet' or the non-human(s), but instead is found in the relation strung over complex spatial, corporeal and normative dimensions. Attempts to unpack, critically examine and pursue the ethics in assemblages (often born out over a set of complex spatial and temporal inter-relations) have often developed an appreciation for technologies role in supporting the transformative potential of matter into various socio-materialities. For example technologies such as Fairtrade food labels ferry the agentive potential of coffee beans that are the product of more fairly-rewarded human labour in the Brazilian coffee-fields, when attached to coffee beans – across multiple sites and great distances in the journey to the Fairtrade coffee drinker (Whatmore and Thorne 1999). However this work has also made it harder to place ethical responsibility within a complex network of human and non-human agencies. Two food-related examples are, firstly, the dangers associated with the assemblage that includes the prion, Bovine Spongiform Encephalopathy and Creutzfeld-Jacob disease (Hinchliffe 2001), and secondly, the technologies of the Fairtrade coffee industry that enables consumers to act at a distance to support fairer trading relations for coffee bean farmers thousands of miles from the street cafés of Europe or America (Whatmore and Thorne 1999). Where normatively the meat industry and the café coffee drinkers may be seen as the ones who held or hold the opportunity to ethically intervene 'to make things right', i.e. to not feed cattle bone-meal to cattle, or to buy Fairtrade coffee, contrastingly this relational ethic instead makes it harder to make that judgement. This is because a whole host of other technologies and people and stuff are active in the network generating the contingencies in which a prion could exist within the UK food-animal meat industry assemblage, or facilitating the emergence of the Fairtrade option in the coffee industry. Critics may say this over-complexifies the situation, but in defence a relational ethic is sensitive to the overlooked subtleties of socio-technical and socio-material arrangements producing contexts where events can take place. It also is a tool for critically examining how normative ethical stances are stabilised from the multiple possible relational ethics that could be addressed, scrutinised and questioned.

Pursuing the food and meat theme, I will consider the implications of new techniques used by farm animal scientists and meat scientists that enables them to apprehend 'sentient materialities' in both the living bodies of animals and in their dead bodies once classified as meat. This empirical example will develop the argument that we should account for the nuances of different kinds of matter and its processes in our studies of assemblages by considering the following questions. Where did 'sentience' come from? Why has it become a 'matter of concern' (Latour 2005, 256-257) now? Why is it being accounted for? And how did it become a 'matter of fact' (ibid.), ordered by, and ordering activities in, sections of the meat industry?

Food-Animal-Meat Industry Assemblage

Food-animal-meat industry assemblages take a variety of forms determined by a number of factors including different species and/or by different cultural cuisines and product branding. Two examples are the assemblages of broiler meat chickens and British welfare-standard pigs. Both animals can be conceived of as assemblages straddling the archaic modernist division between Human and Non-human. For example, chicken grown from the modern-day ancestors of the jungle fowl, now commonly known as table chickens, are housed in an indoor farm shed in the UK and reach culling-weight at 40 days old. Or, a second example is the pig produced at extra costs from a British sow, rather than a Danish sow, which fails to find a commercial market in continental Europe because of the extra costs the UK legal ban on sow-stalls has given the UK pig industry. These pigs and chickens are entities developed by non-humans and humans 'working together' in complex ways to co-generate these 'assemblages'. Assemblage is the English translation of the French term 'agencement' used by Deleuze and Guattari (1988). Some authors (Hardie and MacKenzie 2007) have preferred to continue to use the term 'agencement' because it conveys more closely the notion that subjectivity is enacted through the 'assembling'. However, the term is not used in this chapter, but rather instead the term 'assemblage' because of the inference that agencement as defined by Callon (2005) is made up of human bodies with prostheses, tools, equipment, technical devices, algorithms etc., with, notably, no direct inclusion of other living non-human bodies. Therefore by virtue of this 'assemblages' used with the prefix human/non-human or specific naming like 'food-animal-meat-industry', denotes an interest in what happens when things, bodies, technological devices come together to enact as a subjectivity. By not assuming that a subject is formed around the figure of the human, the term assemblage opens up the possibility for conceptualising subjectivities that may, for example, form around the non-human animal body.

The case study is taken from the meat industry where there is commercial concern for the welfare of animals in the hours before slaughter. This concern has led to the development of a technique that makes visible, or affords the possibility, of reading the material traces of animal sentience in flesh not just ante-mortem but also post-mortem. I propose that this technique is successful at identifying sentient materialities as a 'matter of fact'. I then suggest the implications for our theorisations of matter. These are that we must deduce that there cannot be a singular category of matter, but rather matters because not all materialities can emerge from any kind of matter. Instead materialities emerge from specific kinds of mattering and we must be alert to differences between them. Therefore, to understand the creation of a food–animal–meat–industry assemblage requires a deepened understanding of the *practices* that brought the assemblage into being, the *materialities* evoked by those practices and the *processes of matter* or matterings over space and time that together energise its generation. Consequently, to become alert to a plurality of processes of matter supports a critique of normative non-human ethics as reductionist, and some forms of relational ethics as abstracted. By comparison the approach I outline gestures towards a performative position where specific world-making practices and materialities are actively foreclosing possible expressions of certain kinds of processes of matter and it is in this recognition that an ethical response is made available.

Latour's word to describe the processes of matter is 'plasma', for him it is 'that which is not yet formatted, not yet measured, not yet socialised, not yet engaged in metrological chains, and not yet covered, surveyed, mobilised, or subjectified' (2005, 244). To elaborate, he refers to that matter which is not us, (as it is outside of our perceived boundaries as humans), but yet is tangled up with us extensively, 'providing the resources for every single course of action to be fulfilled' (ibid.); (for example what we eat), generating with us new materialities, new knowledges? What is this 'plasma' like? Are there different kinds of 'processes of matter' that support the generation of different phenomena when met with the affordances of different practices? These questions place attention on the ethical imperative to relate how specific non-human processes of matter are intra-imbricated into the emergence of materialities and specific knowledges.

The background to agro-food geographies interest, and to a significant extent non-representational geographies interest also, in studying the processes of matter can be found in the writings of Margaret Fitzsimmons. She called for the study of 'the matter of nature' (Fitzsimmons 1989) – a matter that requires a need to study the organic, to study life itself. Since then there have been notable efforts to imagine some of the lively, animate capacities of matter, offering alternatives to the often-implied static, solid, inert matter under human control. The political theorist Jane Bennett (2001) writes about what matter does, specifically how it enchants – she draws inspiration from the Lucretian swerve to imagine matter as unexpected, abundant and vital and in this way explores how it exerts influence on moods, dispositions and decisions (Bennett 2007). She builds these ideas to outline an 'ethics of generosity' (Bennett 2001) for how the abundance of matter effervesces to attach human to human, human to non-human and non-human to non-human in the paths we make and cross. Another two geographers Anderson and Wylie (2009) argue for working with a different image of matter, matter as multiplied within a '*thoroughly materialist*' position where all elements i.e. earth, fire, air, and states i.e. solid, liquid, gaseous express a question of materiality, not only obdurate, concrete forms. They fold this principle of multiplication into three modes of how matter exists as turbulent, interrogative and excessive (Anderson and Wylie 2009). By doing so they argue for matter as emergent phenomena 'taking place in multiple different states/elements' (ibid., 332) but they do not consider the ethical imperative to interrogate the practices that afford the possibilities for specific materialities to be generated from these multiple matter states/elements. Anderson and Wylie's imagination of matter is from a different angle to that of philosopher Elizabeth Grosz who calls for the development of knowledges of matter that accommodate duration, change and transformation (Grosz 2004). Grosz's matter is identifiable by how she configures it with spatio-temporal qualities. This engagement with spatio-temporal qualities is also central to Pierre Stassart and Sarah Whatmore's (2003) portrayal of an ethic of matter at work in metabolic processes that cross imagined borders between human and non-human, for example when producing and eating meat. The mobility of matter between bodies in this study on meat is conceived as a messenger for risk through the connectivities and affectivities that exist between the spaces of production and consumption (ibid.). These different studies of matter do not all attend to the ethical imperative in how we imagine matter. Some of the authors (Anderson and Wylie 2009; Grosz 2004) do not directly address an ethic in how they cast matter, whereas Bennett and Stassart and Whatmore do but in quite different ways. For Bennett, the ethic is emergent in the capacities of an abundant matter, whereas Stassart and Whatmore more narrowly define their interest to the metabolic processes of matter. Building on Stassart and Whatmore's specific engagement with a *kind* of process of matter I wish to do the same for the sentient processes of matter that cross imagined borders between life and death during the event of slaughtering an animal. I want to use this example to make the case for more attention towards understanding how within the 'plasma' (Latour 2005) lie processes of matter, the agentive force, or 'matterings' (Barad 2007), charged with diverse energies and capacities, lying beyond our phenomenological experience.

Sentient Matterings Generating Sentient Materialities

'Sentient materialities' are a relatively recent phenomena in the meat industry. Notably, 'sentience' as a widely-used term only found popular understanding in the 1990s and 2000s through the success of animal welfare non-governmental organisations activities lobbying the European Parliament to have the rights of 'sentient beings' to be included within the EU constitution, finally achieved in 1997. We can understand the social recognition of 'sentience' as a result of animal-human assemblages co-generating the scientific evidence to indicate animals had feelings (Appleby and Hughes 1997) and that this message was then mediated effectively to the wider general public (CIWF 2009). The animal's agentive role cannot be under-played and yet the opportunity for sentient beings to gain legal status occurred as recently as 1997 and only in Europe, and was given because of scientific evidence, *not* the experiences of animal-handlers or pet-owners who live and have lived alongside animals for centuries. The impact of the legal recognition of animals as 'sentient beings' has been large. Some parts of the global meat industry have now an increased regard for how animals are handled (Roe and Higgin 2008), especially because of the discovery that bodies, both pre and post animal slaughter carry traces of sentient materialities. Or in other words, the food–animal–meat industry assemblage, that is always more or less stable, is adapting to the emergence of this knowledge. The industry's response has been to code-up, or add this value, that meat is from the body of a sentient being, to meat quality negotiations in the local context via testing procedures in the abattoir, and in the global context via participation in debates on animal welfare standards at world trade talks (Webster 2006). The ethnographic narrative I present in this paper elaborates this empirical context through describing the passage of animals from farm through an abattoir to the packaging of meat products. My analysis considers what this example asks of us conceptually and theoretically in terms of how we conceive the biophysical and biochemical processes of matter that respond to the practices of technologies and humans to co-produce new socio-materialities. Inherent to this process of making an assemblage is an ethical imperative found in the relation between technological practices, materialities and processes of matter, but more specifically in what kinds of matter are worked with. It is here I argue where the role of specific matterings is being overlooked. Social constructionism has been criticised for not including the agency of non-humans, but contemporary accounts of human – non-human relations that do account for non-human agency are failing to be sensitive enough to account for the variegated processes of matter; there is not matter, but matters. Bringing sensitive, specific accounts into our social and cultural geographies will further develop the study of ethics and the non-human.

At this stage it is worth stepping back to remember that deliberately growing a sentient non-human animal for food consumption is a very different human–animal relation to that with zoo animals, domestic pets and wild animals. Historically, livestock, farm animals or food animals (as is perhaps more appropriate to call them) have received little attention by animal geographers, and when agricultural geographers have considered them it has often been as 'units of production' (Symes and Marsden 1985), or cultural icons in post-productivist literature (Evans and Yarwood 2000); they have not been understood as sentient beings. However, there is a small collection of farm animal studies that goes some way to engaging with farm animal bodies (Yarwood and Evans 2006), sentience in hobby farm animals (Holloway 2001) and animal subjectivities (Risan 2005) but none of these pursues the tricky transgression from living animal to meat. They do not attempt to conceive the matterings of sentience as a co-presence through the practices and processes that surround this material transformation and change in meaning in the abattoir. Nor do they really engage with the characteristically huge scale of industrial farm animal production focusing instead on singular animals like Risan's dairy cow and Holloway's hobby animals, rather than the mass production of sentient farm animals.

Karen Barad invented the neologism 'matterings' to detail how the processes of matter 'acquire meaning and form through the realisation of different agential possibilities' (2007, 141). I draw upon her work on mattering to think about *how* the processes of sentient matter, or matter in itself, becomes materialised as sentient in the farm animal-meat-industry assemblage. For Barad, matter is singular but always promiscuous:

> Matter is neither fixed and given, nor the mere end result of different processes. Matter is produced and productive, generated and generative. Matter is agentive, not a fixed essence or property of things ... Matter itself is always already open to, or rather entangled with the 'Other' (ibid., 137, 393).

Whereas natural scientists frequently perceive humans as interference and attempt to understand the world of non-humans in a world without humans, ignoring our and their entanglement with each other, descriptions of matter like that of Barad's allude to matter's 'intra-activity'; its readiness for entangling. In fact she argues that 'the primary ontological units are not "things" but phenomena – dynamic topological reconfigurings/entanglements/relationalities/(re)articulations of the world' (ibid., 141). Into her 'phenomena' of matter I want to conceive different kinds of potential ready for entangling when afforded by Barad's semantic unit 'material discursive practices', notably not words.

Knowledges of sentient materialities are conceived as constituted in the farm–animal–meat assemblage through the entangling of the phenomena of matter with specific material–discursive practices. For us to have knowledge of sentience is dependent upon the tri-partite intra-action between processes of matter, practices and existing materialities. Consequently, there is something about the intra-action of matter, practices and materialities in the farm animal meat assemblage that means animal sentience is now know-able and known. Could they be found anywhere? Or is there a geography to where sentient materialities are likely to be known? How did food animals like sheep, pigs, cattle and chickens become recognised as having feelings that matter to them (Webster 2006) in Europe in 1997 and not worldwide?

The following narrative is based on trips to two different abattoirs and is indicative of the phenomena and material-discursive practices that create the farm animal–meat–industry assemblage, where techniques and values have developed to identify the material affects of 'sentience' in animal flesh post-mortem. The narrative evokes the meat industry world's intra-activity with the matterings of sentience, to demonstrate how the conditions for the generation of material sentience are afforded by specific practices encountering the phenomena of sentient matter. This generation of material sentience by the meat industry occurs through the intra-activity of causation and the effect of measuring and signifying sentience. I visited one large commercial lamb abattoir in West Wales that supplied all the lamb for a major UK supermarket, as well as other retailers in the UK and across Europe. The other was a small research abattoir at a University's vet school where I saw young and mature pigs being slaughtered. What follows is a depiction of what happened on my visits. On both occasions during my ethnographic encounter I was attentive to the 'assemblages that mediate and produce entities that cannot be refracted into words' (Law 2004, 122). Recognising I can't begin to put into words all the 'stuff' that is gathered, that comes together in the assemblage, I relate the event as it appeared to me. This account of two singular events depicts processes of matter at work intra-acting with the social to produce sentient materialities in the agro-food industry.

The Slaughter Event for Sheep and Pigs

Sheep live outdoors for all their life before being slaughtered. The specific embodied sensate capacities of sheep has meant humans have found them harder to intensively rear because their reproductive cycle is incredibly resistant to being kept indoors without true sense of day or season. As the original animal ethologist Jacob von Uexküll (1957 [1934]) might express it – the *umwelt* of the sheep is less suited for indoor intensive housing; it is more difficult to re-create environments, or to re-create sheep, through re-disciplining their bodies by tapping into the affectual topologies of being sheep. (For an extensive discussion on the genealogy of sheep see Franklin 2007).

In contrast, pigs can be reared intensively more successfully, the same embodied resistances are not there that interfere in growth and breeding patterns. However, there are other well-recognised problems that can arise from intensive pig production, including stereotypical behaviour such as tail-biting, and adult boars exhibiting aggressive behaviour to each other which when kept in confined spaces causes what is known as 'boar taint', an unpleasant taste to the meat for human consumers. In some countries the castration of male piglets is the practice for dealing with boar taint: in the UK, where castration is prohibited by the pig assurance scheme, male pigs either are killed when they are younger before their hormones start raging, or they are kept in family groups that inhibit aggressive behaviour. UK pig farming practices have developed to overcome the problems the prohibition of castration places on the edibility of pork meat. Many of these stock-handling techniques for sheep and pigs developed pre-animal science are testament to the durability of knowledges over hundreds of years of animal production; whereas others indicate how intensive housing systems, a development in only the last 50 years, has required changes to knowledges of pig stockmanship.

The sheep arrive at the abattoir after various travelling experiences from being in a small trailer towed from a local farm by a Land Rover or in a large lorry with many sheep travelling down from Scotland. These animals come from farms inspected every 12 to 15 months to ensure husbandry standards comply with industry assurance schemes – known to consumers as logos on food packaging that represent the Assured Food Standard, RSPCA Freedom Food and Organic standards. Assurance schemes are an industry technique to attempt to create repeatability and consistency in terms of farming practice and meat quality, respectively.

The animal may have experienced poor handling and have been fearful during the process of loading as they come in contact again with humans. They may have experienced for some time stress and poor health yet this may not yet be apparent to the untrained eye. Or they may have been quite content and trusting knowing from previous experience how kind and gentle humans are. Yet it can be the first time they have ever been transported by road, so boarding the lorry could be a new experience. Their hooves may slip on the ramp if it is too steep and made of slippery metal, causing them pain. Once inside they may experience travel sickness for the first time and also find themselves around animals they don't know, making them feel anxious, uneasy, and aggressive. Bodily sensibilities intra-act between humans and cows, pigs, spiders to name a few. Those who have a body share in bodily knowledges – the reach of a paw, the playful catching of a trouser leg, the sniff of noses (see Haraway 2003 for further examples).

There is a maximum journey time of 8 hours for live animals under UK regulation. On arrival the animals are put into the lairage; this is a large barn area where they can rest, calm down, and recover from the 'ordeal' of the journey if they have travelled large distances. In smaller abattoirs there may be no lairage and the animals instead are taken straight in to be killed. If the animal's body has no time to recover from the stress of its journey, it is killed with glycogen/glucose having built up in its muscles. The consequences of the presence of glycogen in its muscles will be discussed later.

The livestock-handlers move the sheep through the lairage pens by shaking large plastic bags, they do not physically touch the sheep. The pig handlers push them with their hands to separate out five of them to go through into the stunning space. When the pigs are handled in a similar gesture they make yelping/snorting noises; the sheep are silent. On the wall in the University abattoir is a plaque saying 'sponsored by the humane slaughter society'. Here the slaughter-man separates out one pig and places the two-pronged electrical stunner to its head on either sides. The pig's eyes close and it falls over on its side. The other pigs are standing around it, not seeming to know or realise what has happened. If we humans never knew of this device, would we know what it could do and what the intention of stunning an animal was? Another stock-handler quickly puts a metal shackle around one of its back-ankles and it is hoisted upside-down and through a plastic curtain for sticking and ex-sanguination. Following sticking, blood pours from the throat of the pig and involuntary muscle movement starts –legs shake for up to a minute. Some partial record of the animal's sentient life is now registering in the matter of the carcass. The process and practices of the sheep body's journey from a living and breathing being in the lairage to becoming a dead carcass hanging upside down on a metal conveyance system, is much the same.

The carcass swings round on the conveyor chain into the cutting hall. One person inspects the mouth of the dead sheep for indication of two adult teeth; if present the carcass must legally have its spinal column removed following the tighter post-BSE controls introduced in the late 1990s. Then the carcass swings along ungainly hanging from its feet, and a man (it's hard, tough work I was told – man's work) 'dresses' (or rather undresses) the carcass. The woolly coat of the sheep is pulled off like a banana being unpeeled (stiffer, harder to do later in the year as the sheep gets older).

The pig is put in hot water at 60 degrees Celsius to loosen its hairs. The pig carcass is wet and slippery when it comes out of the hot water bath. Some smaller pig carcasses slip onto the floor and are dragged back up again onto the table. Then the carcass is put in a big tumbler to try and 'rub off' as many hairs as possible. A pig has edible skin, so the hairs are meticulously removed and as least water as possible is used to clean the meat (should any faecal matter slip out of the rectum as the whole of the digestive system is removed). Not feeding the animals 8 hours before slaughter is often considered ideal, yet the lack of food and the imposed starvation can cause additional stress to the animal in its last few living hours, so there is a balance between welfare, meat quality and the cost of feed.

Quick cuts are made first around the anus, then the sexual organs are cut out, and then the cut is continued down the line of the sternum, and the vital organs removed. A machine cuts off the front legs and the sheep's head. The vital organs – heart, lungs, liver – are removed and carefully hung separately for inspection by a vet for signs of ill heath – given the all clear the carcass receives a stamp from the vet representing the Meat and Livestock Commission. Another inspector examines the size and shape, the leanness, fatness of the carcass and gives it a class, as well as identifying which carcasses may be more suited for particular markets. The French favour leaner lamb carcasses. Once this decision is made, the back legs of the carcass are crossed to accentuate the shape of the back end of the animal to make it appeal to the French cultural taste in carcass presentation. Ever since the initial ex-sanguination, small drops of blood can be seen on the floor, marking the journey the hanging carcasses make as they wind their way around the cutting hall. The abattoir processes 60 sheep an hour. A new process of electrolysis is used to speed up the maturation of the meat in the large commercial abattoir, before putting it in cold storage for 24 hours – the carcass judders violently during this process.

It is the following day when *rigor mortis* has set in and the pig carcass is cut up that indications of poor meat quality like Pale, Soft and Exudative (PSE) meat, or Blood Splash, or Pale, Dark and Dry (PDD) meat becomes materialised to the trained eye. The pig, stressed shortly before slaughter has a build-up of glycogen, popularly known as glucose, in its muscles (Velarde et al. 2000). Following death the body is starved of oxygen so cannot break it down and the result is lactic acid causing a sudden drop in the pH of the meat (ibid.). PSE meat is characterised by a pale colour, lack of firmness and fluid (exudates) dripping from its cut surfaces. When cooked, this meat lacks the juiciness of pork meat that the UK consumer expects and the consistency of a good eating experience that the pork industry wants to offer. UK meat processors also find that PSE meat's watery quality makes it unsuitable for processed meat as the product looks too pale and swims around in extra fluid. In a struggling UK pork market where carcass utilisation is the difference between making a profit, or selling at a loss, to have meat unfit for processing is a concern. Likewise for lamb there can be a similar lack of tenderness to the meat if the animal was stressed pre-slaughter. This technique is a widely-used indication of meat quality and records materialities generated and generating sentient matter in the life of the animal. The phenomena of processes of matter intra-act with the pH measuring and interpretation of meat tenderness.

The meat cutting is carried out by rows and rows of predominantly moustachioed Polish male butchers wearing protective chain-mail on arms and fingers, balaclava nets over facial hair, blue hairnets, and blue overalls. Food hygiene is of foremost importance. The carcass arrives with a tag carrying a barcode containing information about which farmer, what kind of production system, which retailer product line it should be prepared for. Some of the cuts are sealed to be re-opened and re-cut to prepare meat for particular retail packs. I saw packets of a top-four UK supermarket retailer's premium quality lamb chops being packaged, labelled and priced before my very eyes. Only the back-end or the hindquarter of the carcass is packaged here. The front-end or forequarter is shipped to Northern Ireland to be turned into processed meat products. The sheep's coat is salted and shipped for a pound a coat to China to be made into shoe-leather among other things. The packaging for the premium retailer brand lamb carries a statement about the retailer's high animal welfare standards; there is a picture of one of the sheep farmers chosen by the abattoir to represent their farm supply-base.

Animal Sentience in the 21st Century

For meat scientists and some, not all, animal scientists, animal sentience can be recognised through a complex biochemical assemblage of indicators; this is a reductionist conceptualisation, but it is this definition that has implications for meat industry practices. The meat industry's primary interest in animal sentience is to alleviate the stress of animals because of how it affects meat quality, although there are some parts of the industry additionally working to provide a life worth living for a sentient animal. It is the intra-action of the processes of matter with techniques for measuring sentience that generate biochemical responses; these biochemical responses are interpreted by animal scientists and meat scientists as relevant for understanding an animal's stress levels. As the narrative indicates, particular events in the final hours of the animal's life are of interest to the UK meat industry because the animal's stress levels are registered in the levels of glucose in the animal's blood up to the point of its death. To explain: as *rigor mortis* sets in, a biochemical change takes place in the flesh because the flesh is being starved of fresh oxygen. This is caused by processes of matter that previously were generating a living, breathing animal, now intra-acting with the practice of killing to transform into a different materiality. Knowledge of the materiality of the fleshy body changes in tandem with the transforming processes of matter; these events co-exist. There are many different forms of knowledge of this post-slaughter materiality that are generated including thickness of fat, the marbling (streaks of fat running through muscles), the distribution of the meat on the carcass, the healthiness of the vital organs. However, the knowledge-practices that focus our attention in this paper are those that can be connected to animal sentience. One of these is the measurement by meat scientists of the changing level of the carcass's pH value (Chambers and Grandin 2001); this is for two connected reasons; firstly, the pH values of the carcass can be interpreted to understand the level of stress the animal experienced pre-slaughter, and secondly, this in turn also indicates the commercial usability of the meat (ibid.).

The material-discursive practice of measuring the pH value of a meat carcass and then interpreting this measurement as evidence of stress in the eight hour period before the animal was slaughtered, tells us something about the processes of matter, the mattering, that constituted this knowledge. What this tells us is that the agency, or mattering, of the process of matter has a temporal-spatial duration (Grosz 2006); its duration crosses the border between life and death of an animal enabling the scientific identification of sentient experiences in the dead animal's carcass. Furthermore, not only do pH-level measurements indicate something of the stress levels of the animal, but also a visual change in the colour of the pork meat occurs. Visibly, meat samples from a non-stressed animal, take on a comparatively different shade of pinkness when compared to those from a stressed animal. A colour-grading scheme showing images of different shades of meat pinkness has become a reference for meat quality assessors (ibid.; Purdue University Animal Sciences 2009). As Morris and Holloway (2008) have remarked, work to improve on meat quality leads to an array of different farming and livestock selective-breeding practices, yet it is the pH measurement, the change in colouration of the meat, commercial meat processing requirements and a good cooking and eating experience, that are relevant meat knowledge-practices that encourage care for the experiences of the sentient living farm animal; not an empathic relation but instead utilitarian. The mattering of processes of matter that cross the border of life and death intra-act with these particular practices, which in turn generate materialities conveying knowledge. This in turn embeds animal sentience into the farm animal meat industry assemblage. What do the knowledge-practices measuring the pH of the meat, and measuring the colour of the meat, tell us about the processes of matter from which they generate?

Firstly, sentience is a meaning given to food animals when cultures of knowledge production intra-act with the processes of sentient matter; therefore *cultural-historical events* determine how and when processes of matter may intra-act with knowledge-making practices. Could animal sentience be present within the farm animal – meat – industry assemblage in another era, in all places? No. Over the last three to four hundred centuries we can chart a cultural development in religious, political and scientific attention towards apprehending and investigating whether animals have feelings, whether they can suffer. Animal sentience has been a philosophical interest since at least the eighteenth century speech of Jeremy Bentham (1996 [1789]) in Parliament on accepting non-human animal suffering:

> A full-grown horse or dog is beyond comparison a more rational, as well as more conversable animal, than an infant of a day, or a week, or even a month old. But

suppose the case were otherwise what would it avail? The question is not, Can they reason? Nor, Can they talk? But, Can they suffer? (Bentham 1996, 283).

However current commercial and legislative activity indicate how only in the last decade has it become an urgent applied philosophical concern as the welfare of 'sentient beings' became legally recognised across the European Union (Anonyme 1997), after decades of non-governmental animal welfare groups working to increase public awareness of the suffering of animals in fur and food farming and animal experimentation. Historically, the sentient animal as 'object of knowledge' (Latour 2005) was produced under conditions originating in Britain, which then spread to other parts of the world. The British writer Ruth Harrison's 1964 book, 'Animal Machines', led the British Government to set up in 1965 the Brambell Commission (Rushen 2008). The following Brambell report (1965) investigated and commented on the welfare of animals living in intensive confinement systems and called for research into farm animal welfare.

Since then, much of the last 40 years of research has focused on reducing the suffering of animals. The 'suffering' animal is a recognisable figure in popular and political society since the widespread appeal of Singer's publication *Animal Liberation* (1974). However, the current depiction of a sentient animal embraces a broader vision to meet the animal needs and desires as well as alleviating suffering (Mench 1998; Veissier et al. 2008) thus broadening the welfare remit from suffering to sentience. For example, species-specific European recommendations start from the biological characteristics of species to consider minimum requirements to ensure the animal's needs for adequate food and water, freedom of movement, physical comfort, social contacts, normal behaviour and protection against physical and psychological stressors are fulfilled (Veissier et al. 2008). Government funding of farm animal welfare science, in recognition of human society's responsibility to care for farm animals, has led to animal welfare law-making. Yet contemporary drives for improvements are through market-led initiatives, encouraging consumers to act ethically and buy higher-welfare products. Its no coincidence then the very recent UK TV and media coverage by celebrity chefs Jamie Oliver (Jamie's Fowl Dinners 2008) and Hugh Fernley Whittingstall (Chicken Out campaign 2008) that draw public attention to farm animal suffering and specifically aim to guide viewer's to buy higher-welfare poultry products.

The act of registering pH values from the carcass and calibrating them as evidence of an animal's stress level in the hours before slaughter, and doing the same with a meat colour grading schema is, against this cultural history, a recent development of putting sentience into the farm animal – meat – industry assemblage. But we might also recognise that the farm animal – meat – industry assemblage never existed in the way we recognise today until the later half of the 20th century. As the mechanisation and mass slaughtering of farm animals increased to factory scale in the later half of the 20th century there were calls to give attention to the suffering of the farm animals, yet it was only in the last couple of decades that the increasing stringency for food safety coupled with greater competition about meat quality consistency focused meat scientist attention to make the link between stress ante-mortem and poorer meat quality. It is a consequence of processes of matter, not only, generating the materiality of meat tenderness that encouraged people to examine whether what the living animal experienced impacted upon meat quality, but also, connecting cultural practices and circumstances identifiable within the contemporary era in the meat industry.

The consequences of a rigorous defence for the existence of animal sentience are profound. The scientific and socio-cultural pursuit of indicators of sentience in animals is often entangled with religious beliefs and debate. The conditions of the assemblage that identifies sentient beings also operates to draw boundaries that for some cultures reveal problematically its proximity to other matterings of sentience, such as those matterings of sentience in humans. In some cultures this possibility, brings us full circle back to the material discursive practices around religious beliefs that often work to constitute boundaries of difference between human and non-human, culture and nature. The matterings of sentience are contentious because of how its assemblage bounds the world differently to challenge the exceptionalism of humans. Yet since science started developing techniques to study and identify sentience, the concept positioned in the language of western science has travelled.

A second point about the processes of matter is that we have to engage with its *spatio-temporal specificities* intra-acting to produce material knowledge-practices. Both of these techniques – the colouring of the meat and the pH values – enable meat scientists to look back into the past, as it were, back into the previous one or two days when the animal was still alive and to evaluate the stress it experienced by the colour-shade of the meat, or by its pH value. It is not the practices, nor materialities, but the processes of sentient matter enabling this glance back in time. The sentient matter has a spatio-temporal durability within flesh that is constitutive of knowledge-making practices post-slaughter. Or to put it another way, sentient matter has a durability that gives it a footprint larger than the life of the body, it maintains its meaningfulness because the bodies of farm animals are exceptional in that after death they have such a meaningful second life as food. The traces of sentience in the carcass of the animal are registers of the processes of matter that happened in the body not when the pH reading was taken, but what happened a day or two before when the animal body was sentient, was breathing, was living, when it was quite other in one respect to the carcass of meat. In other words, processes of matter enable us to access a world prior to the invention of materialities, in this case that of sentience. The animal lived and experienced the journey to the abattoir, the time spent in the lairage, the stunning process. For those taking the pH readings to test meat quality they were not there during this part of the life of the animal. If they had been there they may have struggled to know the stress levels of the living animal; if they had used an empathic sensibility they could have been accused of anthropomorphism. The processes of sentient matter intra-act with material discursive-practices to give more than just the present but also a past, a past of a living, breathing animal body when sentient materialities are registered in meat. In this way the non-relational world of non-human experiences are known.

As we look back at the history of animal suffering and sentience and appreciate the importance of science in making the suffering-sentient-animal assemblage, the charges of anthropomorphism and the problem of Wittgenstein's lion (Wolfe 2003) – if they could talk, would we be able to understand their language? – does reach some resolution. Non-human animals, like human animals, consist of processes of matter that have the capacity for duration, change and transformation. The generation and measurement of sentient materiality offers a new tool to interrogate the sentient animal, marking a turn away from the empathic relation with a fellow sentient, and the charges of anthropomorphism that follow it. It also directly connects the sentient experience with the utility of the animal's body as edible. Incidentally, this becomes only possible because of the ability of science to appear objective, it has of course just invented a different form of human-animal relation through those new scientific practices, but science is rarely accused of anthropomorphism.

Conclusion

FitzSimmons's called us to study 'the matter of nature' (1989). However when we work with non-human processes of matter, practices and materiality as intra-action (Barad 2007) it becomes evident that there is not one matter out there in Latour's (2005) 'plasma' but in fact 'matters'. If there are 'matters' then ethically and politically we must be sensitive to the specific expressions of different kinds of non-human matters. Sentient matter is one kind of matter. Unlike animation, enchantment, and other descriptors for the image of matter, the naming of a matter that can intra-act to generate meanings and materialities of sentience has a greater normative political and ethical charge. This is because by its naming it is establishing as a matter of fact in new places, entities, beings carrying with it a widening remit of those for whom we should care. No longer are sentient materialities located only in human animals but also non-human animals and, who knows, in future may become identifiable in the materialities of other non-human bodies. And so we must recognise that 'sentient matter' is not a generic, universal term at work in all human/non-human assemblages, but instead is a figuring of matter only available in some kinds of mattering. The contingencies of sentient processes of matter intra-acting, urges us to be sensitive to how our material discursive knowledge practices afford or foreclose possibilities of engagement with specific kinds of processes of matter. This develops the relational ethic, central to the study of the non-human and ethics in non-representational theory, into a gestural device for indicating how and when generating and generated practices and materialities may repress other kinds of processes of matter that if intra-acted with would support a different outcome. Introducing the sentient animal, its experiences ante-mortem, and the events surrounding its body post-mortem has enabled the development of the concept of sentient processes of matter. The development of this concept supports social science studies of the huge developments in the natural and biological sciences to engineer sentient materialities, for example cloning, or hybrid human embryos. The identification of sentient processes of matter provides the opportunity to ask ethical and political questions about these developments. Such as, when is 'stuff' a sentient materiality? Are there political and ethical implications to the close entanglement between scientific practices and knowledge of sentient materialities and sentient beings? Thus, we may be with many (Bingham 2006), and have connectivities (Roe 2006) with them, but some of these connectivities, like to those non-humans who share sentient processes of matter with us humans, have arguably a more profound political and ethical consequence for how the human is co-constituted. 'Non-human' processes of matter are generating human/non-human assemblages of materialities when they intra-act with the material-discursive practices of humans. The existence of various kinds of processes of matter demands further specific accounts of the generation of non-human materialities. It equally demands a move away from the reductive category of 'non-human' through a sensitivity for the existence of different kinds of processes of matter, with unknown spatio-temporal capacities and whose ability to generate materialities is defined by cultural-historical events.

References

Appleby, M. and Hughes, B. (1997), *Animal Welfare* (Wallingford: CAB International).

Anderson, B. and Wylie, J. (2009), 'On Geography and materiality', *Environment and Planning A*. 41, 318-335.

Anonyme (1997), Treaty of Amsterdam amending the treaty on European Union, the treaties establishing the European communities and related acts. Official Journal, 340, available at http://eur-lex.europa.eu/en/treaties/dat/11997D/htm/11997D.html (last visit 2 October, 2007).

Barad, K. (2007), *Meeting the Universe Halfway: Quantum Physics and the Entanglement of Matter and Meaning* (Durham: Duke University Press).

Bennett, J. (2001), *The Enchantment of Modern Life* (Princeton, NJ: Princeton University Press).

—— (2007), 'Edible matter', *New Left Review* 45, May-June, 133-145.

Bentham, J. (1996)[1789], *An Introduction to the Principles of Morals and Legislation*. Burns, J. and Hart, H. (eds) (Oxford: Clarendon Press).

Bingham, N. (1996), 'Object-ions: from technological determinism to a geography of relations', *Environment and Planning D: Society and Space* 14, 635-657.

—— (2006), 'Bees, butterflies, and bacteria: biotechnology and the politics of non-human friendship', *Environment and Planning A* 38(3), 483-498.

Brambell, F. (1965), *Report of the Technical Committee to Enquire into the Welfare of Animals Kept Under Intensive Livestock Husbandry Systems* (London: Her Majesty's Stationery Office).

Callon, M. (2005), 'Why virtualism paves the way to political impotence: A reply to Daniel Miller's critique of the laws of the markets', *Economic Sociology: European Electronic Newsletter* 6/2 (February), 3-20.

Chambers, P. and Grandin, T. (compiled by) (2001), *Guidelines for Humane Handling, Transport and Slaughter of Livestock* (ed.) Gunter Heinz and Thinnarat Srisuvan (Food and Agricultural Organisation of the United Nations. Regional Office for Asia and the Pacific. RAP Publication 2001/4).

Chicken Out Campaign (2008), http://www.chickenout.tv/, last accessed 25 October 2008.

CIWF (2009), www.ciwf.org.uk, last accessed 5 April 2009.

Deleuze, G. and Guattari, F. (1988), *A Thousand Plateaus.* Trans. B Massumi (London: Athlone Press).

Evans, N. and Yarwood, R. (2000), 'The politization of livestock: Rare breeds and countryside conservation', *Sociologia Ruralis* 40, 228-248.

FitzSimmons, M. (1989), 'The matter of nature', *Antipode* 21, 106-120.

Franklin, S. (2007), *Dolly Mixtures: The Remaking of Genealogy* (Durham and London: Duke University Press).

Grosz, E. (2004), *The Nick of Time: Politics, Evolution and the Untimely* (Sydney: Allen and Unwin).

Haraway, D. (2003), *The Companion Species Manifesto: Dogs, People, and Significant Otherness* (Chicago: Prickly Paradigm Press).

Harrison, R. (1964), *Animal Machines: The New Factory Farming Industry* (London: Vincent Stuart Publishers Limited).

Hardie, I. and MacKenzie, D. (2007), 'Assembling an economic actor: The agencement of a hedge fund', *The Sociological Review* 55(1), 57-80.

Holloway, L. (2001), 'Pets and protein: Placing domestic livestock on hobby-farms in England and Wales', *Journal of Rural Studies* 17, 293-307.

Hinchliffe, S. (2001), 'Indeterminacy in-decisions: Science, politics and policy in the BSE crisis', *Transactions of the Institute of British Geographers* 26, 182-204.

Jamie's Fowl Dinners (2008), Channel 4. Screened January 2008.

Latour, B. (1993), *We Have Never Been Modern* (Cambridge: Harvard University Press).

—— (2005), *Reassembling the Social: An Introduction to Actor-Network Theory* (Oxford: Oxford University Press).

Law, J. (2004), *After Method: Mess in Social Science Research* (London: Routledge).

McCormack, D. (2003), 'An event of geographical ethics in spaces of affect', *Transactions of the Institute of British Geographers* 4, 488-507.

Mench, J. (1998), 'Thirty years after Brambell: Whither animal welfare science?', *Journal of Applied Animal Welfare Science* 1(2), 91-102.

Morris, C. and L. Holloway (2009), 'Genetic technologies and the transformation of the geographies of UK livestock agriculture: A research agenda', *Progress in Human Geography* 33(3), 313-333.

Philo, C. and Wilbert, C. (2000), *Animal Spaces, Beastly Places: New Geographies of Human–Animal Relations* (London: Routledge).

Purdue University Animal Sciences (2008) *Meat Quality and Safety*, <http://ag.ansc.purdue.edu/meat_quality/>, last accessed 31 October 2008.

Risan, L. (2005), 'The boundary of animality', *Environment and Planning D: Society and Space* 23, 787-793.

Roe, E. (2006), 'Material connectivity, the immaterial and the aesthetic of eating practices: an argument for how genetically modified foodstuff becomes inedible', *Environment and Planning A* 38, 465-481.

Roe, E. and Higgin, M. (2008), 'European meat and dairy retail distribution and supply networks: A comparative study of the current and potential market for welfare-friendly foodstuffs in six European countries: Norway, Sweden, the Netherlands, UK, Italy and France', *Welfare Quality Report Series No 6.*

Rushen, J. (2008), 'Farm animal welfare since the Brambell Report', *Applied Animal Behaviour Science* 113, 277-278.

Singer, P. (1990), *Animal Liberation*, 2nd edition (New York: Avon Books).

Stassart, P. and Whatmore, S. (2003), 'Metabolising risk: Food scares and the Un/Re-making of Belgian beef', *Environment and Planning A* 35(3), 449-462.

Symes, D. and Marsden, T. (1985), 'Industrialization of agriculture: Intensive livestock farming in Humberside'. In Healey, M. and Ilbery, B. (eds), *The Industrialisation of the Countryside* (Norwich: Geo Books), 99-120.

Thrift, N. (1996), *Spatial Formations* (London: Sage).

Veisser, I., Butterworth, A., Bock, B. and Roe, E. (2008), 'European approaches to ensure good animal welfare', *Applied Animal Behaviour Science* 113, 279-297.

Velarde, A., Gispert, M., Faucitano, L., Manteca, X. and Diestra, A. (2000), 'The effect of stunning method on the incidence of PSE meat and haemorrhages in pork carcasses', *Meat Science* 55, 309-314.

Von Uexküll, J. (1957) [1934], 'A stroll through the worlds of animals and men'. In Schiller, C. (ed.) *Instinctive Behavior* (New York: International Universities Press), 5-80.

Webster, J. (2006), 'Animal sentience and animal welfare: What is it to them and what is it to us?', *Applied Animal Behaviour Science* 100, 1-3.

Whatmore, S. (1997), 'Dissecting the autonomous self: Hybrid cartographies for a relational ethics', *Environment and Planning D: Society and Space* 15, 37-53.

—— (2002), *Hybrid Geographies: Natures, Cultures, Spaces* (London: Sage).

—— (2006), 'Materialist returns: Practicing cultural geography in and for a more-than-human world', *Cultural Geographies* 13, 600-609.

Whatmore S. and Thorne L. (2000), 'Elephants on the move: Spatial formations of wildlife exchange', *Environmental and Planning D: Society and Space* 18, 185-203.

Wolfe, C. (2003), *Zoontologies: The Question of the Animal* (Minneapolis: University of Minnesota Press).

Yarwood, R. and Evans, N. (2000), 'Taking stock of farm animals and rurality', in Philo, C. and Wilbert, C. (eds), *Animal Spaces, Beastly Places* (London: Routledge), 98-114.

PART IV
Politics

Chapter 15
Politics and Difference

Arun Saldanha

Introduction

In its most basic form, politics is irrevocably an analytics of difference. Whether enacted by classic liberalism, revolutionary practice or poststructuralist poetry, all politics ultimately seeks ways to reorganize the social for there to be justice and peace, however defined. This at least implicitly acknowledges that there are real social differences to work with. The most enduring differences have operated in binary and oppositional matrices: free/slave, citizen/barbarian, court/aristocracy, bourgeois/worker, city/country, intellectual/mob, metropolis/colony. As we know after poststructuralism, the nature of these large-scale and pervasive differences is that they are constitutive of discursive and subjective processes, including desire.

The enmeshment of discourse and desire means that the relation of existing differences to the political is profoundly *contingent*. On the one hand differences can be so taken-for-granted – sexual difference being the case in point – that politicizing them is a long and arduous process, meeting immediate resistance from the hegemonic order at every turn. On the other hand, their pervasive, seemingly clear-cut existence also means that differences can at least potentially be *represented* in political discourse. In the latter case, the relation between social identity and political subjectivity becomes more automatic, and we can speak of a politics of identity. Though identity politics is usually understood as the bottom-line of 'the new left', in this framework, local unionism, third world nationalism and feminism are also versions of identity politics.

If membership of a social category (say, wage-labourer) does not predict a certain political or moral orientation (communist, internationalist), not everyone is engaged in politics in the same way or degree. For there to be any undoing of present differences, populations implicated in these differences have to secure some form of representation in parliament, party, and public opinion. Hence representation marks the gap between social membership and politics. What is stimulating about the present moment of theorizing political identities, as can be found for example in the work Chantal Mouffe (2005), is that the relation between social difference and politics has been intensely problematized. Political subjectivity has become a creative project of constitution, in which existing identities and differences are reconstituted together with the political discourses they inspire. The critique of the transparency of socioeconomic and everyday 'interests', as well as of any easy recourse to a consensus on what is good and necessary, has entailed the abandonment of old ideas of a people's party and political leadership. The fundamental framework of parliamentary democracy and individual rights protected by the law has also come under anti-foundational scrutiny. Since such poststructuralist politics finds and creates differences, rather than passively representing identities, it is often called the politics of difference.

This chapter will explore some ontological dimensions of representation guided by the question of the politicizability of difference. A first decision is made by following non-representational theory's insistence on *embodiment* (Anderson 2006; Harrison 2000; Thrift 2007). If we look at the differences which have become enduringly politicized in the West during the last third of the 20th century – man/woman, white/black, straight/gay, able/disabled, healthy/sick, human/animal – it is clear that the force of their demand to be taken politically seriously has much to do with their corporeality, i.e., with the double fact that they involve different bodies and *many at the same time* (Saldanha 2007). Difference involves both the fleshy and the aggregate aspects, what I will call with Deleuze and Guattari the molecular and the molar.

How does Human Geography contribute to thinking the ambivalent relationships between politics, representation, identity and difference? Non-representational theory has been at the forefront in alerting the social sciences to the embodied, prelinguistic and precognitive processes that go into gluing groups, spaces, knowledges and economies together, as well as opening them towards their immanent otherness (Thrift 2008). It is precisely the unstoppable, everyday transmogrification of micro-differences (the 'experience of being black'; the *widespread* hardships of labour) that obscures them from political discourse. Corporeality inherently resists becoming known; in particular for non-representational theory, it cannot be exhaustively understood through analysing its capture in language.

I will argue that the critique of mental and semantic representation in non-representational theory can be fruitfully coupled with the critique of representation in the politics of difference. Both these departures from representation – linguistic and political – can be powerfully backed by the theory of human differentiation found in Deleuze and Guattari. More than other poststructuralists, they allow for real social differences to be thought *through* their corporeality and spatiality. Some scenes from the novel *Because It Is Bitter, and Because It Is My Heart* by Joyce Carol Oates and the poetry of Stephen Crane will provide further hints as to the nonrepresentational realm of difference and politics. However, I end by arguing with Alain Badiou that a truly nonrepresentational politics requires stepping away from poststructuralism.

Molar Difference: Identity

Deleuze and Guattari make a distinction between two kinds of order, molar and molecular, which will be helpful to the discussion. The distinction is resolutely materialist, and they associate it with physics: 'the molar direction that goes toward the large numbers and the mass phenomena, and the molecular direction that on the contrary penetrates into singularities, their interactions and connections at a distance or between different orders' (1983, 280). The distinction is in fact more proper to the domain of chemistry, and insofar as the latter investigates the mixtures and transformations of bulk substances, not simply their mechanics or dynamics, it is preferable to locate the origin of the distinction there.[1] Just like in chemistry, it enables analysing and evaluating the mode of *organization* of social, ecological and psychological phenomena. But Deleuze and Guattari are not advocating the reduction of the big to the small:

> Freud was Darwinian, neo-Darwinian, when he said that in the unconscious everything was a problem of population (likewise, in the contemplation of [psychic] multiplicities he saw a sign of psychosis). It is therefore more a matter of the difference between two kinds of collections or populations: the large aggregates and the micromultiplicities (ibid.).

Whether considering the level of the individual brain, the household or the nation-state, its 'molar' aspect is always what binds, repeats, stabilizes, while the level's molecular forces are interspersed, schizophrenic, and potentially disruptive. When considering any phenomenon, its micromultiplicities, its molecular differentials will be physically or phenomenologically 'smaller', more local or askew, than its molar differences (Dewsbury 2003, 1907-1908; Anderson 2006, 736). 'It is crucial for understanding Deleuze and Guattari', Brian Massumi writes, 'that the *distinction between molecular and molar has nothing whatsoever to do with scale*' (1992, 54, *emphasis in original*). Hence Derek McCormack (2007) laments geography's silence on neurotransmission and genetics, which although molecular in the narrow sense, have real effects on societies at large, as exemplified in the dependence of labour on mood-enhancing drugs.

While it is true that every scale is made up of both molar and molecular differences, and also that every scale envelops others, assuming scale is entirely irrelevant to understanding molar and molecular could be misleading, especially to

---

1 In chemistry the distinction is precise and fundamental. The molar mass M of a substance is the mass of one mole of that substance, or Avogadro's number (6.022×10^{23}) of its molecules or atoms. In theory, this same number of uranium atoms would be found in 238 g of $_{238}U$ isotope. The mass of one molecule (or one atom) is of a very different order: one H_2O molecule could weigh $(2 \times 1.00794 \text{ u}) + 15.9994 \text{ u} = 18.01528 \text{ u}$, with the slight deflection due to the prevalent isotopes of hydrogen and oxygen typically found, and u the unified atomic mass unit, taken to be a twelfth of a carbon-12 atom, about 1.6605×10^{-24} g. One mole of such water would have a mass of exactly 18.01528 g. Hence Deleuze and Guattari seem to understand 'molecular' as pertaining to what can only be ascertained experimentally and for particular samples, whereas 'molar' refers to an immensely *larger* bulk of substance and a fixed number that enables a statistical operation.

geographers (cf. the *Transactions* scale debate; Leitner and Miller 2007). Without any scalarity Guattari's chemical quasi-metaphor would lose sense. From his early introduction of the distinction molar/molecular in his clinical practice of the sixties (Guattari 1984), to his activist speeches in Brazil during the eighties (Guattari and Rolnik 2008), the molecular/molar distinction was politically and ontologically clearly homologous to the micro/macro distinction. Guattari opposes creative and grassroots 'molecular revolution' to the sweeping identitarianisms of liberalism, class struggle and psychoanalysis, and speaks

> of a north-south that traverses all countries, a blackness that traverses all races, minor languages that traverse all dominant languages, a becoming-homosexual, a becoming-child, a becoming-plant that traverses demarcated sexes. These are the elements that Deleuze and I group together under the heading of a 'molecular dimension' of the unconscious (Guattari and Rolnik 2008, 103).

Physical scale is therefore certainly one, but not an irreducible, aspect of the distinction between molecularity and molarity. More abstractly, as Massumi explains, molar differences are differences between neatly individuated wholes which rely on a well-integrated organization of their constituent parts, whereas molecular differences are bottom-up, experimental, too rapid or too slow to be perceived from within the molar order. Attuning thought to molecular forces inherently entails breaking open identities and the molar discipline they are based on.

How the reciprocation between molar and molecular gives rise to sexual and racial difference can be fleshed out with *Because It Is Bitter, and Because It Is My Heart*. As in many of her novels, Oates tells of a complex relationship between two individuals ineluctably haunted by racist patriarchy. On the molar level, the level of identities, the census and common sense, the novel's central difference is between a middle class white girl, Iris Courtney, and a working class black boy, Jinx Fairchild. Iris and Jinx grow up in the 1950s in Hammond, a fictitious industrial city in New York state, before national desegregation. The novel ends with the news of the assassination of John F. Kennedy in 1963. Hammond's racial segregation is broadly how it was in the eighteenth century across the United States, and still is today, despite legal desegregation. Jobs, housing, crime, schools, churches and music are all arranged in a black vs. white structure: 'As Hammond eases downward toward the river, as Uptown shifts into Lowertown and the buildings and houses and even the trees become shabbier, there is an increase in dark faces, an ebbing of white faces' (Oates 1991, 22). Because the distribution of privilege hinges on this binarization, transgressing it risks becoming ostracized or worse. As W.E.B. Du Bois (2007, orig. 1903) famously apprehended at the start of the century, this *line* is singularly effective because it is simultaneously materialised spatially and interiorised psychically.

The stability of the colour line – the molar relation black/white – depends on the coherence of the two terms it separates, that is, of black identity and white identity.

These in turn depend on *vigilance*, a constant recognition and reproduction of the elements that are deemed part of those identities. One of the most telling instances of such vigilance, in the 1950s as now, is the routine interpellation of young blacks by white policemen, in which the former are statistically (in molar terms) liable to engage in criminal practices and the latter presume guilt to render their work tractable:

> So John Ritchie is coming home on East Avenue at about 6 P.M. this late-summer day and a Hammond City police car pulls over and two policemen get out yelling to him to 'stop and identify yourself', their billy clubs out and their voices raised, as if there was already some trouble, some threat (Oates 1991, 28).

As going the molar route is always easier, the policemen have already decided that John Ritchie is the 'coal-black Negro' they were looking for, the imagined assailant of a paranoiac white woman. When John Ritchie, whose psyche is tarred by war, acts suspiciously and cannot answer the policemen's questions, they beat him up, with other black people watching. When he finally starts hitting back, the policemen kick him to death, or near-death, it will never be known. Guattari would understand racist, misogynist and homophobic police brutality not as aberration of duty, but as following directly from the molar definitions of populations ('hookers') and places ('the inner city') that policing requires. Aside from incompetence on the policemen's part, the industries of surveillance cameras and tabloids make the molar route even more difficult to resist.

A colour line goes both ways; young blacks will in return instantly and irrationally hate any uniformed body in sight. Oates' novel gives ample indication of facile generalizations about white people among African Americans: whites are careless, selfish, sometimes well-meaning, but naïve and therefore dangerous. The father of Iris warns her, 'They'd peel the skin off us if they could, they hate us so. But they can't. So they're courteous to our faces when they have to be and we're courteous to them, but don't ever confuse it, Iris, for anything else' (Oates 1991, 93). Molar difference thrives through the circularity of etiquette, of affirmed prejudice, of imaginations projected onto the other. They are recognized time and again, and are never confused.

The Topology of Intersectionality

Molar difference is more convoluted than the linear separation of two populations. The colour line does not simply segregate the United States (or any Western country) between disenfranchised and privileged. One important casualty of segregation on the basis of phenotype is a population typically designated by the derogatory term 'white trash'. In capitalism whiteness requires not just a light skin tone but property, education and visibility. By their lack of success as both producers and consumers,

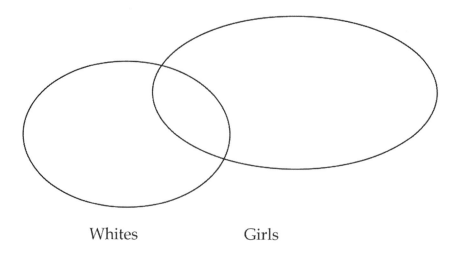

Whites   Girls

Figure 15.1   White girls

white trash live at a cultural and geographical remove from the position their skin is supposed to guarantee them in a racist society (Wray and Newitz 1997). This makes them doubly despicable from the hegemonic middle-class perspective: first in economic terms, and second, as implied explanation, in racial terms, with a lack of gene flow supposedly explaining poverty, stupidity and other deficiencies. The recent websites and festivals attempting to turn the derogatory label into an identity to be proud of only reinforce the distance with the middle class.

If a white girl and a black boy are the main protagonists of *Because It Is Bitter*, the character of the same age that will link them, Little Red Garlock, is white trash. Oates unabashedly repeats all the clichés: the Garlock family is large, ugly, reclusive, ignorant, filthy, and whimsically violent. The white middle class mostly tolerates their dysfunctional presence at the edge of downtown (it was Little Red's delusional mother who called the policemen on John Ritchie), but being white trash, there is already an air of disposability about them. The African American population loathes the Garlocks, partly because the family is brashly racist, and partly because hatred of white trash is a frustrated response to the structural humiliation of being black in America. Through endemic competitiveness, capitalism creates a pecking order in which white trash ranks below working class African Americans. In short, the destabilizing function of the Garlock family in the black-and-white matrix makes Oates' novel a more realistic analysis of molar difference than many social science accounts on race.

Black feminist theory invented the concept of *intersectionality* to account for a social reality in which molar differences (sex, race, class) are not only fundamentally unequal but cross and disrupt each other (Brah and Phoenix 2004; Valentine 2007). Iris and Little Red are both white but unequally so: Little Red is physically

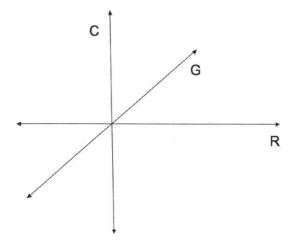

Figure 15.2   Intersectionality in three dimensions

stronger, Iris has better socioeconomic opportunities, etc. Intersectionality quickly gained purchase across the social sciences (Walby 2007), but its descriptive force seems to be not so much empirical as implicitly set-theoretical. For the most basic forms of mathematical set theory, *membership* is the only ontological parameter that matters. Multiple and simultaneous membership is derivatively possible, as is non-membership (Badiou 2005a). For example, a white girl is a member of both the sets Whites and Girls. This overlap illustrates the concept of intersectionality and is easily imagined with a Venn diagram (Figure 15.1), to which sets can be added depending on the molar complexity desired.

It is crucial to remember what abstraction such a visualisation entails. To start, the theory of intersectionality holds that there is no actual body that is a member of only one set. Moreover, membership becomes difficult to determine when faced with the actual range of phenotypes and practices. Most seriously, the theory of intersectionality is concerned not only with debunking the essences of molar identities, but with the unequal structures that their essentialization holds in place, which are not legible in the set-theoretical abstraction.

In a less abstract, more geometrical imagination of intersectionality, a body or group is positioned in an *n*-dimensional space with *n* the amount of polarizing variables of corporeality considered (income, skin tone, gender, health, attractiveness, etc.). These dimensions have a 'slidingness' and directionality to them lacking in sets, so that mutually reinforcing oppressions and privileges can be inferred. In such topology, no element can lie outside the coordinates (with sets, there could be). Neither can an element be understood through one axis alone, which a fortiori means there is no zero-point: there is no body devoid of molar difference. The maximum number of binary molar dimensions that could be easily visualized is three (here class, race and gender), but the 3D figure would have to rotate to properly reveal the intersectional distributions (Figure 15.2).

There obviously remain problems with this reductionist imagination of intersectionality, not in the least to do with to what extent molar difference can be reduced to linear polarization (class to income, race to skin tone, gender to sexual characteristics). Though few theorists of intersectionality advocate this formalism, it can nonetheless be helpful for the philosophical conception of molarity. But even if the topology of intersectionality is better suited for analysing molar difference, without adding the four dimensions of *physical* space-time it remains free-floating and cannot readily explain systemic distributions of power. In stressing the importance of location and connectedness, geographers contribute a rigorous *grounding* of intersectionality. For example, the socioeconomic position of gay Americans is very different among the gentrified elites than it is in poor rural areas (Knopp 1992). Here a population 'gay Americans' is located at the 'gay' end of the 'sexuality' axis but is spread over a range of values on the axes 'social capital' and 'economic capital'. What intersectionality allows for is imagining the interlocking of molar differences between persons and between populations, which is the same as the concomitant formation of unequal identities. However, the concept is only meant for the molar 'level'. As we will see, molecular forces continuously upset the topological localizability of a body.

Representation: Epistemological and Political

In *Ethics and Representation*, Claire Colebrook (1999) calls attention to the reciprocation between two notions of representation constitutive of Europe's breakaway from the Renaissance: epistemological representation of object in subject, and political representation of interests in politics. Modern philosophy *is* basically the epistemological and political situating of subjectivity. Colebrook locates the inauguration of this reciprocation in Kant's critique of Cartesian idealism and Christian metaphysics. For Descartes and Spinoza, representation as such was not the central philosophical problem. In classic rationalism, thinking directly partakes in the great order of being, either by virtue of being divinely endowed (Descartes) or by physically being part of it (Spinoza). Kant's epistemological critique starts with, and from within, a *point of view*, which is nothing but a subject's inner representation of the objects (groups, issues, aims, places), and only those objects, that are relevant to it. Subjectivity and language are therefore always relative and finite, conditioned by geography and history. Kant's revolutionary critique of access to absolute truth also entails the subject is radically self-constituting. Colebrook understands phenomenology and poststructuralism as heirs to this inaugural conception of subjectivity, suggesting that insofar as Heidegger, Foucault, Derrida, Irigaray and Deleuze philosophize the unrepresentable determinations of representation – the limits and chimeras of consciousness, knowledge, writing, speech – they extend Kant's anti-absolutist transcendental project.[2]

Epistemological representation corresponds to political representation. After all, 'it is only because we can think of an "empty" human subject that representative democracy can work: democracy is not just the collection of self-seeking interests and the expressed desires of competing individuals' (Colebrook 1999, 15). Because there is for Kant no transcendent law for what society should be like, whether divine or biological, reason is open to a creation of a free community. Since 'the world' for me is nothing but what I am conditioned to know as represented *to* me, I am responsible for representing it in such a way as to create order for others in my society. Kant's famous categorical imperative – act as if it will become law – is a most elegant, and most modern, injunction, directly twisting autonomous decision-making (hence mental representation) and sociopolitical participation (and by extension parliamentary representation) into a Möbius strip (Kant 1948, esp. 69-73). 'The demand for democratic representation is a demand that modernity break free from its heteronomous past, that it emerge freely into a domain or space or non-interference, communication, recognition and inclusion' (Colebrook 1999, 2). While poststructuralism forcefully decentres this Kantian ego by placing it against its preexisting discursive backgrounds, it remains committed to democracy in that knowledge and ethics remain intractably open and linked.

What does representation mean for the molar arrangement of identities? The idealism of Kant largely bypasses the fact that corporeal, cultural and political-economic differences present huge challenges to the smoothness of representation and the categorical imperative. Poststructuralism, especially third wave feminism, is 'nonrepresentational' to the extent that it uncovers the subtle but inimical work this philosophical bypassing of differences does in maintaining those differences (e.g. Olkowski 1999). It should be added that theoretically speaking, these differences are what Kantian universality and individualism were precisely entitled to overcome. Despite or rather because of the undermining of the firmness of subjectivity and objectivity, there continue to be very diverse elaborations of Kantian ethics, especially via Levinas, which do not depend on a transparent and legalistic concept of representation (Clark 2005; Harrison 2008; Popke 2004).

---

2 If poststructuralism supported non-representational theory's sustained questioning of epistemological representation, it has enabled a turn away from Descartes, not from Kant. Paul Harrison for example notes: 'Representation does not wait for our consent or denial to get underway; rather we find ourselves always already within patterns and regimes of truth as the very resources which allow us to agree or disagree' (Harrison, 2007, 600). This 'always already' of Kantianism has come under trenchant attack from Quentin Meillassoux (2008). Like Colebrook, Meillassoux understands poststructuralism as largely a continuation of the Kantian critical project, but unlike Colebrook, he wants to break with it through a revolutionary return to a kind of Cartesian absolute realism. If there is one philosophical system that is entirely *meant* to demolish representationalism, it is not Derrida's, Foucault's, or even Deleuze's, but Meillassoux' new brand of rationalism.

The political as *public* engagement, as collective action and the discourse on molar interests aimed achieving justice, is no longer a transparent extension of autonomous morality as it was for Kant.

What happens to democracy when the contingency of representation, both epistemic and political, is fully espoused? According to Mouffe (2005) we have 'radical democracy', 'radical' in presuming that there is no ultimate consensus, rationality, and all-inclusive equality achievable under liberal capitalism, also or especially not through socialist revolution (cf. Barnett 2004). Political antagonism about the very foundations of democracy is not only unavoidable, but crucial for democracy itself. For Mouffe it is *through* the multiple conflicts existing in society that political identities are made and remade. In their relating to each other through coalition, activism and debate, political identities constantly mine their own intersectional undecidabilities for dismantling the structures from which they spring. There can be no question of representation in intellectuals or parties of quasi-transcendent categories like 'the worker class', 'the people' or 'the community of *x*' which could prescribe what is to be done. Though poststructuralism is politically nonrepresentational, it is not antidemocratic, endorsing Kant's positive valuation of critique and self-constitutability. Insofar as this engages language (political discourse, the law, rights), representation is in radical democracy not destroyed, but deconstructed and radicalized.

Molecular Difference: Affect

Returning to Joyce Carol Oates' novel, the molecular infrastructures of social identities remain to be examined. Henry Louis Gates notes that Oates' 'basic technique is really quite simple: Find just where it hurts and then press, *hard*' (1990, 28). A writing style can inflect pain or discomfort, conveying much more than the traditional, more or less Kantian concepts of representation allow for (see Doel's chapter in this volume). Oates' writing does not simply 'refer to' diegetic objects like Hammond, segregation, desire or death, but generates myriad and inconclusive affects that smoothly fuse into an unconscious 'canvas', allowing for such objects to be delineated afterward (in criticism for example; cf. Gates 1989). These affects emerge in the complex encounter between the black-on-white print, the reading body, and knowledge of the molar differences spoken of in the narrative. In order to create affect at all, the narrative presupposes the reader's social conditioning in a racist and sexist state. Like a probe, it penetrates this experience of molarity and lodges on its essentialist structure. It is this that hurts.

The story in *Because It Is Bitter, and Because It Is My Heart* is fundamentally one of forestalled love, but we need to first become sensitive to the molecular forces forming this forestallment. When sexual difference is reduced to its utility for the reproduction of what Guattari calls Oedipus (the patriarchal bourgeois family), it operates through its molar dimension. Oedipal molarization occurs at scales from the body to entire cultures and the world economy; an Oedipal version of Oates' novel might have Iris and Jinx transcend segregation to bear happily multicultural offspring. Guattari and Oates insist, instead, that the molar scale-effect of the sexual binary is secondary, merely the institutionalized, oppressive and *bigger* side of an infinity of nitty-gritty and sometimes calamitous erotic corporalities. Beneath or before the molar organization of sexual difference, bodies continuously explore *smaller*, not-yet-structured, even prepersonal intensities. Patriarchy feeds on these intensities but cannot fully discipline them:

> Sexuality is by no means [only] a molar determination that is representable in a familial whole; it is [also] the molecular underdetermination functioning within social and secondarily familial aggregates that trace desire's field of presence and its field of production: an entire non-Oedipal unconscious that will only produce Oedipus as one of its secondary statistical formations ('complexes'), at the end of a history bringing into play the destiny of social machines, their régimes compared to that of desiring-machines (Deleuze and Guattari 1983, 183-184).

The problem in conceiving molecularity lies chiefly in the relationalist, dialectical way most theory has conceives of difference. While molar differences are always relationships *between identities* (the difference between man and woman, between Scotland and England), molecular differences are not relational at all. Molecular differences are not nonrelational in the sense of a negative background of noncompliability for the activity of relationality, as Harrison (2007) has conceived finely in opposition to Deleuzoguattarianism. Empirical and positive, molecular differences do connect, but what they connect is only themselves to where they are going. The 'differences' of molecular sexuality, enacted in erotic games, fantasies and masquerades, are best thought of as not as relations but tensions, gradients, trajectories. Psychiatry and the social sciences operate on the molar level, reductively and predictably, while Deleuze and Guattari attempt to remain true to the physicochemical fluidities of desire.

Three scenes to give an indication of how molecular difference precedes and exceeds molar difference in the U.S.: a school, a street, a bus. As Jonathan Kozol (2005) argues compellingly, racial segregation in schools was never abolished by the *Brown vs. Board of Education* decision of 1954, because it simply runs too deep in American capitalism. But how does federal (de)segregation play out into the *feeling* of being a pupil locally? Writing about the time a little after *Brown*, Oates captures some of the mood:

> In such classrooms in such schools there is a ceaseless drama of wills, as in meteorological crises between contending fronts of atmospheric pressure, and only the benevolent cooperation of the majority of the students allows order to be maintained ... some species of order, however harsh and whimsical (Oates 1991, 33).

A room's *atmosphere* is more than the sum of the senses of place of the people in it. Molecular differences in an integrated school divided by wider institutionalized racism float around in the classrooms like clouds, winds, electricity. These differences are not yet racial; they do not even belong to any body in particular (cf. Tatum 1999). The wills of the students in the quote are not conscious, but instances of Nietzsche's cosmological 'will to power' (1968), unconscious inner forces in student bodies trying to maximize their space, their storminess, shyness, bitterness. Order only arrives if these wills are made to partially converge for some time by some outside force: the teacher. Our analysis of the molecular moves from chemistry to a thermodynamics and meteorology; this inventiveness only signals the extent to which the molecular is not simply more intersections of molar difference, but altogether unrepresentable.

As noted with the death of John Ritchie, the street is a crucial second site for racial, class and gender tensions. The lasting influence on urban ethnography of William Foote Whyte's *Street Corner Society* (1993, orig. 1951) shows that whether in Boston's North End, Newcastle, Johannesburg or Auckland, molar differentiation of populations cannot exist without an overabundance of feelings and practices, most of which are captured in habit and speech, but only gradually and messily.

> *Don't walk too fast* the white girls caution one another, *don't walk too slow*: scared, tremulous, eyes straight ahead, a tingling in the pit of the belly like minnows darting in shadowy water. Sometimes the girls overheard snatches of what these men say to one another; sometimes they have to imagine (Oates 1991, 93).

Although from the molar perspective we are dealing with a *relational* situation of two distinct groups – white girls and black men – it is on the molecular, precognitive level that the gestures, gaits, excitements first occur. Affect is per definition molecular, that is, *almost* molar. Non-representational theory has been at the forefront of examining the molecularity of space, often through a Deleuzoguattarian concept of affect distinguished from verbally expressible 'emotion' (Anderson 2006; Dewsbury 2003; Wylie 2005). Less worked out is how this affectual register gives rise to aggregate difference (Saldanha 2010). Precisely *because* of its doubtful, self-contradictory, animalistic visceralness, the affectual in Hammond is capable of becoming snatched up by habit.

Importantly, place and identity are never formed by one body alone, and the multiciplicities of gut feelings and imaginations are paralleled by the multiciplicites of the bodies which they stir. How many bodies are required for molarity to sprout? For Hammond's sexualized colour line, probably a mere dozen, consolidated over time by hundreds more. We know Hammond is no exception, and that millions of bodies (girls, boys, cops) have over many decades in other similar 'street corner societies' long made the colour line a national fact. For such aggregation to happen there need to be laws and policies, but also an everyday self-disciplining in public spaces. As testimony to the molecular complexity involved, children take many years to master what it takes. In a third example of Hammond's molecular differences, observe how young Iris' naïve urge to share space in a bus is promptly forbidden by her mother Persia:

> 'Why don't they sit with us? – there's room', Iris whispers in Persia's ear. The two of them are together in one of those odd open seats flush with the side of the bus and there is plenty of room beside Persia for a young black mother and her two-year-old, but the woman, hanging from a hand strap, gazes sightlessly beyond them and Persia nudges Iris into silence: 'Just *hush*' (22).

In this scene, affects rapidly interpenetrate and accumulate towards molar solidification: the generosity of little Iris, then the embarrassment of Persia, while the black woman seems tired and indifferent to Iris' inquiry, probably out of habit; then silence. It is always a real possibility *not* to go the molar route, but it requires courage.

On 1 December 1955 in Montgomery, Alabama, Rosa Parks famously defied the racism of the bus system by refusing to give up her seat to a white woman. This simple gesture sparked a mass boycott and constitutional reform, showing *a posteriori* that thousands of African American bodies had long been taking their molar places out of habit, but not without resentment. Martin Luther King, Jr (1958) pointed out the revolutionary Gandhian and Christian aspects in the Montgomery boycott. Parks' story became pedagogical folklore about the fifties: now kids are taught apartheid is wrong. Interestingly, children's books privilege affect not politics, telling of a tired woman who was suddenly fed up, not of a civil rights activist who for many years organized protest in a viciously racist Alabama (Kohl, 2005). In other words, stressing the affectual realm will not by itself instruct on how it can be politicized.

Ben Anderson (2006) contends that *hope* may be the affect par excellence. Through its participating in the excess of the not-yet and its melancholy deferring of accomplishment, hope is what drives politics, consumerism, mourning, but also love. 'Becoming hopeful is marked, therefore, not by a simple act of transcendence in favour of a good elsewhere or elsewhen but by an act of establishing new relations that disclose a point of contingency within a present space-time' (Anderson 2006, 744). To show how contingency and affect constitute love and politics, I will switch from Deleuze and Guattari to Badiou. I cannot elaborate here on how Badiou's position diverges from poststructuralism. In particular, though like Deleuze he attacks representation, Badiou offers no theory of the molecular. This is precisely why his work is more compelling for *embarking on* – as opposed to analysing – such thoroughgoing 'procedures' as love and politics.

For Badiou (1996), love commences with the absolute point of contingency he calls an Event, a singular encounter between two anonymous bodies, to which the lovers will then declare their eternal fidelity. As literature and theatre have always shown, this fidelity itself makes love intrinsically tragic. Returning to Oates' love story, the Event binding Iris and Jinx is macabre.[3] Known in Hammond for his deranged maliciousness, Little Red Garlock follows pubescent Iris around one night just when a friendship has started between her and Jinx. She asks Jinx to walk her home, who fumes, '*I'm* not afraid of that redneck bastard. *I* ain't no helpless girl!' (Oates 1991, 111). Jinx' protection only makes Little Red more reckless. His taunting song reverberates through Iris' life: 'White titties sucks nigger cock – *hey!* White titties sucks nigger cock – *hey!* Big juicy-black nigger cock!' (ibid., 113). The teenage boys get into a short and vicious fight. Dazed with rage, Jinx smashes Little Red's head to pulp with a brick. He orders Iris home and discards the body in the river (the novel begins with its retrieval). The investigation quickly peters out; that the white trash lout would end up murdered is deemed somewhat inevitable.

The strange title of Oates' novel is known in American literary circles and comes from *The Black Riders*, a series of short poems or 'lines' published about a century earlier by the young New York journalist Stephen Crane:

IN THE DESERT
I SAW A CREATURE, NAKED, BESTIAL.
WHO, SQUATTING UPON THE GROUND,
HELD HIS HEART IN HIS HANDS,
AND ATE OF IT.
I SAID, 'IS IT GOOD, FRIEND?'
'IT IS BITTER ——— BITTER,' HE ANSWERED;
'BUT I LIKE IT
'BECAUSE IT IS BITTER,
'AND BECAUSE IT IS MY HEART'.

(Crane 1896, III)

Bitterness, the heart, fleshiness, barren landscapes and God run throughout Crane's lines to conjure the theme of impossible love. Though replete with *fin-de-siècle* Romanticism, Crane's biblical imagery in small capitals is elusive and sparse, prefiguring later American modernism. It is no coincidence that Oates found love's bitterness in Crane's semi-devout broodings.

---

3   While Badiou's Lacanianism requires that the Event is *named* by something like 'I love you', the Event connecting Iris and Jinx is such that it cannot be named, which arguably makes it all the more sacred. The interplay between sacrifice, secrecy and humanity can be further conceived through Georges Bataille, who is remarkably close to Badiou when he writes: 'The world of lovers is no less *true* than that of politics. [It] is constructed, like life, out of a *series of chances that give the awaited answer to an avid and powerful will to be*' (1985, 229). However, for Bataille the unexpected murder of Little Red is what ecstatically unites Iris and Jinx, whereas for Badiou it will forever subjectively divide them.

YOU SAY YOU ARE HOLY,

AND THAT

BECAUSE I HAVE NOT SEEN YOU SIN.

AYE, BUT THERE ARE THOSE

WHO SEE YOU SIN, MY FRIEND.

(ibid., L)

Badiou says love is not union but disjunction. Love generates two irreducible subjectivities which continue figuring out their mutual difference and hence never fuse into a 'couple'. Love is no-one else's business; there is for Badiou no 'third position' who can oversee the disjunction. Little Red Garlock was the only and last to know of Iris and Jinx, and he died when their love was sealed. What momentarily brought Iris and Jinx together is a sin that cannot be known and will keep them apart until they die. Pushing hope to its most sublime, what remains is only the solace that the other, too, cannot escape from this Event and this love. As Oates writes so delicately:

No one is so close to me as you.
No one is so close to us as we are to each other.

(1991, 182)

For Iris and Jinx, the Split of love is literal and spatial. They stay true to each other by carrying a bubble of secret hope of reunion wherever their respective married lives take them to. However stellar (for Iris) or tumultuous (for Jinx) the lives, this melancholy nonplace is precisely where the truth of love patiently lives on.

How to Make a Difference (Yes We Can)

If conceiving love as disjunction in Badiou, Oates and Crane seems pessimistic, it is in fact the most powerful affirmation of the universal capacity to love. Moreover, the rather austere exactitude of this conception has prepared us for the definition of politics. This chapter concludes by contrasting Badiou's theory of politics with Derrida's and returns to the question of representation (cf. Critchley 2007; Dewsbury 2007). As we saw, representation is tied up with the molar differences that politics seeks to rearrange. But even if representation – knowledge of the world and the Kantian imperative – is ineluctably skewed, incomplete, tattered, it would seem we need it as coat against the coldness of absolute absurdity:

IF I SHOULD CAST OFF THIS TATTERED COAT,

AND GO FREE INTO THE MIGHTY SKY;

IF I SHOULD FIND NOTHING THERE

BUT A VAST BLUE,

ECHOLESS, IGNORANT, ——

WHAT THEN?

(Crane 1896, LXVI)

Kant's concept of freedom begins with bracketing the ultimate Guarantor for responsible action by locating it in autonomous reason itself, which represents the (or an) external world to be acted upon. Derrida's freedom, if we can call it that, goes even further, and eliminates representation (both mental and linguistic) as guarantee for correct decisions. *What then?* The genius of Derrida's ethical-political philosophy lies in finding after deconstruction not absurdity, nothingness and nihilism, but as J-D Dewsbury argues (2007, 449-450), Plato's *khôra*, an absolutely Different nonplace without any identity, a receptacle-womb from which new discourse can emerge. This fundamental paradoxicality is nothing but the hope, given down centuries and first formalized in Kant, that democracy *can* deliver in the future through its very emptiness. For Derrida, it is the impossibility, the may-be, the *je ne sais quoi* of a political situation that needs to be sensed philosophically before any decision can be made. Derrida's rediscovery of the *immensity* of responsibility reframes political action as an experience of the outside-reach, the never-enough.

Badiou's generous 'Homage to Jacques Derrida' (2007) focuses on what the two philosophers share, namely the left's quest for the 'inexistent', the vanishing point, or what Badiou (2005a) calls the void. In the 'forest of oppositions' that is a text or situation, deconstruction hunts for the singularity which always recedes further into the darkness while thereby keeping the forest together. Badiou joins Derrida on this continuous hunt for the unrepresentable, except that he *shoots*. However cavalier the metaphor, it helps to discern a difference in philosophical style: literary and gentle on the one hand, mathematical and Maoist on the other. Derrida is Western philosophy's master of theorizing the undecidability lying at the heart of the decision. Instead, Badiou's 'metapolitics' (2005b) insists on the necessity of decision (from *de-caedere*, to cut off) by building on two metaphysical concepts entirely foreign to Derrida: truth and fidelity. Political truths, which Badiou finds throughout the history of popular revolt, do not stop at the *critique* of the representational – laws, rights, elections, polls – but attempt to sidestep democratic institutions altogether. Truths constitute themselves by declaring their allegiance to an Event in the ongoing struggle for equality, like May 1968. This fidelity to an ever-renewing communism goes further than Guattari's micropolitics of the molecular and today's counterglobalization movements, which like historical anarchism, often lack the discipline and clarity for sustained resistance.

Badiou's neo-Maoism is radically nonrepresentational. Identities are not exactly suspended but become irrelevant (even that of 'the worker') to formulating what needs to be done and where the movement needs to go. Despite Badiou's scant attention to how real molecular differences can converge into the procedure of politics, two affects can be inferred that all communism feeds on: surprise in the discovery of a 'we', courage to keep moving. Though not automatically pleasurable (something radical thinkers seldom consider), these affects are definitely and definitively redemptive.

> I WALKED IN A DESERT.
> AND I CRIED,
> 'AH, GOD, TAKE ME FROM THIS PLACE!'
> A VOICE SAID, 'IT IS NO DESERT'.
> I CRIED, 'WELL, BUT ——
> 'THE SAND, THE HEAT, THE VACANT HORIZON'.
> A VOICE SAID, 'IT IS NO DESERT'.

(Crane 1896, XLII)

Unlike in love or poetry, in politics there is no 'I', and this is precisely where its unifying force comes from. The space of the 'we' is placeless, but never a desert: it remains forever a crossroads in which the very 'we' that is moving is continually reconstructed. Within the Event kick-starting politics, the unexpected feeling that 'we' have no *reason* for either nihilistic despair or Romantic solipsism is immediately empowering. Instead of the same-old sorry state without end or beginning, the truly different is affirmed by denying that *despite appearances* ('Well, but ——'), things can change.

Remembering the terror in the French and Cultural Revolutions, Badiou often warns that the suddenness and mysteriousness of political truths can be difficult to bear, though he understands the counterproductive effect violence and fear have on politics. The second affect in political truth, courage, is what guarantees the endurance of the political subject in the face of manifest impossibility.

> MYSTIC SHADOW, BENDING NEAR ME,
> WHO ART THOU?
> WHENCE COME YE?
> AND —— TELL ME —— IS IT FAIR
> OR IS THE TRUTH BITTER AS EATEN FIRE?
> TELL ME!
> FEAR NOT THAT I SHOULD QUAVER,
> FOR I DARE —— I DARE.
> THEN, TELL ME!

(Crane 1896, VII)

The affects of politics share with those of religion a viscerally binding force but are nonetheless of the everyday, and pitched against dogma. They are well summarized in a slogan of Latin American and U.S.-Mexican borderland unionism: '¡*Sí podemos!*', 'Yes we can' (or '¡*Sí se puede!*', 'Yes it can be done'). The slogan found its way to Barack Obama, notably his New Hampshire primaries victory speech, which was turned into a YouTube music video tribute. As with

Martin Luther King, Obama's grasp of the affects required for political truth is impeccable:

> We have been told we cannot do this by a chorus of cynics. ... But in the unlikely story that is America, there has never been anything false about hope. ... And, together, we will begin the next great chapter in the American story, with three words that will ring from coast to coast, from sea to shining sea: Yes, we can.

If Obama's body represents a newly found American urge to transcend the molar differences of race, class, religion, place and gender, this transcendence tends to fall squarely within American patriotism, securely away from the Marxist legacy that Latin American unionists and Badiou pledge allegiance to. As his election proved, Obama's 'audacious hope' for change (2006), and the togetherness he incites, certainly partake in political affect. But to the extent that 'change' is defined on the terms of the same capitalist state that segregates and mis-represents bodies in the first place, it is not yet political. The left has to continue hoping and organizing for fully egalitarian possibilities regardless of who is in power.

When it comes to formulating a politics beyond representation, beyond the hegemonic Möbius strip of knowledge and democracy, we arrive at a triple affirmation. Further than poststructuralism, true politics of difference is communist, overcoming not just molar identity but decisional impossibility and the bitternesses of solipsism. Further than Obama and Martin Luther King, the audacity of hope exists *only* in impeding corporate capital and state mechanisms such as war and patriotism. And finally, further than Badiou, politics is not purely formal. Unearthing courage and surprise outside the platitudes of representation, non-representational theory offers methodologies for politics to both engage and forcefully redirect the molecular forces within the social, so that a real difference can be made.

References

Anderson, B. (2006), 'Becoming and being hopeful: towards a theory of affect', *Environment and Planning D: Society and Space* 24(5), 733-752.

Badiou, A. (1996), 'What is love?'. Trans. J. Clemens. *UMBR(a): A Journal of the Unconscious* 1, 37-53.

—— (2005a), *Being and Event*. Trans. O. Feltham (London: Continuum).

—— (2005b), *Metapolitics*. Trans. J. Barker (London: Verso).

—— (2007), 'Homage to Jacques Derrida', in Douzinas, C. ed. *Adieu Derrida* (Basingstoke: Palgrave Macmillan), 34-46.

Barnett, C. (2004), 'Deconstructing radical democracy: articulation, representation and being-with-others', *Political Geography* 23(5), 503-528.

Bataille, G. (1985), *Visions of Excess: Selected Writings, 1927-1939*. Trans. and ed. Stoekl, A. (Minneapolis: University of Minnesota Press).

Brah, A. and Phoenix, A. (2004), 'Ain't I A Woman? Revisiting intersectionality', *Journal of International Women's Studies* 5(3), 75-86.

Clark, N. (2004), 'Ex-orbitant globality', *Theory, Culture and Society* 22(5), 165-185.

Crane, S. (1896), *The Black Riders and Other Lines* (Boston: Copeland and Day).

Critchley, S. (2007), *Infinitely Demanding: Ethics of Commitment, Politics of Resistance* (London: Verso).

Deleuze, G. and Guattari, F. (1983), *Anti-Oedipus: Capitalism and Schizophrenia*. Trans. R. Hurley, M. Seem and H. Lane (Minneapolis: University of Minnesota Press).

Dewsbury, J-D (2003), 'Witnessing space: "knowledge without contemplation"', *Environment and Planning A* 35(11), 1907-1932.

Dewsbury, J-D (2007), 'Unthinking subjects: Alain Badiou and the event of thought in thinking politics', *Transactions of the Institute of British Geographers* 32(4), 443-459.

Du Bois, W.E.B. (2007), *The Souls of Black Folk* (Oxford: Oxford University Press).

Gates, H. (1989), *Figures In Black: Words, Signs, and the 'Racial' Self* (New York: Oxford University Press).

Gates, H. (1990), 'Murder she wrote', *The Nation*, 2 July, 27-29.

Guattari, F. (1984), *Molecular Revolution: Psychiatry and Politics*. Trans. R Sheed (Harmondsworth: Penguin).

Harrison, P. (2007) '"How shall I say it?" Relating the nonrelational', *Environment and Planning A* 39, 590-608.

—— (2008), 'Corporeal remains: vulnerability, proximity, and living on after the end of the world', *Environment and Planning A* 40(2), 423-445.

Kant, I (1948), *Groundwork of the Metaphysic of Morals*. Trans. H.J. Paton (New York: Harper and Row).

King, Martin Luther, Jr. (1958), *Stride Toward Freedom: The Montgomery Story* (New York: Harper).

Kohl, H. (2007), *She Would Not Be Moved: How We Tell the Story of Rosa Parks and the Montgomery Bus Boycott* (New York: New Press).

Knopp, L. (1992), 'Sexuality and the spatial dynamics of capitalism', *Environment and Planning D: Society and Space* 10(6), 651-669.

Kozol, J. (2005), *The Shame of the Nation: The Restoration of Apartheid Schooling in America* (New York: Random House).

Leitner, H. and Miller, B. (2006), 'Scale and the limitations of ontological debate: a commentary on Marston, Jones and Woodward', *Transactions of the Institute of British Geographers* 32(1), 116-125.

McCormack, D. (2007), 'Molecular affects in human geography', *Environment and Planning A* 39(2), 359-377.

Massumi, B. (1992), *A Guide to Capitalism and Schizophrenia: Deviations From Deleuze and Guattari* (New York: Zone Books).

Mouffe, C. (2005), *The Return of the Political*, 2nd edition (London: Verso).

Nietzsche, F. (1968), *The Will To Power*. Trans. W. Kaufman and R.J. Hollingdale (New York: Vintage).

Oates, J. (1991), *Because it is Bitter, and Because it is My Heart* (New York: Penguin).

Obama, B. (2006), *The Audacity of Hope* (New York: Crown).

Olkowski, D. (1999), *Gilles Deleuze and the Ruin of Representation* (Berkeley: University of California Press).

Popke, J. (2004), 'The face of the other: Zapatismo, responsibility, and the ethics of deconstruction', *Social and Cultural Geography* 5(2), 301-317.

Saldanha, A. (2008), 'The political geography of many bodies' in Cox, K., Low, M. and Robinson, J. (eds), *The Sage Handbook of Political Geography* (London: Sage), 323-334.

—— (2010), 'Skin, affect, aggregation: Guattarian variations on Fanon', *Environment and Planning A*.

Tatum, B. (1999), *Why Are All the Black Kids Sitting Together in the Cafeteria? And Other Conversations About Race* (New York: Basic Books).

Thrift, N. (2008), *Non-representational Theory: Space, Politics, Affect* (London: Routledge).

Valentine, G. (2007), 'Theorizing and researching intersectionality: a challenge for feminist geography', *The Professional Geographer* 59(1), 10-21.

Walby, S. (2007), 'Complexity theory, systems theory, and multiple intersecting social inequalities', *Philosophy of the Social Sciences* 37(4), 449-470.

Whyte, W. (1993), *Street Corner Society: The Social Structure of an Italian Slum* (Chicago: University of Chicago Press).

Wray, M. and Newitz, A. (eds) (1997), *White Trash: Race and Class in America* (New York: Routledge).

Wylie, J. (2005), 'A single day's walking: narrating self and landscape on the South West Coast Path', *Transactions of the Institute of British Geographers* 30(2), 234-247.

Chapter 16

Working with Multiples:
A Non-Representational Approach
to Environmental Issues

Steve Hinchliffe

Introduction: Three Real Gardens

Garden 1: The Women's Garden

A group of women have started to garden on various patches of land that were previously disused and derelict. Allotments that were once covered in bindweed and bramble are now full of herbs and vegetables. A plot of land on the edge of Small Heath Park, in Birmingham, which had been used to store machinery and waste, now contains raised beds in which onions, carrots, rhubarb, potatoes and lettuce are growing. In the corner of the same plot there is a small greenhouse, a poly-tunnel, where seedlings are thinned in the Spring and tomatoes are ripened in late Summer. The women are busy watering, weeding, hoeing, preparing beds, checking soil fertility, staking out plants and chatting. They are growing more organic vegetables than they can eat now. From time to time they cook for open days and festivals. There's a plan, an ambitious plan on the part of the project leader, to turn their attention to forming a cooperative and running a café. Before they started the gardening the women had relatively few contacts outside their families, and few places to go where they could safely enjoy being outdoors. Their health had improved, they say,[1] and they have developed new, or re-discovered old, skills. The gardens were important.

Garden 2: The Urban Garden

Small Heath and Saltley are residential and former industrial areas, just east of Birmingham's city centre. According to the indicators, this is a poor area. In terms of deprivation it's in the bottom 2.5 percent nationally. Health problems and social

---

1   The focus group, interviews and participant observation that informed the arguments in this chapter were carried out in 2003 and 2004 by Matthew Kearnes and Steve Hinchliffe as part of the UK Economic and Social Research Council funded Habitable Cities project at the Open University (Project number R00239283).

exclusion are particularly prominent. It's a majority minority area, with 80 percent of residents described as British Muslim. There's a relative lack of open space and wild space, issues which add to the poor environmental and health statistics and make it an action area in terms of national regeneration budgets and in terms of open space initiatives for the City Council. The gardens are part of a sustainable future for the area – they green the city, they provide healthy activity and good food for residents, they reduce social exclusion. They form part of a more dispersed set of activities that 'garden' the social and the city, drawing on social improvement practices that have a long history and local provenance (from garden cities, and the Bournville-Cadbury 'factory in the garden' in Birmingham, to a more general sense of gardening the state. On the latter see Bauman (1991)).

Garden 3: The Charity Garden

At the funding office of National Lottery Charities Board (latterly the Community Fund and now part of the Big Lottery Fund) the gardens sit on various pieces of paper. There's the application forms on which there are specifications of garden targets (how many gardens will be made over the three year life of the project, how many women will be involved, how visual improvements will be 'dramatic' and how these will be recorded). There's a schedule of payments, marking the release dates of funding for the gardens. There are reports written by the NGO director on the gardening activities, confirming that various targets have been met and justifying others being dropped or re-specified. There are written statements on the purpose of the gardening project. For example, the completed and successful project application form contains hand written entries which speak a language of social, and personal, development. '*The project develops new skills and knowledge which foster personal development, confidence, capacity to affect change, language skills and accessing training*'. Texts like these and the pieces of paper on which they are written help the funding agency to justify their expenditure on gardens to their trustees, to purchasers of lottery tickets and to government ministers. They can also be used to hold the NGO accountable to the funding agency.

The three gardens overlap, indeed, they are in some respects the same garden. But there are also differences. One of the gardens is shaped by hands, seasons, soil fertility and plants. It involves 'timely' events, like planting, thinning, pinching out shoots and harvesting. Another garden is shaped by attempts to provide a greener city, and involves future visions of redevelopment. Yet another is shaped by spreadsheets, returns and numbers that add up over the three year period for which funding has been granted. The garden is done, then, in more than one place and with a variety of things (from trowels to time sheets, from photosynthesis to photographs) and through different times (from events to seasonal cycles and chronological 'development' to the finite time and returns of accounts). These are not three *views* of the same garden, but three ways in which a garden is being made. How can such a tangle or mess (Law 2004a) of things, places and times be understood? My argument in this chapter is

Table 16.1 What's shaping the garden?

Garden	Shaped by ...	Spatial aspects	Temporal aspects...
Women's	Bodies, worms, weather, plants, soils ...	Proximity, working together at the same time and in the same place	Seasons, timely events (like sowing, thinning, harvesting). Rotations, maturing and skills development
Urban renewal	Plans, future visions ...	Co-location for development, then distanciated communication	Long term change, sustainability
Charity	Spreadsheets, numbers, targets, returns ...	Distanciated communication	Annual financial reports, 3 year finite period for funding

that representationalist approaches to this kind of problem offer little purchase on the issues at hand. By representationalist I mean to designate roughly two things. First, I refer to systems of thought which make a firm ontological distinction between the knower and known, the subject world and the object world. Instead, I will prefer a form of connectionism, linking bodies, brains, sensory materials and extended material worlds (Deleuze 1991; Rajchman 2000; Watson 1998), but one where divisions are made and unmade in an ongoing ontological dance (Cussins 1996; Haraway 2008). Second, I refer to a tendency to assume that an event or fashioning of the social is somehow representative of a more fundamental or larger schema. Here, I want to refuse any reading of social worlds which see them as merely local representatives of larger forces (for other refusals which retain a political edge see Law 2004b; Massey 2005). Instead another politics is suggested; a material politics (Law and Mol 2008), where things can and do reverberate beyond their conventional boundaries, but do so in ways that are not inherently structured or necessarily pre-determined.

There are two ways of approaching the problem of more than one garden that seem to stay firmly representationalist as I have defined it above. The first is to treat these gardens as alternative views of the same thing, and look for *the* real garden. In this version of the social, the women's garden, with its plants, events and seasons might be described as the authentic garden. The other two are mere paper representations of gardens, with their plans, chronologies and numbers. Yet, at the NGO and charity offices people will say the *real* work is getting the finances together so that the women's project can be financed, while councillors and urban redevelopment agencies are in the *real* business of combating social exclusion and finding sustainable forms of development. The notion of there being alternative views of the same garden, with one somehow being privileged over the others in terms of its ability to really represent the garden, doesn't capture the ways in which these gardens relate to one another, make one another and so make the garden. Rather than *a* garden that is represented in numerous ways, it may be more interesting to note that the garden is made up of a number of realities. The garden that takes shape will be something of a mixture of the three garden realities, and will depend to a large extent on how those realities relate together.

The second representationalist means of dealing with more than one garden is to treat each 'garden' or set of practices as a single aspect of a coherent garden or social order. That is, while there are different practices involved, an overarching logic or social order ties them together, somehow. Perhaps there's a degree of inevitability about all these practices and the urban garden that results? In this version of the social, the garden adds to and is part of a more or less coherent social process (it could be called neo-liberal governance, romanticism …). So, while there are raised beds to tend and planting days to attend in garden one, public presentations to make and forms to fill in garden two, Ministerial questions to answer and books to balance in garden three, activities that seem worlds apart, all go to make *the* garden. Somehow, all these things come together to make a garden that is to a large extent already pre-scribed and waiting to be realised. Deviations from plans are possible and perhaps inevitable but the garden eventually takes shape as a result of the neat combination of these activity spaces. And yet, as this chapter will seek to demonstrate, it is always more difficult to square all these activities up to a single outcome. These and other activities all garden in different ways. There are different temporalities and spatialities, materials and orderings (see Table 16.1), and they don't necessarily add up neatly. It's obvious in this sense that despite plans and ideas about the future of the garden, the garden doesn't pre-exist all these practices. Less obvious perhaps is the sense that 'social order' doesn't pre-exist the garden either. It too is in the making or at the very least is never finished. There's no pre-established order that makes all the busy activity cohere to a pre-planned scheme. Indeed, as we will see, coherence or even a singular outcome is something that is far from being guaranteed.

How then to talk of non-representational gardens? Instead of an authentic garden, or different aspects on the same garden, there's more than one garden in the making. And more than one garden doing the making, too. To borrow an insight from Mol (Mol 2002), the object/garden doesn't just take shape, it takes shapes. It is multiple. Meanwhile, to emphasise another element of this multiplicity, the object/garden not only takes shapes it is also involved in shaping. It is an actor that is also enacted (this paradoxical phrasing comes from Law and Mol 2007). I return to this in the fourth section, while the second and third sections develop the notion of the non-representational garden and the relational garden through engaging with the trials of the garden multiple. Throughout the chapter, I want to use this field work to emphasise that politics is about more than words and more than representations – it can also be about how things are done in ways that could be otherwise and about struggles between different enactments of reality (Law and Mol 2008). To be sure, words and representations are part of these enactments, but they are not always necessarily the preeminent movers.

More Than Representation, Turning Out Differently and the Multiple Garden

To say that the garden is 'done' in at least three ways suggests that there are at least three groups of doers or actors. The most obvious actors in this telling are the women, the urban development officers and the people at the NGO, each responsible for their respective garden. But the women and the officials are far from being alone, surrounded as they are by plants, computers, strategy documents, ministerial offices, rainfall and seeds. In the more than representational sense that I want to develop here human actors are only part of the story. So what of the other parts and how do they fit together? Are they just context for the people, or is there more to the world of things than background matter?

One shared impetus for the mixed bodies of works that can be given a 'non-representational' label is the following. Thought-as-action is of the world, not prior to it. That is, instead of following the sequence 'thinking then doing', actions enter into thoughts (Ingold 2000; Thrift 2004). Instead of 'thoughts about the world' we have less distinct 'worldly thoughts'. The effects of this seemingly subtle shift of attention are many and often times significant (see other chapters in this volume). There is no attempt to catalogue them here, but one implication is that subjects and objects can become less clearly differentiated (or differentiated by degree rather than in kind), and, more interestingly for this chapter, the traditional fault lines separating human subjects from the rest of the world can become more like folds which bring formerly distant relatives into closer proximity. There are two points to make which follow from the effort to open worldly thoughts.

First, in unsettling subjects and objects, human and non-human being, the world not only becomes more interesting, it also becomes more *and* less malleable. That is to say, compared to some versions of realism, the world is no longer imagined to be one thing that human representations get more or less right. There is no real version of the garden from which the spreadsheets and plans somehow deviate. And, contrary to some versions of social constructivism, that world can no longer be anything that gets represented. There aren't three or more separate gardens, each going on independently. Instead of several aspects on a single reality, or of fragmented plural worlds, we gain a more practical and political task of dealing with more than one world but less than many (Mol and Law 2002, 11). We gain a garden multiple. The question shifts from being 'which is the true garden', or 'which garden should we prefer', to how do these gardens work or not work together to make a garden, and, can we make a better garden? This is an ontological politics, an engagement with the making of realities, their distributions, their effects and the possibility that things could be improved (for a background see Law 2004a; Law and Mol 2008; Mol 1999; Mol 2002).

Second, shifting from thoughts and plans *about* the world to worldly thought suggests a trickier world. As I have already suggested, no longer are matters one thing and one thing only (with the only question being more or less accurate representations), nor are they malleable to thought and therefore simple delivery devices of human or divine intention. Rather, things make matters complex. Once all the bits and pieces, soils and spreadsheets, women and weather, government ministers and gardening tools have been added to the gardening process then, to put it crudely, things can go awry. This is an old point,[2] one that Actor Network Theory in particular has championed in its version of the making of the social (the body of work is large, but see for example Callon 1986; Latour 1988; Law 1991). By adding social and material complexity to accounts of how the social is made, and by demonstrating just how 'badly behaved' people and things could be, Actor Network Theory effectively demonstrated the limited force with which plans and programmes could be put into practice. Far from an overarching discourse running affairs, or there being successful disciplining of elements of a network, there were continual failures, misfits as well as creative appropriations and re-deployments in almost any set up. Things don't re-present plans, just as plans don't make things present in and of themselves – there are translations, circulations and movements. As numerous empirical studies have demonstrated, to get things done often requires a demonstration of things working in practice (on the material and political importance of demonstrations see Barry 2002). From scientific verification in laboratories, to finding ways to beat anthrax on French farms (Latour 1988) to getting a bush pump to work in Zimbabwe (de Laet and Mol 2000), to getting the gardens going in Small Heath (Hinchliffe et al. 2007) – getting things done involves more than convincing others of logic or rationality. It is practical, heterogeneous engineering (Law 1987), a process of continual trial, error and repair (for the importance of the latter in making the social see Graham and Thrift 2007).

One aspect of this misbehaving world is the inevitable way that things 'turn out' differently than planned. Another way of characterising this is to envisage something the Epicureans called 'swerve', the inevitable deviation of matter from a straight path or simple trajectory (Bennett 2001; Latour 2003; Lucretius 1951). To understand swerve, or turning out differently, it's useful not only to recognise the limited force of orders and programmes but also their limited extent (Mol 2002). That is to say, not only did the garden evolve along lines that no one could have fully envisaged at the start of the project, it was already a complex object, made up of more than one garden. The project itself was already multiple. Which meant that numerous orderings were being practised and that they stretched and pulled the project in a number of ways. Redevelopment, social inclusion, primary health care, improving diets, urban environmental improvement, making friends, enjoying being with others, religious and gender sensitivities, making policies, moving money to good causes – these were just some of the concerns that coursed through and so shaped the gardens. And they each spoke to a variety of orderings, from care to entrepreneurship, to bureaucracy (Hinchliffe 2007). I'm following Law's (1994) lead in stressing that these are orderings rather than established

---

2   As Robert Burns' 18th-century poem, 'To a Mouse' attests, most famously in the lines 'The best laid schemes o' Mice an' Men, Gang aft agley'.

orders. They are, in other words, born of the world and not visited on it from some imaginary outside (like rationality, pure consciousness, altruism or other non-worldly alternative). In being of the world they are in process, incomplete and therefore best expressed as verbs rather than nouns. They are attempts at ordering rather than orders that are imposed on the world. The broader point here is that things don't just turn out differently, they are already different from themselves, or better, they are multiple, and they are far from being alone (indeed, they are convivial (Hinchliffe and Whatmore 2006), made together, and not simply made through exclusions, a point that Barnett is right to raise with respect to what he calls generic post-structuralist approaches to identity (Barnett 2005)). Orderings exist in mixes and their effectiveness will depend on how they relate to other orderings. As I have already suggested, orderings don't necessarily cohere, neatly, into a single object. 'This is because various "orderings" of similar objects, topics, fields, do not always reinforce the same simplicities or impose the same silences. Instead they may work – and relate – in different ways' (Mol and Law 2002, 7). The question becomes how do the gardens and orderings interact with one another, and affect one another? How do they relate?

Gardening Relations

In the previous section I highlighted the non-representational elements of a garden that was done in more than one place and through more than one practice. The garden was, in this account, not taking shape as a result of some blueprint, but was actively being shaped by a number of orderings which didn't necessarily cohere or add up to something neat and tidy. The garden had frayed edges and was far from a settled object. In this section I want to describe in more details the relations which made the garden/s. And I want to highlight the different ways in which these gardens co-existed with one another. For the gardens weren't always mutually supportive, or even blissfully indifferent to one another. At key moments, there were tensions to deal with and to manage. This is a key issue for an ontological politics – how do we deal with more than one reality being made at the same time, especially but not only when one of those realities can threaten the others and, in turn, threaten the multiple?

A common enough claim of work in a non-representational register is that things are not simply of or for themselves, rather they are made through their relations. Relations secrete realities (Mol 2002). The more a garden takes shape, the more entangled it becomes with gardeners, who are of course not only human (in the simplest of lists, insects, micro-organisms, wind, plant catalogues, fertilisers and so on, garden the garden). And, of course, the garden makes the gardeners. Nothing stays the same as relations are made. Following this 'to be is to be related' move, another move is to say that not only do things take shape or become in relation, but they take *shapes*. That is, in becoming more real, things also become multiple (again Mol 2002 provides the clearest example). For if we accept that things are done through practices, and that practices are heterogeneous, involving different places, people and many different things, then it follows that things will not be entirely settled matters. They will be pushed and pulled in different ways. An important point to stress is that the reality of the garden is dependent on it being more than one thing. The garden depends on the women, the soil, the plants, the spreadsheets, the plans – it depends on there being accounts, entrepreneurial activities, on care for a green city and for this part of Birmingham. To be a garden, it needs to be a charity garden, a women's garden and an urban garden. It needs to be a multiple. Simply put, the garden would not be if it wasn't all three gardens. Without the women, the garden couldn't exist in this form. Likewise, the garden would not be a garden, or would be a very different garden, without the NGO, the urban developers, the funding agency, the application forms as well as the seed potatoes and garden forks. So the garden *involved* many people and things – application forms, audits, drawings, voluntary labour, tools, plants, weather, gardeners, insects and so on. By using the term 'involve' I want to move away from any sense of a simple inclusion, or seamless mix. Rather, I am more interested here in the diversity of involutions (Deleuze and Guattari 1988, 238) that are possible and that make the social. To be sure the garden is a hybrid, a crossing involving the three gardens I have highlighted, but more than this it is a complex gathering. How elements come together, in concert, in conflict or perhaps indifferently, is a matter that can vary, and can change as the garden multiple grows. I will develop this point by looking in more detail at the various relations of the garden multiple, starting in the offices of the NGO.

The director of the NGO involved in securing the money for the gardening project hints at some of the problems he faces when he talks about the process of raising money for these kinds of project.

> It's a bit of a juggling act … at the end of the day I'm obviously interested in the quality of the work that we actually do and how responsive that is to community need. But obviously I need to recognise that in terms of getting funding I'm going to have to actually satisfy the needs of the funder. So it's a bit of a balancing act really. To some extent it's actually playing the game or at least the funder's game but trying to come up with quantitative outputs that are not actually going to cripple the project. Because I think that's the great danger. There's always a temptation that the more figures you can write in, the more chances at the end of the day that that will provide you with the money.

For the NGO Director, figures and targets can interfere, in negative ways, with the quality of work that the NGO does. They can subtract, according to this, from the work of responding to community need. Too much emphasis can be placed on quantity and not enough on the quality of work. However, all is not lost, and there is a 'game' to be played (one the funding people know too) wherein targets help to secure a project, and help to make it accountable, at the same time as allowing good things to happen. Nevertheless, there's a risk. So, as the NGO director also says:

> The constraints on funding have been that [it] tends to be for visible actions on the ground. The funders expect physical outcomes for their money, whether it's the number of trees planted or whatever. You know there needs to be some sort of quantifiable output with a lot of funding. And I think that has meant that we've generally employed staff who have been very focussed on the delivery of whatever piece of work it actually is. And what we've not been able to do is to actually develop an infrastructure that actually supports those project officers.

So the garden is being pushed and pulled a number of ways. One pull is responding to a community need, the other is responding to something like a need for audit, a need for making certain things present that can easily circulate on forms (like trees, numbers of gardens and so on). Another pull is from the NGO where there is a felt need for an organisation with an infrastructure that can do more than simply deliver time-limited funded projects. Those paid to deliver projects (project officers) find themselves stretched between the auditable project and the NGO, with the result that the organisation can lose a capacity to see a bigger picture. So the requirements for calculable entities and returns seem to detract from the need to respond to a community and to service an organisational infrastructure. Again, though, these are risks, and partly condition the ways in which the director plays the funding game, juggling orderings as business is conducted, forms are filled in and funding agencies are lobbied.

If it's merely a risky game for the Birmingham-based NGO, for others the gardening enacted by funding agencies is a more serious problem, particularly so when the finite temporality of the funding contract interferes negatively with the longer time frames of urban re-development and environmental sustainability, and downplays the recurrent and timely or event-full practices that go into something like a garden. The Black Environmental Network, another NGO who had indirect involvement with the gardens (seeing them as exemplars of environmental action with a social justice component), make the general point in a communication on the problems of funding:

> Funders tend to consider their commitment in the short term and challenge small organisations with providing an exit strategy which would make no further demand on the funder as proof of their being fit to have a grant. This is experienced as an enormous and unfair burden, especially when a high proportion of projects from ethnic groups are ones which enable ethnic groups to get a first foothold on the problems to be addressed (BEN 2000, 31)

What this 'exit strategy' signals is that funding is often a one off, temporary, affair and that by the time money runs out the project should have found a way to either carry on under its own steam, or fold, with participants taking new skills and capabilities to other (economic) activities. Once funding ceases, for the project to be regarded as a success, the funding agencies and the NGOs should be able to walk away, withdrawing the funding relation's life-support structures, without fear of a collapse of the 'patient'. The patient should have been cured, or at least be on the way to a full recovery. A term that is often used in this social sustainability model is 'social capital' which broadly suggests that the funding should have created networks, reciprocities and associations which can form resources for longer term mutual benefit. The language is from social science, and most notably from the work of sociologist Robert Putnam (though with important antecedents in Jane Jacobs and Pierre Bourdieu, see Bourdieu 1986; Putnam 2000), and has moved through policy circles like wildfire (Fine 2002). It speaks of social (people to people) bonds, treating those bonds as assets that can provide a stream of utility to individuals. If sufficient social capital has been generated, the funds invested should have created the opportunity for self-sufficiency and 'sustainability' (which means in this instance, an ability to continue in some form, even a garden-less form, without further outside funds). Any requirement for more charitable funding is regarded as a failure, for insufficient capital has been made.

In considering the post-funding future of the charity garden, four possibilities arise. First, there is a transformation from charity garden to a social enterprise garden which makes enough money to enable the purchase of the transport and tools that are needed to keep things going. For this, the women would need to start selling their harvests. A second possibility is transformation to a hobby garden, requiring more of an investment in terms of time and money from the women. While both of these are achievable, they are difficult given the circumstances of the women. Many of the women have domestic and work commitments that take up much of their time and some are reliant on social welfare payments (which may be jeopardised if the group developed their economic activity in particular ways). Many are relatively elderly and close to retirement age. Few if any of the women can afford to take on being organisers of the group (and therefore filling the role of the NGO), and even the extra work of getting themselves and their tools, water and other essentials to the sites without help from the NGO will be a hurdle. There's no money to buy a vehicle or replace lost and broken tools. Any loans would have to be arranged in accordance with Islamic law, which would mean sharing the risk with an investor. The continual process of repairing the garden and the gardening group as things go wrong, people leave and events need organising, requires investments that are costly in terms of time, energy and materials. A third possibility is that the charitable garden transforms into a set of transferable skills. In this case the women's garden fades into the background and even closes, and the 'gardeners' go on to other things. Here the responsibility for moving on is firmly with the women, as socio-economic agents rather than as gardeners *per se*. The inconvenience of losing the gardens is made up for by the convenience of storying the gardeners as newly equipped social capitalists (something that is more difficult to measure and thereby difficult to deny). This storying is another way in which the 'accountability' of the project can have effects. To be sure, people need to be able to tell stories about projects and the resulting narratives need a beginning and if possible a happy ending. The social capital stories overlap with and may reinforce but may not always sit neatly with the charity garden with its returns and finite times. The fourth 'exit' possibility to mention is that the project leaves little lasting trace on the landscape of east Birmingham and the women's group disbands with no real sense of ongoing achievement.

Perhaps it is the third option, where social capital is made but not necessarily demonstrated, that is the easiest exit strategy? It seems to allow all the funding parties and NGOs to leave the scene with a sense of achievement. It is certainly the option that circulates in policy documents and colours a good deal of the funding landscape, with its' one-off, opportunity, 'kick start' ethos (Hinchliffe et al. 2007). It's a mode of accounting, a storying that fits the temporal horizons of the return, but also fits the narratives of self-improvement and self help. In practice, though, things are less straightforward. To be sure, the charity garden has a framed and limited time-span and it tends to focus on social (in the narrow sense, people to people) relations, disentangling the women and the gardens. Nevertheless, given its own heterogeneous make up as a charity garden (formed you could say of a mix of orderings, including administration, care and entreprise, see Hinchliffe (2007)), it is far from being immune to the demands of the other orderings and gardens. One demand is from some of the urban developers who need ways to story people *and* environments as ongoing, stretching into the future. Another pull is from the women's garden which will soon be overrun with bindweed, knotweed and brambles if the hoeing and planting stops. Meanwhile, many of the women are gardeners now, and taking the women out of the garden might not be so easy. Just as importantly the NGO is called CSV Environment[3] and has a mission statement that incorporates environmental sustainability. The women's social capital is therefore but one of a number of concerns. CSV Environment also prides itself on being located in this part of the city, and caring for the future of east Birmingham (I return to care in section 4). It has other projects to run and good relationships with all manner of people living in the area to maintain. Its mission is ongoing and the gardens form an important part of its success as an organisation. They are a symbol of that success. Finally the project has a wonderfully evocative name, 'Concrete to Coriander' which again speaks to urban environmental improvement, and evokes another storying of the project which speaks to a social that is different from the social capital story. It speaks to urban greening and to more permanent changes to the urban fabric. In short, to let go of the project and the gardens is difficult.

So the charity garden is pulled by its own heterogeneity, by the women's garden, the urban garden and the NGO towards a number of exits. In turn, of course, the women's garden and the urban garden are being pulled and shaped. Throughout the project the women's garden is focused on meeting targets and

---

3   CSV stands for 'Community Service Volunteers'. CSV Environment operates as one of seven major national programmes of the parent organisation, CSV.

on being entrepreneurial. The women and the project officer are continually looking for opportunities to promote the garden, to find other sources of funding and to organise themselves into a cooperative that can outlast the current grant. Urban developers meanwhile are continuously reminded of the reality of social enterprise models, of finding ways to garden the city that enable participants to be contributors to the urban economy. Such matters are rarely resolved and it becomes a practical question as to how matters go on.

Multiple Relations and Their Limits

As I have already noted, it's a convenience and somewhat a commonplace to say that 'to be' is to 'be related'. Certainly, there would be no garden if it wasn't a multiple. Relational geography has become one way of doing a kind of non-representational theory. The refusal of relational styles of working to cede an origin to the making of the social, particularly perhaps an origin that invokes a conscious human subject, has chimed rather well with non-foundational aspects of non-representational theory. Hybrids, mixtures, crossings, filiations, all speak to this relational turn. But what the multiple can suggest is more than a claim that things are made in and of relations, it is also a claim that the relations themselves are complex and differentiated, and form part of an ontological politics – a making of more than one reality which can then be contested, re-made or made subject to a normative sense that things could be done in different and possibly better ways.

The three gardens are not then simply related, they are related or can be related in differing ways. There are two ways that I want to develop this notion. The first is to look at the kinds of relations that exist between gardens and orderings. The second is to start to foreground the multiplicity of relations and to explore what work a notion of 'partial relations' can do in developing a sense of the non-representational 'thingness' of matter.

'Political Ecologies'

Notwithstanding this short telling of a garden it is relatively clear that there is more than one relation at work. So, for example, while the charity garden seems to run affairs, governing the life and shape of the women's garden, this may be more an effect of a certain kind of telling (this chapter has tended to start from the offices of the NGO rather than from the council office or from the women), and is in any case hardly having things its own way. The gardening that takes place at the funding agency is crucial, but it can't be allowed to dominate the scene totally. If it was the only garden, or even dominated the scene, nothing much would happen outside the office. The charity garden is not dominant in part because the other gardens have effects but also because the charity garden is not one thing either. So while there is a certain amount of hierarchy in the way I have told the gardens here, this hierarchy is but one kind of relation. Sometimes, for example, the administrative need for accountability and neat returns runs up against longer term care, and two orderings seem to be in competition, with neither one able to exist if the other triumphs. At other times, the administrative ordering of the charity can only work if a game is played whereby the quantitative administration of a project is juggled with an entrepreneurial, opportunist ordering (see the quotes from the director of the NGO cited earlier). Too much administration and the NGO and its projects will cease to function. Too little and all the entrepreneurship in the world will not secure projects. The orderings seem to depend on one another (Law 1994). Without various forms of caring the NGO would not be in this part of Birmingham and the women would not turn up every week to tend the garden. But that is not to say that caring alone is sufficient, or even fundamental, or even located within the people at the NGO and in the women. Caring, like the other orderings, is distributed across the gardens. It is in the forms, in the offices, in the polytunnel, and it depends on other modes of ordering (for more on the complexities of care see Mol 2008). Meanwhile, beyond these overlapping and mutually constitutive relations, there's also indifference. Some plants tend to grow regardless of the spreadsheets. The women turn up to garden with little interest in government policy on green spaces.

Relations include hierarchy, competition, co-dependency, distribution and indifference. There will be other possibilities, but the point is that the garden multiple is not simply a hybrid of things, people, numbers and orderings, it is a complex of matters that are arranged through a dynamic set of different and differentiating relations. It is an ecology of action (Hinchliffe et al. 2007). So, there's no sense here of one garden or one ordering being dominant or there being certain forms of resistance to this single order. The romance (a defeatist romance, see (Gibson-Graham 1996)) of there being a single order in charge of the social to which people may more or less successfully resist is too easy and too limiting and returns us to a representational version of the social (where reality represents an instance of a bigger picture – for alternatives see (Kwa 2002; Law 2004b; Mol and Law 2002)). Gardening the social is a more complex, multiple affair, requiring social scientists to jettison their representationalist mind set for a project that is both more difficult (in the sense that there is more than one thing and one kind of relation at work) and more promising (in that the future is not foretold).

The Multiple as Actant

The second point builds on this sense of there being more than one kind of relation in order to start to open up a space for the garden to do things. So far I have, to summarise the argument, moved from saying that things are made in relation, to saying things are made in relation*s*. They have complex histories and geographies. They are not only shaped but pulled into shapes. But what of the garden multiple? What can it do? It isn't just made (though it is clearly made), it also makes and contributes to its mattering. In the traditions of Spinoza and Deleuze, we shouldn't be asking only what a body is, but what it can do. And following Latour (Latour

2004), we can take this to ask not what is an environment, but what can an environment do?

In Mol's detailed ethnography of the body multiple we learn that the body did not precede 'the various coordination strategies' which succeed 'in reassembling multiple versions of reality' (Mol and Law 2002, 10). Clinical consultations, pathology results, surgery and palliative care are related together in ways that can produce a useful treatment for a patient. There are many kinds of relation within and between these practices, but the end result of finding a way forward for the patient overrides any sense of there being a single body, or a truth which grounds all decisions. There isn't a single body that can be accessed and represented and which can provide *the* right answer. There are instead lots of words, numbers, blood clots, feelings, care practices and so on that can be brought together, more or less coherently, to provide a body for which treatment and care can be arranged. In the garden case there are lots of knowledges and practices in play too, and finding a way forward similarly involves, from time to time, coordinating matters in ways that can work effectively. This kind of ontological politics has much to recommend it in terms of moving away from a representationalist/epistemological politics. Instead of trying to find the garden, the task is to experiment with ways of intervening in the continual reassembling of the social. Here though I want to extend the remit of this politics in such a way that can underscore a more-than-human element. To be clear, anthropocentrism and speciesism is something that those working in science and technology studies and Actor Network Theory have long been concerned to excise from social theory, so my job is not a critique, it is an attempt to find openings and consider further possibilities for onto-politics.

In order to talk not only about how garden multiples are done, but also about what they can do, we need to emphasise the partiality of relations. Instead of a garden that is made from its relations we need to be able to understand the garden as a thing that is far from being exhausted by those relations. Moreover, this needs to be done without drawing us back to a pre-existing garden or social order. The philosopher Graham Harman argues this point very nicely. He suggests that if we reduce 'the being of objects to their relational situation' (Harman 2002, 229), if we privilege the network of negotiations between things (something he accuses Latour of doing), then we end up with a single world, where nothing really happens (a point made convincingly by Lee and Brown some time ago now: see Lee and Brown 1994). His argument is that things are never simply of their current set of relations, there is always something in reserve, something that withdraws. 'If an entity always holds something in reserve beyond any of its relations, then it must exist somewhere else. And since this surplus or reserve is what it is, quite apart from whatever might stumble into it, it is actual rather than potential' (Harman 2002, 230). Now, as Harman is at pains to show, this is not to suggest that there is some core or essence, some old fashioned unchanging Aristotelian substance to things. Rather there is a philosophical challenge to 're-establish the firewalls that protect every entity from its neighbours ... without relapsing into a conservative version of substances' (Harman 2002, 256-257).

Harman's philosophical project to get at things in themselves is one way to remind ourselves of the partiality of relations. Taking a slightly different tack, one that is less interested in withdrawal and more concerned with the creativity involved in a multiplicity, Law and Mol also suggest that 'the reality of an entity is never exhausted' (Law and Mol 2007, 14). In their case inexhaustibility is not due to a holding back in reserve but to the shear number and complexity of relations which mean that any thing is already more than one thing. Things are always involved in more than one relation, making it inevitable that they will go awry, or do things that are not quite expected of them. Callon puts it like this, any element of a relation 'at the very same time as it is helping to structure and frame the interaction of which it more or less forms the substance, is simultaneously a potential conduit for overflow' (Callon 1998, 254). The multiple is already indeterminate by virtue of its being made up of things that are making other things, and that are therefore elsewhere too.

Despite the differences in these approaches, the point is that things, like a garden or a body, never lend themselves fully to a relation, a network or to an association. They are as Strathern has put it, partially connected (Strathern 1991). Things are made by more than one practice, a multiplicity that produces a potential for new configurations. Meanwhile, rather than this being a simple ontology of force, the multiplicity of things throws up another dimension, one where things are not only shaped but also may have to cope with more than one shape. The point is that an element in a relation will be the subject of and subject to many other relations. The garden multiple that emerges is not then exhausted by the charity, the women and the city, it adds to these and many others besides. It is therefore an 'actor-enacted'. As Law and Mol explain: 'an actor-enacted acts in collaboration with others to such an extent that it is not always clear who is doing what ... But this is not to say that an actor-enacted is determined by its surroundings. It has its own stubbornness and specificities: it is full of surprises' (Law and Mol 2007, 14-15).

The empirical point is that the gardens won't necessarily go away quietly once the funding stops. To be sure, they won't last for ever, they are not self-sufficient and they do not act alone. But they are not easy for even the most committed social capitalist to move to the background, for they are not passive. Meanwhile, not being exhausted by relations is an important reminder that, first of all, things don't simply re-present their situation; they are not simply part of a bigger picture. Rather they engage with and alter their multiple situations, in ways that are complex and may be likened to ecologies. Second, a crucial aspect of this ecology is the requirement to live with a becoming otherwise. It is to recognise that things are indeterminate, can always relate to something else, and, no matter how domestic or tame a garden, a sheep or even a human body seems to be, they can do something else.

Conclusions

The garden multiple is a complex matter, one that takes a lot of work to keep going. That work takes in many different worlds and orderings, and cannot be reduced to one thing or another. To garden and to garden well is to be attendant to many matters, to their differences and their changing complexions. This is the kind of work that goes on in an NGO that manages to keep things going.

So how do non-representational geographies help us to understand this kind of work? My suggestion in this chapter has been that non-representational geographies require a sensitivity to the multiple makings of things, where things don't represent their situation or relational set up, but are instead involved in many set ups. It is through these multiple involvements that they can be and do become otherwise. Let me add to this with three closing comments which emphasise the main points. First of all, things, like gardens, are more than one thing. This may not matter too much, as the women happily garden without much need to bother with the ways in which the funding agency structures its support. But often this multiplicity can generate effects. Some will be difficult, even contradictory, others will be matters to juggle and to get on with. The details are always to be worked out. Second, things are not simply produced from this complex present, they have other complex time-spaces (histories and geographies) and this multiplicity leads to both stubbornness and to creativity. They are actors-enacted as Law and Mol term them (Law and Mol 2007). Third, this stubbornness and creativity suggests that making things happen, through policies, activities or otherwise, means that any successful shaping of the world, any assembling of the social, needs to respond to other shapings that are going on with different orderings, motives and trajectories. In short, learning to affect the world involves learning to be affected by others. And this ontological politics is therefore a complex process which needs to articulate or join together many kinds of doing.

Acknowledgements

Thanks to Nick Bingham, John Law, Annemarie Mol, Ingunn Moser and Kristin Asdal for leading me into multiplicity, and for conversations on the topic of multiplicity and on the case study

References

Barnett, C. (2005), 'Ways of Relating: hospitality and the acknowledgement of otherness', *Progress in Human Geography* 29, 5-21.
Barry, A. (2002), *Political Machines: Governing a Technological Society* (London: Athlone).
Bauman, Z. (1991), *Modernity and Ambivalence* (Cambridge: Polity Press).

BEN (2000), 'Funding issues affecting ethnic communities'. Black Environment Network http://www.ben-network.org.uk/resources/downlds.html (last accessed 22 July 2004).

Bennett, J. (2001), *The Enchantment of Modern Life: Attachments, Crossings and Ethics* (Princeton and Oxford: Princeton University Press).

Bourdieu, P. (1986), 'The Forms of capital', in *Handbook of Theory and Research for the Sociology of Education*, edited by J. Richardson (New York: Greenwood Press), 241-258.

Callon, M. (1986), 'Some elements of a sociology of translation: domestication of the scallops and the fishermen of St Brieuc Bay', in *Power, Action and Belief*, edited by J. Law (London: Routledge and Kegan Paul), 196-223.

—— (1998), 'An essay on framing and overflowing: economic externalities revisited by sociology', in *The Laws of the Markets*, edited by M. Callon (Oxford and Keele: Blackwell and Sociological Review), 244-269.

Cussins, C. (1996), 'Ontological choreography: Agency through objectification in infertility clinics', *Social Studies of Science* 26, 575-610.

de Laet, M. and A. Mol. (2000), 'The Zimbabwe Bush Pump: Mechanics of a fluid technology', *Social Studies of Science* 30, 225-263.

Deleuze, G. (1991), *Bergsonism*. Trans. H. Tomlinson (New York: Zone Books).

Deleuze, G. and F. Guattari. (1988), *A Thousand Plateaus: Capitalism and Schizophrenia*. Trans. B. Massumi (London: Athlone).

Fine, B. (2002), 'They f**k you up those social capitalists', *Antipode* 34, 796-799.

Gibson-Graham, J.-K. (1996), *The End of Capitalism (As We Knew It): A Feminist Critique of Political Economy* (Oxford: Blackwell).

Graham, S. and N. Thrift. (2007), 'Out of order: Understanding repair and maintenance', *Theory, Culture and Society* 24, 1-25.

Haraway, D. (2008), *When Species Meet* (Minneapolis: University of Minnesota Press).

Harman, G. (2002), *Tool-being: Heidegger and the Metaphysics of Objects* (Chicago and La Salle, Illinois: Open Court).

Hinchliffe, S. (2007), *Geographies of Nature: Societies, Environments, Ecologies* (London: Sage).

Hinchliffe, S, M. Kearnes, M. Degen, and S. Whatmore (2007), 'Ecologies and economies of action: sustainability, calculations, and other things', *Environment and Planning A* 39, 260-282.

Hinchliffe, S. and S. Whatmore (2006), 'Living cities: towards a politics of conviviality', *Science as Culture* 15, 123-138.

Ingold, T. (2000), *The Perception of the Environment: Essays in Livelihood, Dwelling and Skill* (London: Routledge).

Kwa, C. (2002), 'Romantic and baroque conceptions of complex wholes in the sciences', in *Complexities: Social Studies of Knowledge Practices*, edited by J. Law and A. Mol (Durham and London: Duke University Press).

Latour, B. (1988), *The Pasteurisation of France*. Trans. A. Sheridan and J. Law (Cambridge, MA: Harvard University Press).

—— (2003), 'What if we talked politics a little', *Contemporary Political Theory* 2, 143-164.

—— (2004), *Politics of Nature: How to Bring the Sciences into Democracy*. Trans. C. Porter (Cambridge, MA: Harvard University Press).

Law, J. (1987), 'Technology and heterogeneous engineering: the Case of Portuguese Expansion', in *The Social Construction of Technological Systems: New Directions in the Sociology and History of Technology*, edited by W. Bijker, T. Hughes, and T. Pinch (Cambridge, MA: MIT Press), 111-134.

—— (ed.) (1991), *A Sociology of Monsters* (London: Routledge).

—— (1994), *Organizing Modernity* (Oxford: Blackwell).

—— (2004a), *After Method: Mess in Social Science Research* (London: Routledge).

—— (2004b), 'And if the global were small and noncoherent? Method, complexity, and the baroque', *Environment and Planning D: Society and Space* 22, 13-26.

Law, J. and A. Mol (2007), 'The Actor-Enacted: Cumbrian sheep in 2001', in *Material Agency: Towards a Non-anthropcentric Approach*, edited by Knappett, C. and Malafouris, L. (New York: Springer), 57-78.

—— (2008), 'Globalisation in practice: on the politics of boiling pigswill', *Geoforum* 39, 133-143.

Lee, N. and S. Brown (1994), 'Otherness and the actor network: the undiscovered continent', *American Behavioural Scientist* 37, 772-790.

Lucretius (1951), *On the Nature of the Universe*. Trans. R.E. Latham (London: Penguin).

Massey, D. (2005), *For Space* (London: Routledge).

Mol, A. (1999), 'Ontological politics, a word and some questions', in *Actor Network Theory and After*, edited by J. Law and J. Hassard (Oxford and Keele: Blackwell/ Sociological Review), 74-89.

—— (2002), *The Body Multiple: Ontology in Medical Practice* (Durham, NC.: Duke University Press).

—— (2008), *The Logic of Care: Health and the Problem of Patient Choice* (London: Routledge).

Mol, A. and J. Law (2002), 'Complexities: an introduction', in *Complexities: Social Studies of Knowledge Practices*, edited by J. Law and A. Mol (Durham and London: Duke University Press), 1-22.

Putnam, R. (2000), *Bowling Alone: The Collapse and Revival of American Community* (New York: Simon & Schuster).

Rajchman, J. (2000), *The Deleuze Connections* (Cambridge, MA: MIT Press).

Strathern, M. (1991), *Partial Connections* (Savage, MD: Rowman and Littlefield).

Thrift, N. (2004), 'Movement-space: the changing domain of thinking resulting from the development of new kinds of spatial awareness', *Economy and Society* 33, 582-604.

Watson, S. (1998), 'The new Bergsonism', *Radical Philosophy* 92, 1-23.

Chapter 17

Events, Spontaneity and Abrupt Conditions

Keith Woodward

Introduction

Politics and events circle each other like twin stars, generating an immense theoretico-gravitational field that hangs constellations of questions above patchworks of speculative, critical, and pragmatic landscapes: what is the political event? What, if any, epistemological or ontological access might we gain to it? What is the status of our actions – or, for example, those of gathering storms, circling vultures, or crashing markets – relative to its emergence? Like a night sky scattered with black holes – each of which, though hidden from sight, pulls us in different directions – this field of problems orients many of the key approaches in current social theory. In the past several decades, it has transformed our understandings of the performance of identity (Butler 1990), the practice of self-exploitation (Foucault 1977; Deleuze and Guattari 1983), and the struggle to envision and realize social change (Pignarre and Stengers 2005; Shukaitis et al. 2007). Beyond its more anthropocentric impacts, it has also transformed key debates in a range of topics from political ecology (Vayda and Walters 1999) to forestry management and lawn care (Robbins 1998, 2007), and from particle physics (Anderson 2007) to cosmology (Trotta 2007).

Within Human Geography, non-representational theorists are likewise gradually revisiting the dyadic relation between politics and events (Dewsbury et al. 2002). Focusing upon its everyday connections to embodiment, affect and perception, they have explored the more- and less-than-rational components of witnessing (Dewsbury 2003), listening (Anderson 2006b), walking (Wylie 2002, 2005), touching (Paterson 2007), and even 'being still' (Harrison 2009). These interventions often focus upon their indebtedness to the work of a relatively small group of recent figures in continental philosophy – from Levinas, Foucault and Derrida to Deleuze, Butler and Badiou – who currently cast long shadows across our theoretical imaginaries. At the same time, the non-representational has long been a source of speculation within Western and non-Western thought (Derrida 1981, Deleuze 1990b), predating even the dialogues of Plato (2000, 220-223). Seen in this light, querying the non-representational may be less a matter of *whether* certain thinkers take account of it than *how* it works its way into their descriptions of ontology, epistemology, representation, and so on.

Mindful of such subterranean currents, this chapter examines the roles played by spontaneity and the event in non-representational understandings of politics. Recalling the centrality of spontaneism for Kant's critical philosophy (1987, 1996, 1997), it questions the recent fetish for the 'new' in political theorizing and the accompanying belief that such novelty constitutes a fundamental shift from earlier liberal and radical philosophies (Amin and Thrift 2005). By making these challenges, I am not espousing Kantianism. However, I am arguing that it continues to exert a disproportionately large influence upon current understandings of thought, ontology and political practice; so much so that it is not uncommon to find thinkers assuming Kantian positions despite pronouncements to the contrary. This is a recurrent difficulty, as Brassier (2007a, 2007b) notes, even in recent challenges to Kantianism launched by non-representational thinkers of the so-called 'speculative turn' (Meillassoux 2008, Bryant et al. forthcoming). At the same time – and paradoxically – radical thought *also* drives these theoretical trajectories, particularly where they connect with the many non-representational theorists of the 1970s and 1980s, whose political ontologies grew out of engagements with the radical philosophies of Marxism and anarchism. These are a key influence, for example, in the work of Foucault (1977, 163-164) and Deleuze (1988b, 70), whose understandings of affect draw upon Marx's descriptions of force – 'labour power' – in capitalist social relations (Marx 1976; see also Hardt and Negri 2004, Lazzarato 1996, Read 2003, Woodward and Lea 2010).

There is an ever-present temptation to succumb to the anxiety of influence and thus reduce these earlier contributions to outdated political strawmen against which we juxtapose (and oversell the novelty of) our contemporary moment. This chapter holds that we are better positioned were we to avoid such a reflex and, instead, complicate and enrich our portraits of their crisscrossing continua. Doing so, I suggest, highlights intersections, divergences and influences that enable us to glimpse emerging political alternatives. Turning to the tradition that braids together thought, embodiment, spontaneity and politics, I begin by tracing two influential dimensions of 'non-representational Kantianism'. The first of these surfaces in the tendency for conceptions of political 'liberation' and 'play' to reproduce representationalist versions of spontaneism. The second arises through Deleuze's reversal of Kantian 'synthetic' (or, representational) understanding; a manoeuvre that opens the door for theorizing material syntheses of non-*presentational* forces. This intervention transforms the importance of events for the politics of collective action. Finally, drawing upon Marxist philosophy's suspicions of spontaneism, I turn to the politics of popular protest, along the way hijacking an undeveloped, one-off concept from Althusser – the 'abrupt condition' – as a tool for considering *counter-* or *anti*-representational politics.

Thought, Spontaneity and Embodiment

The most frequently revisited of modernity's theoretical landscapes has been Kant's 'land of pure understanding', a realm founded upon rationality and surrounded by a chaotic and unknowable external world. Human cognition, he explains, 'is an island, and is enclosed by nature itself within unchangeable bounds. It is the land of truth ..., and is surrounded by a vast and stormy ocean, where illusion properly resides and many fog banks and much fast-melting ice feign new-found lands' (Kant 1996, 303). While classical versions of ontology and epistemology had populated the external world with objects whose *essences* we could only struggle vainly to comprehend, Kant's 'Copernican revolution' installed thought as the organizer of an otherwise inconceivable 'outside' (Kant 1996, 21-22). By this he did not mean that thought projects order *onto* the world. Rather, he contended that, aided by a set of intuitive, transcendental rules that are knowable prior to experience (or, *a priori*), understanding *spontaneously* moulds erratic sensory data into recognizable object-images. According to Kant, for example, *your* knowledge of *this book* as an object is the product of a synthesizing, rational act that unites numerous fragmented, differing and otherwise disorderly perceptions (one thinks of a cubist painting) and re-presents them to consciousness as a singular, spatially situated thing. As Deleuze puts it: 'Representation means the synthesis of that which is presented' (Deleuze 1984, 14). In such representation, 'space' is neither an empirical perception nor a manifestation of an 'outside'. Rather, it is an *a priori* component of rational intuition and a basic condition for human understanding, organizing the clamour of sensations according to the logic of 'the pure form of all outer appearances' (Kant 1996, 88): extensivity. With Kant, space becomes a universal and necessary condition for making sense of 'my' perceptions, and thus 'my' understanding does not represent the world as it 'really is' (that 'stormy ocean' *in itself*), but is a synthesis of the specific ways it appears *for-me*. As Grant notes, 'Kant defines nature as "the sum total of all things", before adding the familiar Copernican caveat, "insofar as they can be objects of our senses"' (2006, 7; see also Toscano 2006, 25-27).

At the same time, Kant's reversal does not imply that we can simply *choose* how we will represent the world to ourselves. Representation is a spontaneous, transcendental act of *thinking*, independent of individual agency. (Indeed, rather than being the *source* of representation, the subject is one of its *products*). Nevertheless, while thought organizes objects, the appearances from which these arise are *immanent* to a subject's embodied situatedness. Objects of thought do not sit somewhere randomly 'out there', waiting to be perceived, but are constructed by cognition out of specific presentations of the world *for-me*. To every such 'synthetic' intervention, representation attaches an 'I' – as in 'I think' – that is simultaneously an object of understanding *and* a reference point for the organization of sensory data. This intersection of transcendental reason and spontaneous cognition generates several conditions for the work of representation. First, it initiates a *situating function* that places the thinker simultaneously *in the world* and *at a mediated distance from it*. Second, it furnishes the world – *for-me* – with objects fished from a sea of otherwise indeterminate, chaotic sensations. And third, it is *singularizing, spontaneous* and *spontaneously subjectivizing:* 'only because I can comprise the manifold of the presentations in one consciousness, do I call them one and all *my* presentations. For otherwise I would have a self as many-coloured and varied as I have presentations that I am conscious of' (Kant 1996, 179). In turn, these conditions have consequences for reconsidering thought, subjectivity and politics in light of recent non-representational theories.

On the one hand, *the politics of spontaneity arise out of representation.* In Kantian philosophy, representation is a *spontaneous,* object-assembling act *for-me* that simultaneously engenders 'me' *as* a subject. The resultant '*I thinking*' of this process does not conjure its 'I' from thin air, but *liberates* it *as a position* from the undifferentiated chaos of appearances, making it express its situation and impose a perspective upon that which had heretofore been non-representable. That is, Kant modernizes the political correlation between liberation and self-expression. Thus, where contemporary theories invoke spontaneous action to exemplify non-representational politics, it is only with difficulty that they disentangle it from the legacy of Kantian liberalism. More often, such approaches reintroduce representation through the back door. The following section will discuss one such case of 'non-representational Kantianism'.

On the other hand, *Kantian critical philosophy establishes several of the crucial points upon which non-representational theories continue to pivot.* By placing representation at a remove from the external world – mediated by situated, bodily perceptions and synthesized by *a priori* categories – the uncertain character of Kant's 'Nature *in-itself*' tends to occupy the place of the non-representation. Concerning the metaphysics of Nature, Kant suggests that such an outside is not representable by virtue of its being a mess of unfolding, infinitely complex forces (Kant 2002, 225). Here, the *in-itself* is an *absolutely disaggregated something else* seated at the limit of the reasons capacity to *object*ify the world: the producer of *objects* of thought, synthetic representation can make little sense of non-objects such as force. This account resonates strongly with the experience of the 'many-coloured and varied' self that Kant speculated was the product of the pure force of sensation if unmediated by representation. Though he attributes very little philosophical value to the *less than representational* character of the raw data feeding appearances, Kant clearly acknowledges a connection between thought, experience and non-representation. Crucially, theorizing synthetic cognition simultaneously sets the formal conditions for making claims about the non-representational: what it *requires* to situate something outside or beyond representation. Looking to just such an outside, in the third section, I discuss a second site of non-representational Kantianism: Deleuze's non-*presentational* synthesis.

Non-Representational Kantianism, No. 1: The Politics of Spontaneism

Deleuze and Guattari (1983) enlist several wildly dynamic styles when formulating their break with representationalism. Coupled with Deleuze's acknowledged admiration of Bergson (1988a), it should come as little surprise that this stylistic energy inspired some, re-thinking the political in his wake, to import spontaneism to Deleuzian ethics of liberation and resistance. In spite of his having rejected such an interpretation (Boutang 1996; Valentin 2006, 187), it nevertheless remained common even amongst his students. Though most prevalent during the years immediately following the publication of *Anti-Oedipus*, it continues to resurface today. Consider, for example, Braidotti's recent description of the politics of 'becoming-minoritarian':

> This specific sensibility combines a strong historical memory with consciousness and the desire for resistance. [It favours the] production of joyful acts of transformation. The spontaneous and rather anarchical aspects of this practice combine with a profound form of aesceticism that is today the determination to focus on and build upon micro-instances of activism, avoiding over-arching generalizations (Braidotti 2008, 24).

Much of this passage gestures to strong possible linkages between Deleuze and contemporary political activism. At the same time, it reproduces recent tendencies to treat spontaneity as an unqualified virtue, to link it exclusively to 'joyful' acts and yet to simultaneously and paradoxically grant it the capacity to avoid 'over-arching generalizations'. Such accounts are often less clear about how to distinguish joyful versions of spontaneism from the widespread *conservative,* minoritarian acts, such as the racist violence that sprung up across the US in the xenophobic wake of 11 September 2001.

From time to time, geographic linkages between spontaneity to resistance generate similar ambiguities. Merging non-representational and performativity theories, Thrift (1997, 125) discusses dance as a resource for individual, embodied liberation. Such an activity, he contends, provides escape routes from representation by disrupting the body's disciplined and routinized participation in power networks. Nash (2000) has challenged this formulation, noting that dance is already a thoroughly representational art. This critique is cogent, both *in its own right* and in terms of the disciplinary frameworks against which Thrift frames resistance. However, given the picture of representation drawn in this chapter, Nash (and, occasionally, Thrift) mistakes *discursive-disciplinary* representation – sign systems – for acts of cognitive re-presentation – the proper object of non-representational critique (though these, admittedly, need not be *entirely* distinct). While his is not the only voice of non-representational politics – they have grown both in diversity and complexity during the past decade (Anderson 2007; Lim 2007; Saldanha 2006; Popke 2009; Woodward and Lea 2010) – Thrift's approach echoes the popular desire to clothe the ethics of liberation in the neutral colours of Deleuze's 'pure' event and its corresponding 'will "of indifference"' (Deleuze 1990b, 100). But such accounts stumble where they make the subject – political or otherwise – the agent of non-representational materialities. By accessing regimes of expression set outside more familiar *modes* of representation, Thift suggests that experimental playfulness spontaneously gives rise to potential alternative worlds: 'dance is clearly using the body to conjure up the "virtual", "as-if" worlds by configuring alternative ways of being through play' (Thrift 1997, 147). Conjuring such worlds 'is not self-evidently about discourses of power and control. It is about play ... Play is, in other words, a process of performative experiment' (Thrift 1997, 145).

The 'play' that guides Thrift's intervention does not concern choreographed performance: his dancer is not a mimic, but a *conjurer of worlds,* an *agent of the event.* However, by anchoring it to the *dancer's* body, liberation becomes a self-objectivizing act that, to the contrary, *reproduces* Kantian representation and subjectification (that is, liberation *for-me*). Play, or 'free play', is a key aesthetic concept for Kant that describes a complex of non/relations between different modes of cognition: 'since bringing a presentation of the imagination to concepts is the same as *expounding* it, aesthetic ideas may be called *unexpoundable* presentations of the imagination (in its free play)' (Kant 1987, 217). Resonant not only with Thrift's usage, but with contemporary non-representational theories more broadly (Deleuze and Guattari 1994), the unexpoundability of play describes aesthetic ideas that cannot 'move from an intuition to a concept' (Longuenesse 1998, 98). However, Kant continues, 'both kinds of ideas, rational as well as aesthetic, must have their principles, and both must have them in reason: the principles of rational ideas must be objective principles of reason's employment, those of aesthetic ideas subjective ones' (Kant 1987, 217). In the effort to found a non-representational politics upon play, in other words, 'liberation' becomes a representational mediator of *as-if* worlds, spontaneously synthesizing *this* dancing body *as* a liberated object *for-me*. Here, the very *idea* of non-representational politics commits itself to spontaneous acts of representation. May suggests that:

> Any political intervention, if it is to be successful, must discard all projects ... that work through representation; instead, such intervention must embark upon a program of subverting the pretensions to completeness of the representational structure. It must open up other possibilities for action that cannot be reduced to representation and its negativity, but that instead allow for non-representational realizations of the libidinous. Since all political action involves representation, this will necessarily be a paradoxical project (May 1994, 83).

Further, recognizing the presence of representation in politics is, in many ways, helpful for developing approaches to political struggle and critical work. For instance, once you have decided to squat a politician's unoccupied second home, changing the locks on the doors is – recalling Nash's critique – an important *discursive*-representational act that helps legally protect you from being immediately, forcibly evicted (Squatters 2009). By the same token, within *critical/ cognitive* representational perspectives, it takes but a few steps through Hegel or Heidegger, for example, to move from the *for-me* of Kantian representationalism to the ethics of the Other found in Levinas. For it is in part through the subjectivizing dimension of representational cognition and its later variants (e.g. otherness, alterity) that the oppressive character of exploitation, privatization and discrimination presents itself. This holds even in the work of the master of free-play – Derrida – who, in early 1968, accepted an invitation to deliver a talk in New York 'only when I was assured that I could bear witness here, now, to my agreement, and to a certain point my solidarity with those, in this country, who were fighting against what was then their country's official policy in certain parts of the world, notably in Vietnam' (Derrida 1982, 113). The 'play' at work between representation and non-representation in Derridian deconstruction is concerned not with the invention of other worlds, but with searching out, exposing and resisting the modes through which *this* world constructs regimes of *impossibility* – the impossibility of equality, of peace, of resistance – as a strategy for control and exploitation.

Non-representational Kantianism, No. 2: The Non-Presentational Synthesis

Nietzsche echoes Kant where he insists upon the existence of an inherently non-representable external world (Clark 1990, 81). In so doing, he *bends* the relationship between the 'I' and cognition, making spontaneity a characteristic of non-subjective, extra-rational forces, thus generating a 'wilder version of Kant's transcendental faculties' (Braver 2007, 146). Amongst other things, this sets the groundwork for identifying false consciousness as a symptom of organizational understanding, a problem against which Kant saw little practical recourse outside of appeals to moderation (Ross 2000, 77). Nietzsche explains that:

> life itself has been defined as an increasingly efficient inner adaptation of external circumstances ... But this is to misunderstand the essence of life, its *will to power*, we overlook the prime importance that the spontaneous, aggressive, expansive, re-interpreting, re-directing and formative forces have, which 'adaptation' follows only when they have had their effect; in the organism itself, the dominant role of these highest functionaries, in whom life will is active and manifests itself, is denied (Nietzsche 2007, 52).

The target here is the reductive dimension of Kantian syntheses. By contrast, 'the world consists not of things, but of quanta of force entangled in something on the order of "universal power struggle" ... with each centre of force having or being a tendency to extend its influence and incorporate others' (Clark 1990, 206). Replacing Kant's world of objects with fields of *imperceptible* forces, Nietzsche shifts the question of appearance from the *for-me* to the *in-itself* (Henry 1993, 5), and radically re-conceptualizes the relation between synthetic cognition and the world. Ontologically speaking, this unshackles spontaneity from its enslavement to a transcendental consciousness, returning it to the complicated, self-differentiating forces of materiality.

Following this, Deleuze closes his book on Kant with a discussion of *The Critique of Judgement* wherein he describes 'sensible nature' in terms of its 'pure relations of forces, conflicts of tendencies which weave a web of madness like childish vanity' (Deleuze 1984, 75). The excitedly Nietzschean form of the analysis arises from an inversion that subjects 'Kantian thought to *the heterogenesis of its unthought* in order to carry it off toward an outside' (Alliez 2004, 97). Deleuze accomplishes this, according to Alliez, by way of 'reversals' of worldly time, intensive time, the Law and the sublime (Alliez 2004, 98), each a different aspect of the synthesis Kant develops across his three *Critiques*. Further, Nietzsche's ontology of forces enables Deleuze to transform synthetic cognition into a disjunctive synthesis that dissolves the inside-outside distinction. Adapting Leibniz's (1989, 112) description of the sea to illustrate this dissolution, Deleuze explains:

> The idea of the sea ... is a system of liaisons or differential relations between particulars and singularities corresponding to the degrees of variation among these relations – the totality of the system being incarnated in the real movement of the waves. To learn to swim is to conjugate the distinctive points of our bodies with the singular points of the objective Idea in order to form a problematic field. This conjugation determines for us a threshold of consciousness at which our real acts are adjusted to our perceptions of the real relations, thereby providing a solution to the problem. Moreover, problematic Ideas are precisely the ultimate elements of nature and the subliminal objects of little perceptions. As a result, 'learning' always takes place in and through the unconscious, thereby establishing the bond of a profound complicity between nature and mind (Deleuze 1994, 165).

Here, the work of sensation-affect moves all the way 'in' to establish the problems that are neither *a priori* nor *a posteriori* for the mode of intuition that is neither passive nor omniscient, but instead *engaged, participatory*. Synthesis, no longer a spontaneous rendering, becomes genetic, gradual, something that *learns* and *readjusts* to the world it encounters – not as its object – but as dynamic series of movement and force. As a result, the manifolds it maps remain *incomplete*, becoming more complicated, more elliptical, fleeing what it was, becoming what it is not, subject to constant revision by virtue of continuous changes at the liaising thresholds (points) of the communicating unconsciousness-consciousness-body-waves. Nor, finally, does noise get resolved through synthetic positionality. Making the *unconscious* a relay for the image, Deleuzian intuition dissolves the *for-me* in the infinite, localized singularities of worldly forces, fostering what might be called a '*non-presentational synthesis*' of the 'many-coloured selves' – incompossible selves, becomings-other – that had marked the conceptual limit for Kant's synthetic representation. This in no way constitutes a *some place else* conjured by a body-self. It gestures instead toward a situatedness that is infinitely more complex than the paradigm of Kantian space-time, but rather than anchoring this to the body, it is more helpful to explore the non-presentational synthesis in relation to the event.

The View From the Event

When approached as a resource for political liberation, spontaneity falls back upon cognitive modes that, though useful in other capacities, are not capable of carrying politics *beyond* representation. Rather, the political is articulated within the representational position, the manifold viewpoint, even when such a perspective is subaltern (as Spivak's fieldwork indicates, see: Wainwright 2008, 230) or minoritarian (Deleuze and Guattari 1987), or when it corresponds to acts of absolute refusal (preferring 'not to'; see Hardt and Negri 2000, 203-4), or remaining still (Harrison 2009). The political cannot assume a non-relational point. To expect otherwise – by, for example, positing 'pure' politics, metapolitics or political totalities housed in non-representational outsides – evacuates the power of situated engagement and critique, only to replace it with tacky sorcery promising 'new' politicalities, new ways of being, and other such subjunctive snake oil, almost inevitably conjured from the uneven positions of old-school, vanguardist knowledge-power: a seductive spectacle of non-representable possibilities preying upon the desire for accessible, realizable novelty.

At the same time, neither can the entirety of materiality fall under the purview of the political. The continua of forces that underscores Deleuze's passage on 'swimming' – for example, those emerging from interaction between waves and legs – are *non*-political. Destabilizing the borders that rope in Kantian subjective consciousness, such forces are not available (i.e., 'present') to the understanding in a way that might be subsequently synthesized as representational objects *for-me*. Cognition can neither make manifest sense of the specificity of such forces, nor describe a politics appropriate to them (in this regard, even Foucault and Deleuze must turn to the *routines* established in forceful repetitions, or 'diagrams'; see Deleuze 1988b). There are doubtless a thousand ways to represent the politicalities that might place *this* body in *these* waves crashing upon the shores of *this* beach resort. We might, for example, identify systems of very real social and economic exploitation that grant this individual to access such spaces while simultaneously barring others, including their local populations (Kingsbury 2005). The politics of such situations

– particularly with regard to organizing, decision-making and resistance – arise by taking a representational stand, as it were, *identifying* distributions of bodies *as* people, *objectifying* dynamic processes *as* capital, tourism and exploitation, or *recognizing* routine practices *as* disciplinary. At the same time there are specificities in such situations – interactions of varying aggregates and trajectories of force, their activities and passivities – that are not presented to sensation, perception or consciousness in ways that can be made politically meaningful. Turning towards these, representation finds only non-objects, non-spaces, non-presentations in such a hazy mess of forces. Still, although the non-presentational synthesis does not enable such referential politics, the forceful inter-relations and orientations articulate problems that offer two important challenges to representation's routine solutions. First, beyond the logics of the spontaneous, synthesizing subject, it offers a collectivizing or aggregative view from the event. Consequentially, it offers a rereading of political events in terms of their 'abrupt conditions' (to be detailed in the next section).

The problem expressed in the swimming example gestures to what Deleuze (1990b) calls the 'pure' event. By 'pure', he is suggesting that the event is something complete unto itself, the sole resource for its own manifestation, and in need of no external or supplemental causes, designers, managers, drivers or transcendental organizers. Because Deleuze is not a phenomenologist (Alliez 2004, 89; Williams 2008; Anderson and Wylie 2009), his account does not require affirmation by a human observer (that is, it is non-presentational). Nowhere is this more evident than in the centrality he grants to the *immanence* of the event. Kant reduced immanence to a characterization of the connection between experiential understanding and representational synthesis (Kant 1996, 371). Pure immanence, by contrast, bears immediately upon the self-organizing processes – for example, the 'auto-affective' forces driving everything from the division of cells to the sudden surge of a crowd – that compose a continuum irreducible to the actions of its specific members nor to the representation of a thought-object (Deleuze 2001). As Badiou explains, the Deleuzian 'event is what composes a life somewhat as a musical composition is organized by its theme. "Variety" must here be understood as "variation", as variation on a theme. The event is not what happens to a life, but what is in what happens, or what happens in what happens, such that it can only have a single Event' (Badiou 2007, 39; see also Badiou 2009, 383-4).

To illustrate this, consider the distinction between the *points* of view *of* figures on a battlefield and the view *from* the event of a battle *itself*. On the one hand, representation constructs a unique perspective for the soldier (amongst other perspectives) by selectively assembling the surrounding happenings. The battle, on the other hand:

> is not an example of an event among others, but rather the Event in its essence, it is no doubt because it is actualized in diverse manners at once, and because each participant may grasp it at a different level of actualization within its variable

present ... [while the event of] the battle hovers over its own field, being neutral in relation to all of its temporal actualizations, neutral and impassive in relation to the victor and the vanquished, the coward and the brave (Deleuze 1990b, 100).

Although Deleuze deploys a discourse of 'hovering', he is not suggesting that the event presents a view from above or a perspective from a singular point. Rather, it is a distribution of problems that the world 'solves' (i.e., 'actualizes') by way of localized determinations of materiality: the perspective of the running soldier, the *specific* redistributions of soil by an exploding shells, and so on. What gets re-presented to consciousness is not the event (the problem), but the ways it gets worked out in matter (the solution). For the swimmer, such problems populate the continuum of forceful bodies (limbs, waves), the solutions to which were the *actual* forceful bodily responses to its situation. Further, the swimmer's viewpoint – and those of the soldiers on a battlefield – is also one such material solution. These should not be confused with the view from the event, which concerns an emerging situation's 'making-available' a multiplicity of viewpoints (potentially) to the bodies (humans, bits of matter, animality, languages, and so on) that compose it. While the perspective of each participant (like the many points of articulation in bodies of water and swimmers) is presented with localized fragments of unfolding actuality (if not *for-me*, then certainly *for the perspective, for the figure*), the view from the event is the aggregate view, the worldly perspective, of divergent perspectives – viewpoints of subjects, yes, but also blades of grass, screaming bullets, phantom limbs and countless, unthinkable others – a manifold of changing perspectives, forces and relata. The event is not simply non-representational, it is non-*presentational:* while the event forges complex and specific singularities – a material synthesis – what gets presented to a thinker and subjected to *re*-presentation are only its fragments and material traces. Thus, Deleuze is being playful when he echoes Spinoza's contention that we do not know what a body can do (Deleuze 1988c, 17-18; 1990a). Only the event can 'know' what a body can do: the complex field of problems that charts of bodies' immediate orientations toward their own situatedness – the non-presentational synthesis – inevitably leaves us slightly in the dark, playing catch-up. However, it is also by virtue of its being imbued with an edge of unknowability that the event becomes fruitful for exploring the nature of political intervention.

Abrupt Conditions

Speculation about the event has long been key to theorizing political practice. After all, what is radical social change if not an material transformation that confounds oppressive political representations? The vague logics of such disruptions fuelled numerous discussions within the First International – particularly those between the anarchist Bakunin and the communist Marx, who pitted the ethics of spontaneity against the philosophy of conditions in an increasingly ugly disagreement over revolutionary ontology (Robertson 2003). This dichotomy continues to resurface in political actions and academic debates today, and underscoring, for example, the recent *Antipode* exchange between Amin and Thrift (2005) and Smith (2005). An individualist, Bakunin links liberation to spontaneous, 'revolutionary hardihood, and that troublesome and *savage energy* characteristic of the grandest geniuses, ever called to destroy old tottering worlds and lay the foundation of new', all grounded in the '*power of thought*' (Bakunin 1970, 31, *my emphasis*). Marx, on the other hand, speaks from onto-political perspective that rejects individualism and spontaneism in favour of collective politicalities conditioned by modes of 'work' whose force relations form social-natural aggregates:

> Really free working, e.g. composing, is at the same time precisely the most damned seriousness, the most intense exertion. The work of material production can achieve this character only (1) when its social character is posited, (2) when it is of scientific and at the same time general character, not merely human exertion as a specifically *harnessed* natural, spontaneous form, but as an activity relating to all the *forces of nature* (Marx 1973, 611-612, *my emphasis*; see also Woodward and Lea 2010).

Still, although they trace different roots to political mobilization, both agree that resistance interrupts representable conditions of oppression and that radical political change is tied to relations of force. It is not the subject-thinker, but a 'savage energy' – the force proper to thought – that Bakunin claims is a resource for liberation against existing social conditions. Meanwhile, Marx's political potentiality swims in the very sea of forces from which Kantian syntheses 'harness' objects of representational understanding. These forceful dimensions, I will suggest, offer a regime for aligning politics with non-presentational syntheses. Before exploring this, however, it is helpful to look more closely at the Marxist critique of spontaneism.

Suspicious of spontaneous mass movements, Lenin links them to past failures in revolutionary uprisings and a key source for the reproduction of bourgeois representationalism (i.e., Kantianism). These criticisms centre on the tendency – *in the wake of a radical event* – for social practices to revert to familiar, pre-revolutionary routines (capitalist social relations) and for thought to fall back upon reflexive, common notions (liberalist ideology). Couching the revolutionary event in explicitly representational terms, Lenin explains:

> There is much talk of spontaneity. But the spontaneous development of the working-class movement leads to its subordination to bourgeois ideology ... for the spontaneous working-class movement is trade-unionism ... and trade-unionism means the ideological enslavement of the workers to the bourgeoisie (Lenin 1988, 107).

Whereas Kant anchors spontaneous representation to individualized acts of cognition, Lenin makes the 'outside' a condition for representational understanding, subjecting it to the gravity of history and stratified social relations.

This has important implications for Althusser, who, while sharing many of Lenin's targets, launches his arrows from a considerably different historical position. A witness to the slow implosion of the Soviet experiment beneath the weight of statist violence, exploitation and power-mongering, Althusser considered himself to be levelling the 'first *left-wing* critique of Stalinism' (Althusser, qtd in Elliott 1990, xviii). Of this era, he notes, 'The denunciation of "the cult of personality" [i.e., Stalinism], the *abrupt conditions* and the forms in which it took place, have had profound repercussions, not only on the political domain, but on the ideological domain as well' (Althusser 1969, 10, *my emphasis*). But what are (these) 'abrupt conditions'? Mindful of Lenin's attacks upon post-revolutionary regression, Althusser is spotlighting the left's rediscovery of forms of liberalism that were waiting in the wings in the aftermath of Stalinism, particularly the rapid softening of Soviet policy toward the West and the sudden rise of Marxist humanism. But more broadly, by imbuing materiality with abrupt conditions he also formalizes the ontological shift in Marxist-Leninism that detaches spontaneity from individualism and realigns it with broad social forces. Doing so avails them of descriptions of the entire working political field – radical, liberal or otherwise – and, further, makes them a potentially productive resource for resistance and political theorizing. However, while it seems fairly clear how abrupt conditions can help illuminate the liberalist imaginaries that haunt the margins of Marxist politics, it is less certain how they might be a useful tool for the transformation of really existing, situated modes of material oppression and exploitation.

The answer, I suggest, rests in seeking out the connection between abrupt conditions and non-presentational syntheses. Recall that radical representations of class relations negotiate conflicting and aggregating material forces and affects (Marx 1976; Read 2003; Woodward and Lea 2009). That is, they address not only the exploitative tendencies that stabilize into familiar, bourgeois conditions, but also a simultaneous proliferation of counter-forces: black markets, collective organizing, localized practices of mutual aid and other such underground currents of the everyday. The latter of these offer insight into abruptness as a characteristic of radical resistance, but in a way that differentiates them from individualistic spontaneism.

Consider, for example, the political solidarities that emerged on Seattle's city streets during World Trade Organization's Third Ministerial Conference in late 1999 (Wainwright et al. 2000). Prior to the disruption of the WTO, North American activist groups, having broadly different – even conflicting – politics, tended to organize and mobilize independently of each other. Their success in Seattle, on the other hand, depended significantly upon the sudden collectivization of several groups, particularly union members, environmental activists and anarchists. But we pass over political complexity when we characterize the suddenness of its mobilization as spontaneous, just as we reduce it to cartoonishness where we assume that spontaneity makes it somehow 'more' authentic (Shepard and Hayduk 2002, 5). Rather, the Seattle protests were animated by situated, pragmatic acts – a radical bricolage – drawing heavily upon the recent transformations in collective action that had been a growing feature of Zapatista solidarity campaigns. While these strategies were to varying degrees unfamiliar or well-rehearsed to the activists employing them, they were sufficiently new and overwhelming to those policing the protests to have had the *appearance* of spontaneity. That is, in mobilizing dynamic forces, activists engendered abrupt conditions that were not, recalling the language of Kant, representable as an object of knowledge *for-me*. Such fragmented forces gather unpredictably, gesturing toward an uncertain aggregate, a non-objectifiable representation, *a movement* that gains velocity by simultaneously writing and concealing it own logics (Derrida 1978, 240). In 1999, these non-presentable aggregations were the events called the 'Battles in Seattle' (Wainwright et al. 2000). The abrupt conditions that drew them into an event were the many and varied happenings, witnessings, doings and confusions that its situatedness enabled.

Seattle introduced not something *new*, but something *different* to the enactment and governance of the political event. Particularly instructive in this regard are the dynamic, radical potentials of emerging political events through the *becoming disruptive* of abrupt conditions. By mobilizing in ways that resisted or deflected the localized policing representational reflex to organize, objectify and reduce complex forces to common-sensical relations (i.e., union members, environmental activists, anarchists, citizens and so on), these activists fostered a general (and fruitful) confusion that was subsequently *enlisted* as part of the political force of the protest. That is, the politics of the event arose out of forceful work that was, for a time at least, *anti-* or *counter*-representational. In Seattle, this had many progressive and positive results; however, it cannot be assumed that such politicalities are available to be coopted or controlled by a subject – or a gang of subjects, such as a vanguard. Though it inevitably emerges from any number of localized decisions, guesses, accidents and treacheries, when it works, the political event is a lucky mangle of tendencies, a grand aggregate, that re-contextualizes and re-situates its components. Like the countless perspectives on Deleuze's battlefield we bear witness only to the pieces, traces, edges and aftermaths of such mobilizations. So, too, was the case for Seattle riot police, for whom the protests, while not entirely localizable, were yet capable of leaving entire policing systems momentarily flabbergasted. Abrupt conditions *intrude* upon relatively stratified lines and trajectories of policing and protest by refusing to submit to representation, and while perhaps having *some* sense, grow from an active refusal of the discipline of sense-making.

Conclusion

In their discussion of the contemporary alter-globalization movement, Pignarre and Stengers (2005) suggest that radical politics has, to a significant degree, 'inherited' much of its innovation from Seattle's abrupt conditions. In an important variation on Deleuze and Guarttari's famous claim that 'Philosophy's sole aim is to become worthy of the event' (Deleuze and Guattari 1994, 160), they acknowledge a political inheritance from Seattle that enjoins us to 'become a child of the event' (Pignarre and Stengers 2005, 11, *my translation*). This rests the relation to the political event not simply upon our being theoretically or conceptually *inclined* toward its immanent character, but *also* upon our being historically, materially, *genealogically* situated *with* it. Resonant with discussions of utopianism that have marked one entryway for more recent accounts of politics and non-representational theory (Anderson 2006a), they explain that the event in Seattle 'created a "now" in response to the question of a certain "acting *as if*" that is unique to children when they fantasize and create' (Pignarre and Stengers 2005, 11, *my translation*). Accordingly, events such as Seattle do not create new worlds (the *as-if*), but forge interventions – responses, counter-representations – in the supposed impossibilities built into the world we inhabit.

The intersections of radical philosophy and non-representational theory should thus be repeatedly revisited to find more engaged, participatory approaches to changing and adjusting politicalities. In closing, I would like to suggest that these concerns cross paths at critical points. (1) *Non-representational theories require accounts of representation.* Although non-representational theory gestures to that which sits outside or beyond representation, a specific, localized account of representation is required to reflect the interventions described by an equally specific non-representational theory. Neglecting to do so mortgages one account of non-representation upon another thinker's (often Kant's) account of representation. The problem, of course, is that nuances in non-representational thinking have implications that re-situate and reframe representationalism, in turn asking for a new articulation of each. Nowhere is this more true than in the non/ relation between non-representation and the political. (2) *Because it politicizes representation (and counter-representation) radical theory is a key subject for non-representational politics.* Non-representational theorists are uniquely situated to explore nuances that abrupt conditions introduce to political events. Although the non-representational cannot be a domain of the political – nor can it have a politics exclusive to it – political events such as Seattle draw important, impermanent *counter-* or *anti-*representational trajectories illuminated by the light shed from the edges of the non-representable.

Acknowledgements

Ben Anderson and Paul Harrison provided helpful commentary and boundless patience as versions of this chapter traveled through Germany, France, Spain, Portugal, Canada, Cuba and several points in the US and the UK before finally finding its way up to Durham. In the meantime, it cluttered the vacation tabletops and ruined the holiday appetites of John Paul Jones III, Cindi Katz, Eric Lott, Sallie Marston, Jennifer McCormack, Sarah Moore, and Paul Robbins. At Exeter, it would every now and then pull John Wylie down the hall to deliver a sympathetic reminder of its tenuous and tardy existence. Like the politics of social change, the metaphysics of writing are founded in friendship. My many thanks to all of you for yours.

References

Alliez, E. (2004), *The Signature of the World*. Trans. E.R. Albert and A. Toscano (New York: Continuum).

Althusser, L. (1969), *For Marx*. Trans. B. Brewster (London: Verso).

—— (1990), *Philosophy and the Spontaneous Philosophy of the Scientists*. Trans. B Brewster, et al. (New York: Verso).

Amin, A. and Thrift, N. (2005), 'What's left? Just the future', *Antipode* 37, 220-238.

Anderson, B. (2006a), 'Transcending without transcendence: utopianism and an ethos of hope', *Antipode* 38(4), 691-710.

—— (2006b), 'Becoming and being hopeful: Towards a theory of affect', *Environment and Planning D: Society and Space* 24, 733-752.

—— (2007), 'Hope for nanotechnology: Anticipatory knowledge and the governance of affect', *Area* 19, 156-65.

Anderson, B. and Wylie, J. (2009), 'On geography and materiality', *Environment and Planning A* 41, 318-35.

Badiou, A. (2007), 'The event in Deleuze', Trans. J. Roffe. *Parrhesia* 2, 37-44.

—— (2009), *Logics of Worlds*. Trans. A. Toscano (New York: Continuum).

Bakunin, M. (1970), *God and the State* (Minneola, NY: Dover).

Boutang, P.-A. (dir.) (1996), *L'Abécédaire de Gilles Deleuze*, G. Deleuze and C. Parnet (perf.) (Paris: Éditions Montparnasse).

Braidotti, R. (2008), 'The politics of radical immanence: May 1968 as an event', *New Formations* 68(1), 19-33.

Brassier, R. (2007a), 'The enigma of realism: On Quentin Meillassoux's *After Finitude*', *Collapse* II, 15-54.

—— (2007b), *Nihil Unbound: Enlightenment and Extinction* (New York: Palgrave Macmillan).

Braver, L (2007), *A Thing in the World: A History of Continental Anti-Realism* (Evanston, IL: Northwestern University Press).

Bryant, L., Srnicek, N, and Harman, G. (forthcoming), *The Speculative Turn: Continental Materialism and Realism* (Melbourne: re.press).

Butler, J. (1990), *Gender Trouble* (New York: Routledge).

Clark, M. (1990), *Nietzsche on Truth and Philosophy* (New York: Cambridge University Press).

Deleuze, G. (1984), *Kant's Critical Philosophy*. Trans. H. Tomlinson (Minneapolis: University of Minnesota Press).

—— (1988a), *Bergsonism*. Trans. H. Tomlinson and B. Habberjam (New York: Zone Books).

—— (1988b), *Foucault*. Trans. S. Hand (Minneapolis: University of Minnesota Press).

—— (1988c), *Spinoza: Practical Philosophy*. Trans. R. Hurley (San Francisco: City Lights Books).

—— (1990a), *Expressionism in Philosophy: Spinoza*. Trans. M. Joughin (New York: Zone Books).

—— (1990b), *The Logic of Sense*. Trans. M. Lester (New York: Columbia University Press).

—— (1994), *Difference and Repetition*. Trans. P. Patton (New York: Columbia University Press).

—— (2001), *Pure Immanence: Essays on a Life*. Trans. A. Boyman (New York: Zone Books).

Deleuze, G. and Guattari, F. (1983), *Anti-Oedipus: Capitalism and Schizophrenia*, Volume 1. Trans. R. Hurley, et al. (Minneapolis: Minnesota University Press).

—— (1987), *A Thousand Plateaus: Capitalism and Schizophrenia*, Volume 2. Trans. B. Massumi (Minneapolis: Minnesota University Press).

—— (1994), *What is Philosophy?* Trans. H. Tomlinson and G. Burchell (New York: Columbia University Press).

Derrida, J. (1978), *Writing and Difference*. Trans. A. Bass (Chicago: University of Chicago Press).

—— (1981), *Dissemination*. Trans. B. Johnson (Chicago: University of Chicago Press).

—— (1982), *Margins of Philosophy*. Trans. A. Bass (Chicago: University of Chicago Press).

Dewsbury, J-D (2003), 'Witnessing space: "Knowledge without contemplation"', *Environment and Planning A* 35, 1907-1932.

Dewsbury, J-D, Harrison, P., Rose, M. and Wylie, J. (2002), 'Enacting geographies', *Geoforum* 33, 437-440.

Elliott, G. (1990), 'Introduction', in Althusser, L. *Philosophy and the Spontaneous Philosophy of the Scientists*. Trans. B. Brewster, et al. (New York: Verso), vii-xx.

Foucault, M. (1977), *Discipline and Punish: The Birth of the Prison. Trans. A Sheridan* (New York: Random House).

Grant, I.H. (2006), *Philosophies of Nature After Schelling* (New York: Continuum).

Hardt, M. and Negri, A. (2000), *Empire* (Cambridge, MA: Harvard University Press).

Hardt, M. and Negri, A. (2004), *Multitude* (New York: Penguin).

Harrison, P. (2009), 'Remaining still', *M/C Journal* 12(1) <http://journal.media-culture.org.au/index.php/mcjournal/article/viewArticle/135>.

Henry, M. (1993), *The Genealogy of Psychoanalysis*. Trans. D. Brick (Stanford: Stanford University Press).

Kant, I. (1987), *The Critique of Judgment*. Trans. W.S. Pluhar (Indianapolis, IN: Hackett).

Kant, I. (1996), *The Critique of Pure Reason*. Trans. W.S. Pluhar (Indianapolis, IN: Hackett).

Kant, I. (1997), *The Critique of Practical Reason*. Trans. M. Gregor (New York: Cambridge University Press).

Kant, I. (2002), *Theoretical Philosophy after 1781*. Trans. G. Hatfield, et al. (New York: Cambridge University Press).

Kingsbury, P. (2005), 'Jamaican tourism and the politics of enjoyment', *Geoforum* 36, 113-132.

Lazzarato, M. (1996), 'Immaterial labor', in Virno, P. and Hardt, M. (eds), *Radical Thought in Italy: A Potential Politics* (Minneapolis: University of Minnesota Press), 133-147.

Leibniz, G.W. (1989), *Philosophical Essays*. Trans. R. Ariew and D. Garber (Indianapolis, IN: Hackett).

Lenin, V.I. (1988), *What Is To Be Done?* Trans. J. Fineberg and G. Hanna (New York: Penguin).

Lim, J. (2007), 'Queer critique and the politics of affect', in Browne, K., Lim, J. and Brown, G. (eds), *Geographies of Sexualities: Theory, Practices and Politics* (Aldershot: Ashgate), 53-68.

Longuenesse, B. (1998), *Kant and the Capacity to Judge*. Trans. C.T. Wolfe (Princeton: Princeton University Press).

Marx, K. (1973), *Grundrisse*. Trans. B. Fowkes (New York: Penguin).

Marx, K. (1976), *Capital, vol. 1,* Trans. B. Fowkes (New York: Penguin).

May, T. (1994), *The Political Philosophy of Poststructuralist Anarchism* (University Park, PN: The Pennsylvania State University Press).

Meillassoux, Q. (2008), *After Finitude*. Trans. R. Brassier (New York: Continuum).

Nash, C. (2000), 'Performativity in practice: Some recent work in cultural geography', *Progress in Human Geography* 24, 653-664.

Nietzsche, F. (2007), *On the Genealogy of Morality*. Trans. C. Diethe (New York: Cambridge University Press).

Paterson, M. (2007), *The Senses of Touch: Haptics, Affects and Technologies* (Oxford: Berg).

Pignarre, P. and Stengers, I. (2005), *La Sorcellerie Capitaliste: Pratiques de Désenvoûtement* (Paris: Éditions la Découverte).

Plato (2000), *The Republic*. Trans. T. Griffith (Cambridge: Cambridge University Press).

Popke, J. (2009), 'Geography and ethics: Non-representational encounters, collective responsibility and economic difference', *Progress in Human Geography* 33(1), 81-90.

Read, J. (2003), *The Micro-politics of Capital: Marx and the Prehistory of the Present* (Albany: State University of New York Press).

Robbins, P. (1998), 'Paper Forests: Imagining and deploying exogenous ecologies in arid India', *Geoforum* 29, 69-86.

Robbins, P. (2007), *Lawn People: How Grasses, Weeds, and Chemicals Make Us Who We Are* (Philadelphia: Temple University Press).

Robertson, A. (2003), 'The philosophical roots of the Marx-Bakunin conflict' *What Next?* 27. <http://www.whatnextjournal.co.uk/Pages/Back/Wnext27/Marxbak.html>.

Ross, A. (2000), 'Introduction to Monique David-Ménard on Kant and madness', *Hypatia* 15(4), 77-81.

Saldanha, A. (2006), 'Reontologizing race: The machinic geography of phenotype', *Environment and Planning D: Society and Space* 24, 9-24.

Shepard, B.H., and Hayduk, R. (2002), 'Urban protest and community building in the era of globalization', in Shepard, B.H., and Hayduk, R. (eds), *From ACT UP to the WTO: Urban Protest and Community Building in the Era of Globalization* (New York: Verso), 1-9.

Shukaitis, S., Graeber, D. and Biddle, E. (eds) (2007), *Constituent Imagination: Militant Investigations Collective Theorization* (Oakland: AK Press).

Smith, N. (2005), 'Neo-critical geography, or, the flat pluralist world of business class', *Antipode* 37, 887-899.

'Squatters Occupy MP's "Main Home"' (2009), *BBC News*, 29 June 2009. <http://news.bbc.co.uk/2/hi/uk_news/8124763.stm>.

Thrift, N. (1997), 'The still point: Resistance, expressive embodiment and dance', in Pile, S. and Keith, M. (eds), *Geographies of Resistance* (New York: Routledge), 124-151.

Toscano, A. (2006), *The Theatre of Production* (New York: Palgrave Macmillan).

Trotta, R. (2007), 'Dark matter: Probing the arche-fossil', *Collapse* II, 83-170.

Valentin, J. (2006), 'Gilles Deleuze's political posture', in Boundas, C. (ed.) *Deleuze and Philosophy* (Edinburgh: Edinburgh University), 185-201.

Vayda, A.P., and Walters, B.B. (1999), 'Against political ecology', *Human Ecology* 27(1), 167-179.

Wainwright, J. (2008), *Decolonizing Development: Colonial Power and the Maya* (Malden, MA: Blackwell).

Wainwright, J. Prudham, S. and Glassman, J. (2000), 'The battles in Seattle: Microgeographies of resistance and the challenge of building alternative futures', *Environment and Planning D: Society and Space* 18(1), 5-13.

Williams, J. (2008), 'Gilles Deleuze and Michel Henry: Critical contrasts in the deduction of life as transcendental', *Sophia* 47(3), 265-79.

Woodward, K. and Lea, J. (2010), 'Geographies of affect', in Smith, S., Pain, R., Marston, S. and Jones, J.P. III (eds), *The SAGE Handbook of Social Geographies* (London: Sage), 154-175.

Wylie, J. (2002), 'An essay on ascending Glastonbury Tor', *Geoforum* 33, 441-454.

—— (2005), 'A single day's walking: Narrating self and landscape on the South West coast path', *Transactions of the Institute of British Geographers* 30, 234-247.

Chapter 18

Envisioning the Future: Ontology, Time and the Politics of Non-Representation

Mitch Rose

Introduction

Without question the advent of non-representational theory has introduced into the discipline an array of topics and concerns that, before now, were not seriously considered academic much less geography. Questions about movement (Dubow 2001; Wylie 2002; Dubow 2004; Wylie 2005; Spinney 2006), the emotional (Jones 2005; Patterson 2005; Saville 2008), the sensible and the material (Alexander 2008; Wilford 2008; Anderson and Wylie 2009) and the affective (McCormack 2003; Anderson 2005; Anderson and Harrison 2006) have not only broadened Human Geography's purview, but have re-worked some of the discipline's cherished concepts such as spatiality (as ecology) (see Bingham 1996; Thrift 1999; Whatmore 2002; Simpson 2008), practice (as sense, affect and becoming) (see Dewsbury 2000; McCormack 2003; McCormack 2005; Simpson forthcoming) and method (as witnessing and listening) (see Dewsbury 2003; Harrison 2007). Yet, despite this expansion of perspective and this dispersion of concerns, non-representational theory has nonetheless remained powerfully gravitated around the question of the political. No doubt non-representational theory has always been political. It has endeavoured, from its earliest articulations, to open Human Geography's conception of what the political means – i.e., what counts as a proper political question – by supplementing the epistemological logic of traditional forms of social/political theory (Thrift 1983; Thrift 1997; Hinchliffe 2000; Thrift 2000; McCormack 2003; Thrift 2004; Amin and Thrift 2005; Anderson 2006; Anderson 2007; Hinchliffe 2008; Jones 2008). Given that non-representational theory has been around for over 10 years and that it promises so much more than new modalities of political practice, one wonders why the question of the political has remained so central (indeed restrictively so) to non-representational theory's concerns (see Wylie this volume). I, for one, would like to see non-representational theory do more than justify its political potential or at least get on with the business of performing its political commitment in the creative ways it espouses. But I can also see why it has not yet managed to escape from a set of somewhat narrow political debates. Despite the unique and creative ways it has rearticulated how we think about ethical/political practice, there is something deeply unsatisfactory about its notion of political commitment. Indeed, while I myself have great sympathy with the ontological positions of non-representational theory, and little sympathy with the social/theoretical reductionisms of structuralism in its various lefty guises, I cannot help but find myself aligned *not* with non-representational theory's critics but with the sentiments their critiques often suggest. In other words, while I do not agree with what the critical banners say, I am sympathetic to why they were raised. Non-representational theory gives us a powerful sense of how much can be gained by looking beyond representational politics. But it does not acknowledge what has been lost, that is, what is necessarily forsaken once representational politics are abandoned.

The problem, it seems to me, is about the future. Politics, after all, is about change. It is about desiring to bring about change – about attempting to create or produce a difference and, significantly, a difference for the better. While these terms (change, attempt, create, better) carry within them a density of problematic conflations, they nonetheless signal a commitment to something that is lacking in non-representational theory, that is, a vision of the future. Indeed, in non-representational theory it is precisely this long-standing tradition of envisioning the future (and equating that vision with the good, the ethical and the just) that stultifies politics. By pre-establishing a representational economy that not only situates subjects but the desires and capacities *of* subjects, traditional notions of political progress condemn the good politics to a limited tactical universe with miserly criteria for measuring success. In contrast, non-representational theory offers a far more dispersed and dynamic ethical/political field – a field defined *not* by the agency of subjects but by a set of pre-subjective (and a-subjective) affectivities whose various configurations give rise to intensities whose effects reverberate at diverse and multivalent registers (McCormack 2003; Thrift 2004; Anderson 2006). To put it another way, non-representational theory molecularises Human Geography's common sense notions of practice, will and hegemony by focusing on agencies, rather than agents, as the primary figures of political action. Unlike a representational economy that views political action through the lens of causality and exchange (as well as their measurable relations), non-representational theory explores the effects of composite intensities which cannot be readily judged as progressive, just and/or exploitative. Indeed, it is precisely the inclination to reduce these effects to such pre-established positions that limits their political potential. In taking a set of complex multivalent affective configurations and securing them to available political positions that can be identified, analysed and acted upon – what Thrift (2004) calls a 'know and tell' politics – the levers for ethical/political action are reduced to a blunt political calculus – e.g. ranked political priorities, hierarchies of suffering and notions of the 'greater good' (see Bassi 2010 for a description of how these rankings are operating in left-wing academic circles in response to 9/11 and the 'War on Terror').

The problem for non-representational theory comes when it attempts to figure out what constitutes ethical/political action given this dispersed, complex, unknowable and, hence, un-judge-able political field. The answer, it seems to me, can be summed up through non-representational theory's notion of *generous pragmatism.* As a number of commentators suggest (McCormack 2003; Thrift 2004; Anderson 2005), the ethics of non-representational theory involve an awareness of and devotion to, the various forms of life that exceed experiential consciousness. In this sense, it endeavours to foster, increase and foment affective energies in order to extend and enliven the capacities by which we (and others) live. Its pragmatism is defined by its willingness to see things up close. To measure a situation not in relation to a pre-theorised utilitarian code, determining what is 'best', but in relation to skilful judgement, determining what is best for now. Its generosity is defined by its commitment to extension – using thought to supplement and multiply the affective energies by which the world prospers. In sum, generous pragmatism cultivates what Thrift (2004, 14) calls 'an ethos of awareness working experimentally upon virtualities' in order to encourage and promote unique forms of flourishing appropriate to the situations we face.

The aim of this chapter is not to level a critique of non-representational theory or its conception of politics. I do, however, want to raise a question. A question that begins, not by taking sides in some of the long-standing debates that have surrounded non-representational theory, but by making a comment about the quality of the debate itself. A debate, it seems to me, where both sides are guilty of a certain deafness and, thus, a lack of real engagement with the issues put forward. On the one hand, critics of non-representational theory (Nash 2000; Cresswell 2002), particularly from the Marxist left (Castree and MacMillan 2004; Smith 2005; Harvey 2006), accuse its proponents of retreating from the political sphere and of approaching political action as a matter of play rather than purpose. But as Amin and Thrift (2005) suggest, there is a poor equation here between a recognised inability to provide a full account of an event and an unwillingness to say or do anything about it. Indeed, for many in non-representational theory, generous pragmatism massively expands the levers of ethical political action, allowing for modest adjustments to situations as well as broad-scale coalitions of convenience. In this sense, the critics of non-representational theory have placed too much emphasis on the pragmatism, founded as it is on a retreat from abstract theoretical damnation, and too little emphasis on the generosity – the hopeful forms of ethical/political practice it puts at our disposal.

Yet, non-representational theory is equally guilty of a certain deafness. As compelling as an ethics of generosity sounds, it is understandable why sceptics might question what all this liveliness actually does for life. What does this production of new energies actually energise? While non-representational theory provides us with an incredibly versatile and positive image of political potential, it does not provide any sense of who or what that potential should serve. In short, why be generous? Amin and Thrift (2005) are correct when they suggest that there is nothing within non-representational theory that precludes the traditional orientations of left-wing politics *but* there is also nothing within non-representational theory that encourages it (also see Popke 2009). It is a question, once again, about the future. What in non-representational theory determines progress? What allows us to claim certain forms of life as better than others?

As Smith (2005, 231) suggests 'a host of traditional liberal sentiments turn up [in non-representational theory]: they are against too much corporate power, but handwringingly unsure what determines "too much"'. No doubt within this question lies precisely the figure of the pre-established political ideology non-representational theory seeks to escape. But there is more to this critique than a simple demand to erect a utopian teleology. While pragmatic generosity seeks to free and open life from the immediate structures of a representational logic, there is a temporal implication to this logic that extends beyond the immediacy of events and the new immediacies those events situate. An implication whose interrogation cannot be reduced to a desire to know and tell. It is a question, I would suggest, about responsibility. A question that is ultimately a question about time.

The aim of this chapter is to think more carefully about the politics of non-representational theory by thinking more carefully about time. Indeed, I hope to show that it is only by thinking through the temporality of non-representational theory that we can understand the necessary limitations of its political commitments. Yet, as I have already said this is not a critique. My purpose, rather, is to illustrate that non-representational theory's ethical/political commitments are a product of its ontology and that its ontology, like any ontology, is burdened with certain limitations. Exposing these limitations comes by way of describing another ontology that, while harbouring its own limitations, nonetheless achieves some of the indeterminacy of non-representational theory while managing to put forward a strong responsible vision and a certain kind of commitment to the future. I conduct this comparison by examining these ontologies in relation to their differing conceptions of time and the temporality of their ethical/political horizons.

Before moving on, however, it is important to recognise that non-representational theory is not a singular theoretical enterprise. Rather it is an expanding trajectory of thought that already has many facets developing in their own unique directions (which this volume itself attests to). Thus, when I say I am going to explore the temporal implications of non-representational theory's ontology, I am referring specifically to the ontology of Gilles Deleuze. While I understand this is a somewhat problematic choice (Deleuze cannot be said to be the ultimate ground for non-representational theory as a whole), I believe it is justifiable since Deleuze grounds so many of non-representational theory's initial positions and continues to play a significant role in the development of the field. The second ontology to be discussed is that proposed by Emmanuel Levinas. In some ways the idea of comparing the ontology of Deleuze to that of Levinas is a bit absurd. The former espouses an ontology of immanence and the latter transcendence. Thus, they begin from fundamentally different starting points and, thus, unsurprisingly evolve in very different directions (even as there are many points of overlap in-between). The aim of this chapter, however, is not to undermine non-representational theory's notion of pragmatic generosity by way of Levinas. Indeed, Levinas' ontology has its own baggage and I would not want to argue that it represents something better or even more progressive. The aim of this chapter is more modest. It is simply to introduce what might be thought of as a check on non-representational theory's enthusiasm – its enthusiasm for politics, for practice and for life. Non-representational theory's ethical/political challenge to live life differently is thrilling for its exuberance – its joyous affirmation of and encouragement for supplementation, multiplicity and vitalism. But I worry sometimes about what might be called its secularism. Life itself is not the end all and be all of living. Indeed, in the spirit of non-representational theory I would like to suggest that there is *more:* more to living than life, more to life than life itself and more to ethics than living ethically. There is, as I will suggest, responsibility. A responsibility to the future, to the dead and the not-yet-living. Thus, even as I am encouraged and excited by non-representational theory's faith in human (and non-human) potential, I also think that we need to stop being so enamoured by what we *can* do in order to start considering what we *should* do. The question of responsibility is a serious one – and while it is tempting to chalk the question up to an anachronistic left-wing utopianism, that would simply add to the already tedious dialogue of deaf political posturing.

The remaining chapter is divided into three parts. In the first two I discuss the ontologies of Deleuze and Levinas respectively, illustrating how each one establishes a unique perspective on time – a perspective that does not simply 'follow from' the ontology but is inherent to the ontology and its founding concepts. The third section moves on to analysing how each of these temporal ontologies situate a particular set of political concerns and commitments. Specifically, I argue that while a Deleuzian ontology poses the question of politics as 'worthiness in the service of freedom', Levinas, poses it as 'waiting in the service of responsibility'. I conclude by arguing that while it is necessary to surpass a representational politics in order to find new political capacities, it is nonetheless the representational economy that situates and orients political ends. I use Derrida's (1992) discussion of justice in *Force of Law* at the end of the chapter to illustrate and exemplify this position.

Time as Risk: Deleuze

As previously suggested, the aim of this chapter is to explicate the temporality of Deleuze's ontology. In many ways it is odd that this question of time has not been a central concern for geographers interested in Deleuze (Dewsbury 2002). Thus far, the geographic implications of Deleuze's work in the discipline have been thought primarily in terms of space (Thrift 1999; Dewsbury, Harrison et al. 2002; Dewsbury and Thrift 2005; McCormack 2005; Wylie 2006; Thrift 2008). Possibly this is no surprise as time, though ostensibly the second of geography's two core concepts, remains an under utilised term. Yet, Deleuze's concept of time is a far more central and necessary component of his ontology than space – a peripheral concept at best, taking shape primarily in his collaborations with Guatarri (1983, 1987) (which have an obvious political and, thus, spatial resonance) and in a different way, in his work on cinema (see Deleuze 1986). In his earlier work, it could be argued that Deleuze's ontology is fundamentally steeped in a particular conception of time. As an ontology of difference, it relies upon temporal processes of differentiation, rather than categories of distinction. We can see this from the earliest chapters of *Difference and Repetition* (1994), where Deleuze lays out his concept of pure difference in temporal terms (see Boundas 1995). For Deleuze, relations of difference constitute the essential quality of the world, an elemental fabric of variation that invites the world, in all its diverse resonances, affects and modalities of sense, to be expressed by degrees via abstract infinitives existing in and through multi-dimensional relations of intensity. In addition, it is a process that works through what Deleuze calls series, repetition or, more broadly, time.

Deleuze's concept of time begins with the following premise: that while all events involve a contraction of differences in the present (a unique synthesis of infinitives), the source of those differences are, in their essence, timeless. For Deleuze, events happen and are happening all the time. Indeed, this is the nature of the present – the presence of events happening all the time in time. But those present events are nonetheless marked by a temporality that exceeds the present. They would have to or there would be no memory or anticipation. Thus, Deleuze needs a way to illustrate how the past and future reside in the present without recourse to human memory or inherited reflexes. His answer is to illustrate how the present is permanently informed by a generalised temporality. Understanding the nature of this temporality, and its relation to the present, means understanding time as something separate from, and yet related to, our experience of the present. Deleuze illustrates this relation by discussing what he calls three syntheses of time, all of which explicate the temporal nature of the present event.

The first synthesis is that of anticipation. Here Deleuze illustrates how previous conjugations of events create an expectation of similar or the same conjugation in further events. As with all things Deleuzian this expectancy should not be conceived in purely human terms. The sun's rising is an expectancy anticipated by numerous non-human processes. The point is that such expectations create habits – events that synthesise the past and future in the present. These habits can be regular (e.g. I expect the sun to rise) or intermittent, such as a song that provokes a memory that foments a desire ('listening to that takes me back, lets go drink margaritas on the beach'). Thus, expectations reside in events via their capacity to sense and synthesise the resonance of past events.

The second synthesis, which Deleuze terms archiving, explains how past events return or remain in the present. Thus, while the first synthesis reveals how past events create expectancy, this synthesis reveals how past events are always present or, more accurately, how present events are always past, always passing away, always becoming past, even as they are presently experienced. Deleuze explains this by discussing how we experience the present as something that passes, that moves away from us (just as we experience it at as something familiar, as something we expect). Experiencing the present as something passing, as something that moves away, means we experience the past as something present (as something that is always in the present). The implications here are quite extensive.

In suggesting that the past is always already in the present, Deleuze is saying that the past and present co-exist, as if the past were not a loss or distancing but an archiving, a living record present in our moving lives. In addition, it is not simply our lives that are part of the passing present but all lives and all events. It is for this reason that Deleuze refers to this past as an *a priori* past, that is, a generalised and abstract archive whose resonance is available and necessary for all events in the moving present. The *a priori* past is the permanent presence of all past events – living stock of the whole past present. In addition, as a living record the *a priori* past *gives* the present the means to make sense of the now, the past and the future, as Deleuze (1994) puts it:

> It is in this present that time is deployed. To it belong both the past and the future: the past in so far as the preceding instants are retained in the contractions; the future because its expectation is anticipated in this same contraction. *The past and the future do not designate instants distinct from a supposed present instant, but rather the dimensions of the present itself* (71 my emphasis).

The passing present is, thus, never lost, but on the contrary, always with us, informing and anticipating in the movement of temporal experience.

The final synthesis of time resides in the syntheses engendered by and in a temporal subject. Here Deleuze is concerned with illustrating how the capacity to synthesize time is a fundamental condition of human subjectivity. Specifically, he argues two things. First, he suggests that before an 'I' has consciousness she *has* time, that is, she has the capacity to contract the *a priori* past into a present that anticipates the future. This is a synthesis that precedes all the other diverse and varied syntheses that engender subjectivity. Second, Deleuze argues that the subject not only has the ability to contract time, she has the ability to *sense* the various ways time can be contracted. Thus, before the subject is aware of herself, she is aware of an archived past that she can contract into an infinite array of anticipated futures. Why is this embedded capacity so important to Deleuze? Because it is the subject's capacity to conjugate the pure past towards a unique future that keeps time radically open. The past, for Deleuze, is not infinite. While its dimensions are expansive, it is ultimately an archive of numerable past events. Yet, the past's capacity for making sense of the passing present *is* infinite since the past conjugated in the present occurs in relation to a wholly unique correlation of intensities. Thus, the present is, in this sense, infinite. If we understand every present as staging a unique set of conditions, than the temporal sense of that present will be based upon the unique way its own distinctive intensities resonate with the pure past. This is why Deleuze refers to the present as a form of 'chance' or 'risk'. The risk is the risk of the incalculable, the risk of not knowing how the unique conditions of the present will resonate with the pregnant magnitude of the pure past. The risk of trying out a joke (it was funny when I told it yesterday), of putting up a barricade (the police chief is too old and jaded to care), of searching for mass graves long denied and forgotten (the discovery re-ignited ancient resentments).

For Deleuze, the risk of the present (passing into the past and anticipating the future), is one that puts the whole of life at risk, not just the moving present, but the incalculable relations that infinitely tie that present to an *a priori* past and to an expected tomorrow. This is why the past can make the future radically different (the discovery that changed the world) and why the future can radically change the past (it made us rethink everything we thought we knew). Temporal being is not only quintessentially complex, it is quintessentially precarious.

There are a number of interesting implications that come out of this conception of time as risk. The first is that time, conceptualised in these terms, operates as a kind of ontological ground for Deleuze's vitalist ontology. By this I mean that it is precisely the risky nature of the present that precipitates the event (and the powers that engender the event). Time, in this rendering, is the engine for the event, i.e., that which calls the event or compels the event to take place. Chance or risk is, thus, the first imperative. It is a problem whose appearance beckons to be taken hold of – to be met with bravery or cowardice, with generosity or meagreness, with energy or frailty. The risk of the present cannot be turned away from or otherwise avoided or parried. It is a demand, indeed, the first demand, and as such it exacts not so much a will to power nor even a desire to power but *a desire to will*. It is that which first claims an action, a difference, a synthesis, i.e. an event.

The second implication is that time (as risk) situates the question of power as an irresolvable problem, or in Deleuze's (or Nietzsche's) phrasing, a problem that eternally returns. For Deleuze, to be in the present is to be at risk. Risk cannot be evacuated or mitigated. As temporal beings we are, by definition, risking. This means that while events engender responses to the problem of risk, such responses can never properly resolve the problem. They can offer solutions, possibilities, forms of management, ways of getting from A to B, but they can never resolve, dampen, administer or direct the ontological condition of risk. Indeed, Deleuze takes this point further by suggesting that every new solution to the problem of risk simply multiplies the problem, thus, perpetually reworking risk into ever new permutations as solutions are posed. Power and empowerment is a problem that remains problematic; it not only exceeds any and all solutions but enrols those solutions into ever new multiplicities and extensions of the problem.

This leads to what I take to be the third implication of time as risk which is that it guarantees the present of the infinite in every event. If the problem of risk is infinitely extended through the posing of responses, than those responses must also be infinitely extensive. This is not only true by logic but true by the relationship Deleuze establishes between the three syntheses of time. As previously suggested, the present is constituted by conjugating an *a priori* past in a moving present that anticipates the future. In addition, because the *a priori* past is extensive and the present is always new, the potential modalities for enacting the present are infinite. In a later language, Deleuze will refer to this *a priori* past as the *virtual* and the various creative enactments that cut the past into the present *the actual*. In many ways this is an obvious point since it is this same infinite potential inherent in conjugating the present (making the virtual actual) that makes the present risky.

The potential for risk and the potential for creative willing/desiring/empowerment are the same. They are both founded on the infinite potential (the virtuality) of the *a priori* past in the present, facing the future.

Time as Intrusion: Levinas

Unlike Deleuze, Levinas does not articulate an explicit philosophy of time. Rather, his ideas on time emerge in a semi-haphazard fashion, surfacing, for example, in his discussions of the future as mystery (*Time and the Other*), of the impossible possession of the present (*Existence and Existents*) and of the unsayability of the past (*Otherwise than Being*). Even as time is a persistent theme for Levinas, it appears as a set of semi-disconnected musings rather than as a central concept. And yet, I would argue that without understanding time in Levinas, you cannot understand his conception of subjectivity and, thus, his conception of ethics and the ethical relation. Indeed, while we are often told that Levinas is a philosopher of ethics, his conception of ethics is founded on an explicit ontology of the subject – an ontology that, I hope to show, is fundamentally temporal. From his earliest articulations of subjectivity, Levinas thinks the subject in terms of *presence*. Specifically, he asks what does it mean for a subject to be present? What does it mean to be *in the present*?

For Deleuze, the question would be non-sensical. As previously suggested, Deleuze understands all events, and the beings (or quasi beings) that emerge from events, as occurring *in* time. We are all in time. There is nothing outside time. Life is temporal. Such a position is a logical consequence of (or asset to) developing a philosophy of immanence, that is, a philosophy that accounts for beings and events through its own internal mechanisms rather than external non-empirical structures. But as compelling as Deleuze' ontology is, one cannot help wonder if there is anything beyond those mechanisms? The genius of Deleuze's ontology is its endless capacity to account for (and bring under scrutiny) the sheer variety of forms of life that life itself begets. But what about those forms of life that life does not beget? Those forms of life given by something beyond life, for example, those forms of life that death begets, such as mourning? For Deleuze mourning would be another form of life, as he suggests every death is double and represents the cancellation of large differences in extension as well as the liberation and swarming of little differences in intensity (1994, 259). In every death we find a releasing of new affectivities. The role of the dead is, thus, positive in the sense that they engender new complexes of affect, even as those affects can be characterised by suffering, vertigo or tragedy. Mourning, in this rendering, is not an emptiness or a fallen memory. It is not a phantom limb – an architecture of sensibilities roaring along in a dimension suddenly absent. It is a new architecture, a new syntheses of intensities made operative by the dislocative event of death. Suffering, in Deleuze, is its own modality of being. It is a form of life given by death.

For Levinas (1969), however, suffering is precisely that mode of being that brings subjects to the edge of their being – to the threshold of what it means to be. Suffering is not another mode of being but is a reckoning with that which is outside being – that which cannot be recuperated for or by being. It is through events like death where subjects face another dimension that is not of this life – a dimension that subjects have no purchase on, control over and can provide no inspiration or reason for existence. This is a dimension that fundamentally questions the subject's sense of self-ownership. It is precisely this dimension 'outside being' that Deleuze attempts to eradicate from his philosophy and he would have no truck with it as a concept or with the modes of subjectivity it engenders. Yet, in raising the possibility of an exterior transcendent dimension (a dimension that looks very different than it did in Hegel, Kant and Husserl), Levinas gives us another concept of the subject, of time, and ultimately, of politics, that cannot be so easily dismissed.

Perhaps the best place to begin explaining how Levinas conceives the relationship between subjectivity and time is by explaining where, for Levinas, the subject is thought to begin. In *Existence and Existents*, Levinas (1978) reworks the classical concept of hypostasis from being a noun (the essence of individuated being) to being a verb. Hypostasis marks what Levinas calls an arising from anonymous being. It signifies a primordial movement – a coming forth out of a smooth plane of uncharacterised non-delineated existence. Even as this rising is not characterised in terms of consciousness or self-ownership (it is not an appropriation or naming of one's *own* being), it is a breaking free – a generalised coming forth from the anonymity of presence. The question that Levinas then raises is what precipitates this claim – what invites this initial movement out and away from anonymous being? For Levinas this is the work of the transcendent. While the transcendent dimension that Levinas describes is non-present (it resides beyond being), its emptiness nonetheless performs a powerful gravitational effect. An example of this effect is the calling performed by the Judeo-Christian God, a being who, by virtue of being infinite and transcendent, is radically exterior to our being, and yet, calls on us to come forth and account for ourselves. Another example is the call of the future, a realm that is infinite in its totalising opaqueness and, yet, solicits us into various preparations, predictions and calculations, none of which assert any mastery or control over what will come. The point is that the event of hypostasis is precipitated by a calling, a summons that arrives from another dimension of existence.

Given this description of hypostasis we can immediately see that the origin of subjectivity involves a conundrum. The rise from anonymous being is an arising preceded by a call. And in responding to this call the subject first takes hold of itself as a being capable of answering. Thus, the capacity to have a voice (to answer) is predicated not on one's own internal abilities but on a summons from elsewhere – and therein lies the conundrum. The voice with which the subject answers is never properly its *own*. The subject is not the origin of its own arising but is dependent on that which calls it – an Other beyond being whose summons first gives the subject its voice. This is precisely the problem of hypostasis. Subjectivity, from the beginning, is never truly possessed. The event of hypostasis (the event of subjectification) does not signal a permanent state of affairs but a recuperation – a gesture of appropriation. Hypostasis marks a retrieval from anonymous being that is never entirely fulfilled since we can never take possession of that which we retrieve.

So how does this conception of hypostasis relate to the question of time? At various points in Levinas' work (1969, 1981, 1987, 1996; also see Critchley 1999; Hutchens 2004) he refers to two dimensions of time. The first is synchronous time which we can think of as the time of the present. *Presence*, here, marks both a spatial and temporal immediacy. Beings that appear in the present appear in the 'here' (spatial immediacy) and the 'now' (temporal immediacy) of the world. They are not only available and 'on-hand' but ready to be appropriated and distributed by a subject. It is for this reason that we can talk of synchronous time as situating a temporal economy. In synchronous time we 'have' time, we 'run out of time', we 'waste time' and 'give our time' to others. Time, in this rendering, is not only something we possess but something we divide, measure and circulate according to various priorities and desires (time for my work, my family, me, etc.). The second concept of time is diachronic time. For Levinas, synchronous time is an imaginary idea (supported by a long philosophical history) rather than a proper description of time. To understand time properly, i.e., outside the idea of synchronous time, we need to see it as something that cannot be owned or possessed by a subject nor assimilated or distributed through an economy. Subjects, for Levinas, never *have* time, they are never *in* time and they never fully inhabit the present. In Levinas' description of hypostasis, subjectivity is rendered as a *movement towards* self-possession rather than an event of actual self-mastery. Diachrony is the term Levinas uses to explain the disjuncture between the temporal economy of synchronous time and what he calls the time of the infinite – a *total time* that transcends and disrupts the time of lived experience.

So what does Levinas mean by this concept of total time? Primarily it needs to be understood as transcendent, a time that can never be of being. Total time stands for a future that can never be anticipated and a past that is impossible to retrieve. In this sense, total time exposes the feeble-ness of prediction and the impossibility of history. Like Deleuze, Levinas understands the past as a past of past events that extend beyond the events that present themselves to an individual's experience or consciousness. Thus, it is comprised of pasts that can never be remembered because they were never part of our present. Yet unlike Deleuze, Levinas does not conceive this pure past as a resource for being or becoming. It is not an archive that can be appropriated or used. On the contrary, it exposes precisely the limits of such endeavours. The same goes for the future. Unlike Deleuze, the future in Levinas is not a form of risking. It is not an opportunity to play one's hand but rather, is a site of utter vulnerability; it cannot be positively met, 'taken hold of' nor risked towards destiny. For Levinas (1987), the future is precisely what eludes all meeting, all holding and all risking, 'the future is what is not grasped, [it is] what befalls us, what *lays hold of us*' (76, my emphasis). The example Levinas repeatedly uses to illustrate the intrusion of diachronic time is the arrival of the other person. As another person presents themselves a new array of possibilities come forward which we cannot anticipate, predict or manage. As Levinas (1987) suggests, 'the other is the future. The very relationship with the other is the relationship with the future' (77). The other person brings with them not only a measure of utter unpredictability but a past that is unique and outside a subject's capacity to assimilate or know. In this sense, other people lay claim to the subject's time. They do not simply ask for the subject's time – as if that time where already owned – they take it. They undermine the very terms of possession. In addition, this taking is performed not only by the other people around the subject but by other people everywhere and always – alive, dead and not-yet-born. Time, as an infinite transcendent dimension, intrudes into synchronicity at every angle, robbing the subject of yesterday, today and tomorrow.

As we can see Levinas provides a very different conception of time than the one we find in Deleuze. While a closer, more extended examination, would speak more readily to the numerous points of similarities between their temporal ideas, their divergent ontological paths ultimately drive them to distinct conclusions. The aim of the final section is to bring these distinctions back to the question of politics – that is, the ethical/political trajectories for change that each of these philosophies establish.

Freedom and Responsibility

So how do these differing temporal ontologies lend themselves to particular conceptions of politics? This is the question that ultimately drives this chapter: how does each ontology situate a unique comportment towards ethical/political issues? The question here is not about the nature of politics or political action but about the orientation of politics: what does *being ethical* mean within the remit of these different temporal worlds? Once gain the question is about the future: what change would Deleuze and Levinas recognise as 'for the better'? What kind of future does each ontology value?

In addressing these questions in relation to Deleuze one would think there would be a ready-made response given how much ink has been devoted to the relationship between Deleuzian thought and political theory (Connolly 2002; Hardt 1993; Honig 1993; Patton 2000, 2005, 2007; Protevi 2001). In particular there has been much interest in what might be called the *machinic quality* of power – the way power relations propose particular trajectories of desire which become embroiled in temporary pragmatic coalitions producing various power effects (DeLanda 1991; Goodchild 1996; Massumi 1992; Patton 2000). But the question of power is not the same as the question of politics. Indeed the former is an effect of the latter. Machines account for *the how* of power but do not explain *the why*: why do machines emerge, why do they take the forms they do, why do they fail or succeed? Machines are a way of explaining the technologies of power – the techniques by which desires seek fulfilment. In this sense, they are an effect of a more primary demand – a solution to a more problematic problem. The question of politics concerns not the methods but the meaning of politics – the demands inherent in Deleuze's temporal ontology.

In line with much of what has already been said in non-representational theory, I would argue that Deleuze's ontology demands worthiness in the service of freedom. Unlike traditional forms of politics, this conception of worthiness is not a worthiness to 'a future' or 'our future' but a worthiness to time itself and all that time puts at risk. It is a worthiness to the immemorial past and the infinite possibilities of the future. It is a worthiness to risk and the risky nature of the event. For Deleuze, this kind of worthiness can best be understood as a valuing of *all* that happens. Not just that which happens to *us* or those happenings judged 'good' but everything. The temporality of the event means that every event produces effects and intensities that reverberate in the past and future, filling the present with potentials whose outcomes cannot be seen. Worthiness, in its most generous reading, is an ethical commitment to the potentialities created by risk. It is a stoical facing of all that the future gives.

There are a number of implications to this concept of worthiness. First, in terms of comportment, worthiness asks us to view events in terms of what they give and add to the world – the pasts they set free and the futures they open – rather than in terms of the immediate effects they engender. As Deleuze (1990, 169) suggests, 'to grasp whatever happens as unjust and unwarranted ... is ... what renders our sores repugnant – veritable *ressentiment*, resentment of the event'. Being worthy involves honouring both the totality and inescapability of risk as well as the untimely events that risk precipitates. It means regarding risk from the perspective of risk itself rather than judging the consequences that risking engenders. This leads to the second implication which concerns the question of agency. While non-representational theory celebrates the different kinds of things that an ethical sensibility can do, Deleuze's notion of risk keeps the actual capacity of these doings quite modest. Indeed, risk (like the event itself) is double-edged. While it unleashes infinite possibilities, such possibilities actualise within a context of infinite possibilities, thus, making any cause and effect attribution incalculable.

Taken together these implications render the notion of envisioning a future patently untenable. While Deleuze's work is suffuse with particular values and ethical orientations, its unending elevation of indeterminacy prevents the philosophy itself from determining any abstract criteria for ethical/political judgement – bar possibly one. Throughout Deleuze's work there is a consistent endorsement of and commitment to what Foucault (1983, xiii) calls a 'non-fascist life', an ethical commitment to develop an 'art of living counter to all forms of fascism, whether already present or impending'. Central to this position is a devotion to freedom: 'to liberate for each thing "its immaculate portion"' (Deleuze 1990, 172). Or as Foucault (1983, xiii) puts it, to 'free political action from all unitary and totalising paranoia'. This is why I characterise Deleuze's politics as worthiness *in the service of* freedom. While worthiness signals a form of ethical comportment – a commitment to particular values – 'in the service of' signals an orientation. Deleuze tells us to be worthy of risk *in the service* of freedom and in opposition to fascism. While it is the riskiness of the event that takes priority here, riskiness is nonetheless valued for the kinds of freedoms it sets forth.

Even in writing this I find myself, once again (as I have been before) attracted to Deleuze's notion of ethics and politics. Its got chutzpah: managing to be nuanced but not reticent; passionate but not arrogant; endeavouring to change the world but content with a modest supplement. And yet, it is precisely this daring positivity and affirmative character that is discordant with Levinas' ontology. The problem of power in Levinas is the problem of a temporality that arrives from nowhere, disrupting the synchronous economy of the present, and thus, disrupting the means by which subjects pursue change. While we could certainly talk about machines in Levinas, such techniques would be feeble mechanisms working in the face of a transcendent time. Indeed, the problem of power in Levinas is the problem of not ever *having* power. The problem of time cannot be faced because the subject has nothing to face it with – she can never risk the future nor conjoin the past because the subject never *has* time. She is never *in* time. She never fully inhabits a temporal economy. The problem of time, therefore, is not one that returns, but rather, one that never leaves. It never allows for solutions to problems. It is not a love of fate (*amor fati*) nor even a stoic comportment in the face of fate (Harrison 2007). In Levinas, facing the future is not a mode of empowerment or being but a threshold of exposure. It is facing a temporality that *disrupts* rather than *gives* possession, as Levinas suggests, time cannot taken be taken hold of – it takes hold of us.

So what does the question of politics mean given this Levinasian ontology of temporal vulnerability, this exposure to time and its power to deny? In a move not dissimilar to Deleuze's injunction to be worthy, Levinas suggests we wait – we wait in the service of responsibility. Responsibility for Levinas, is not a 'taking responsibility for ourselves' (our self is never ours to commit) but a responsibility to the other: 'this is a responsibility that lies outside the category of choice. It has not been chosen because is not something that can be chosen. This responsibility was always already – it is a past more ancient than history and a passivity more passive than any passivity. This is an extreme passivity, which is not assumed but was already there' (Katz 2003, 17). The question here is not whether to be or not to be responsible (there is no danger here of *ressentiment*), but of facing a responsibility already situated in the subject's desire for presence. It is, as Katz (2003) suggests, a position of extreme passivity – of hanging on for something that can never be taken. It is a matter of *waiting*. Waiting in a manner that perpetually defers and delays self mastery. Waiting as a vigilant denial of self in the service of the other (see Levinas 1989).

At first glance it is difficult to see what kind of politics can be gleaned from this position. Waiting in the service of responsibility signals an infinite suspension – a radical passivity to be sure. Yet, Levinas (1989) also tells us that within every waiting is a forgetting, meaning a retreat back into the representational economy of synchronic time. Indeed, waiting as an ethical comportment requires this retreat since it is only by existing in synchronic time that one's subjectivity can be interrupted. In this sense, we can see within Levinas the possibility for a political project and a political vision. While Levinas endeavours to stage the relation between the subject and time in ethical terms – i.e. the subject's orientation towards the future and the radical alterity therein – the advent of retreat is a necessary violence that carries with it certain political implications.

The question, therefore, is whether this violence can be accepted in order for the political implications to be exploited. For Levinas, it is only once the subject retreats into synchronous time that she both discovers and loses freedom. In other words, the subject discovers responsibility from the standpoint of being in a representational economy of synchronic time. Thus, it is only through existing in synchronic time (and the 'as if' dimension it situates) that she discovers her time as a gift given to her by an other – a gift she can never fully possess. While Levinas' conception of ethics demands a forsaking of that which is not ours in the service of responsibility, one can nonetheless see within this ontology the possibility of not forsaking and not being responsible. There is, in other words, the possibility for neglecting the demand to be ethical in the service of being political. For Levinas, it is not as if the future will go away. As a transcendent dimension it exists beyond being. Whether we recognise it or not, the future interrupts and demands recognition. The same cannot be said for Deleuze. If we are not worthy of the event, and choose to embed ourselves in a representational economy, the risky nature of time recedes from view. This is why machines are such a powerful force in Deleuze and Guattari – they miraculate the world, and in doing so, have the capacity to subsume alterity to its pre-established dimensions, what they call 'machinic enslavement' (Deleuze and Guattari 1987). In Levinas' representational economy, machines are a paper fortress, perpetually vulnerable to and ineffectual before transcendent time. Given this we can trust that our various political projects will always be vulnerable.

Thus, there is, within a Levinasian ontology, a means to envision the future – even if we recognise that vision as always already undermined by a primordial vulnerability. In this fashion, Levinas gives us a means to imagine and pursue a future political horizon, even if we must simultaneously acknowledge that that horizon perpetually recedes. This paradoxical approach to politics, this affirmation of change in the face of the impossible, is best exemplified not in Levinas' own work but in the writings of one of Levinas' most significant interlocutors – Jacques Derrida. In a number of his later essays – on justice (1992), friendship (1997) and the messianic (1994) – Derrida sketches out both the impossibility of arriving at an imagined future, and yet, the necessity of moving towards that future nonetheless. For example, in *Force of Law*, Derrida discusses the difficulty of defining justice in abstract terms. While we have laws which are born from a desire for justice, the laws themselves do not create justice nor are they obeyed because they engender a just world. Laws are indicative of a juridical calculation, the operative denominators of a just economy. Justice itself is what exceeds the law; it is what transcends the ambition or desire *to be just*, thus marking 'an experience of the impossible: a will, a desire, a demand for justice the structure of which ... would have no chance to be what it is' (Derrida 1992, 244). And yet, Derrida suggests, justice cannot exist, as a concept, if it is not (failingly) pursued. Thus, even as Derrida deconstructs the means by which we calculate and articulate justice (as law) *and* positions the attainment of justice itself as a necessary impossibility, such manoeuvres do not make justice disappear, but on the contrary, make it appear more forcefully upon our political horizon. Justice, for Derrida, is a call or demand that appears *in perpetuity* because articulations of justice always fail. The constancy of justice is secured by its unattainability – it proceeds by falling short. Thus, while Levinas' primary interest is in exposing the ethical relation itself (and finding an appropriate comportment towards that relation), within this project lies the means to envision the future – even if it is a future we can never have.

Conclusions: A Defence of More-Than-Representational

The idea that non-representational theory does not pre-establish an ethical political vision is not exactly news. As Thrift (2001) and McCormack (2003) suggest, affect is not inherently progressive. And much of their project can be understood as an attempt to develop counter-forces to engage an affective dimension already infiltrated by, what they take to be, problematic capitalist agencies. The reason they reject representational politics is not because it is a bad politics *per se* but because it ignores a set of more immediate political concerns – concerns that lie outside the purview of cultural geography's traditional political vision. In this sense, non-representational theory presents the field with a choice. Either it can choose a politics of 'knowing and telling' and, thus, ignore a series of problematic political activities operating outside that lens, or it can choose a politics of supplementation, a 'ways of enacting and actively manipulating space and time [in a manner] that provide[s] for the potential extension and multiplication of the very field of the ethical' (McCormack 2003, 490).

The problem with promoting this choice in non-representational theory is not with the options themselves. The political world that non-representational theory has brought to our attention is significant and should be treated, explored and engaged, with the utmost seriousness. The problem, I would argue, is about the choosing – i.e., the choice non-representational theory presents and the choosing that it promotes. Non-representational theory has given us a refreshing escape from representational modes of thinking, and the political strictures such modes impose, primarily by arguing that, despite long-held assumptions, we *can* escape from representation. The problem is that in making this claim they simultaneously come to assume that we *can choose* to escape. That we can choose a politics of non-representation and that we can choose a change of political terrain. It is here that non-representational theory reaches its limit as a coherent political trajectory. In the very process of presenting us with a choice, in suggesting that *there is* a choice that *we* have, non-representational theory reconstitutes a representational political terrain

– i.e., a terrain where *we have* choices that are *ours* to make. I say this choosing marks the limits of non-representational theory political thought since it marks the threshold at which non-representational theory, as a political choice, *must* become representational. To be political is to choose and to choose is to represent. If nothing else, the aim of this chapter has been to envision this limit more carefully. In a Levinasian ontology, the representational political terrain (the terrain of synchronic time) is the only terrain a subject has – it is the only terrain where a subject can be a subject. And yet, it is a terrain without terra firma, perpetually undermined by the call of the Other. While synchronic time allows me as a subject to live *as if* I have a past, present and future, *as if* I have time to lose, give and commit to a cause and *as if* I have the capacities to choose one future over another, this time is lent to me by the Other and it is a lending I cannot refuse. I cannot choose to leave synchronic time since it is precisely in the moment of leaving that my I-ness (and my capacities as an I) is disrupted. It is in having the synchronicity of a representational life destroyed, that my capacity for politics (for choosing and for being) are obliterated. We cannot escape a representational economy *and* be an I or an agent. It is only in time that we are subjects capable of choosing futures.

Choosing a representational politics, therefore, is not a denial of the non-representational world but a recognition of its limits for choosing a future. It is also an investment in one's capacity to choose, and ultimately, an investment in oneself as a subject. Such investments are not safe havens. They necessitate accepting a paradox, indeed, *the* paradox, of responsibility in a synchronic representational world: an immediate responsibility to a vision of the future *and* an infinite responsibility to that which destroys that vision. This is why I would endorse thinking about politics as *more-than*-representational rather than *non-representational* (Lorimer 2005). We need representation not only to have a vision of the future but to have an 'I' for choosing that vision. We need to recognise our limits for choosing – a recognition of the thresholds we face, Janus-like, infinitely responsible to a future we can and should pursue but can never expect to possess.

Acknowledgements

The author would like to thank the editors Ben Anderson and Paul Harrison for their engaged reading and insightful comments.

References

Alexander, J.C. (2008), 'Iconic consciousness: the material feeling of meaning', *Environment and Planning D: Society and Space* 26, 782-794.

Amin, A. and Thrift, N. (2005), 'What's left: just the future', *Antipode* 37, 220-338.

Anderson, B. (2005), 'Practices of judgement and domestic geographies of affect', *Social and Cultural Geography* 6, 645-660.

—— (2006), '"Transcending without transcendence': utopianism and an ethos of hope', *Antipode* 38(4), 691-711.

—— (2007), 'Hope for nanotechnology: Anticipatory knowledge and governance of affect', *Area* 19, 156-165.

Anderson, B. and Harrison, P. (2006), 'Questioning affect and emotion', *Area* 38 333-335.

Anderson, B. and Wylie, J. (2009), 'On Geography and Materiality', *Environment and Planning A* 41, 318-335.

Bassi, C. (2010), '"The anti-imperialism of fools": A cautionary story of the revolutionary left vanguard of England's post-9/11 anti-war movement', *ACME: An International E-Journal for Critical Geographies* 9(2), 113-137.

Bingham, N. (1996), 'Object-ions: from technological determinism towards geographies of relations', *Environment and Planning D: Society and Space* 14, 635-657.

Boundas, C. (1995), 'Deleuze: serialization and subject-formation', in C. Boundas and D. Olkowski (eds), *Gilles Deleuze and the Theater of Philosophy* (London: Routledge).

Castree, N. and MacMillan, T. (2004), 'Old news: representation and academic novelty', *Environment and Planning A* 36, 469-480.

Connolly, W.E. (2002), *Neuropolitics: Thinking, Culture Speed* (Minneapolis: University of Minnesota Press).

Cresswell, T. (2002), 'Guest editorial', *Environment and Planning D: Society and Space* 20, 379-382.

Critchley, S. (1999), *The Ethics of Deconstruction: Derrida and Levinas* (Edinburgh: Edinburgh University Press/Marston).

DeLanda, M. (1991), *War in the Age of Intelligent Machines* (New York: Zone).

Deleuze, G. (1986), *Cinema 1: The Movement Image* (Minneapolis: University of Minnesota Press).

—— (1990), *The Logic of Sense* (London: Althlone Press).

—— (1994), *Difference and Repetition* (New York: Columbia University Press).

Deleuze, G. and Guattari, F. (1983), *Anti-Oedipus: Capitalism and Schizophrenia* (Minneapolis: University of Minnesota Press).

—— (1987), *A Thousand Plateaus: Capitalism and Schizophrenia* (London: Athlone).

Derrida, J. (1992), 'Force of Law: "the mystical foundation of authority"', in D. Cornell and M. Rosenfield (ed.).

—— (1994), *Spectres of Marx: The State of the Debt, The Work of Mourning, and the New International* (New York, London: Routledge).

—— (1997), *Politics of Friendship* (London: Verso).

Dewsbury, J-D (2000), 'Performativity and the event: enacting a philosophy of difference', *Environment and Planning D: Society and Space* 18, 473-496.

—— (2002), 'Embodying time, imagined and sensed', *Time and Society* 11(1), 147-154.

—— (2003), 'Witnessing space: "knowledge without contemplation"', *Environment and Planning A* 35(11), 1907-1932.

Dewsbury, J-D, Harrison, P., Rose, M. and Wylie, J. (2002), 'Enacting geographies: editorial introduction', *Geoforum* 33(4), 437-440.

Dewsbury, J-D and Thrift, N. (2005), '"Genesis eternal": after Paul Klee', in I. Buchanan and G. Lambert (eds), *Deleuze and Space* (Edinburgh: Edinburgh University Press).

Dubow, J. (2001), 'Rites of passage: travel and the materiality of vision at the Cape of Good Hope', in B. Bender and M. Winer (ed.), *Contested Landscapes: Movement, Exile and Place* (Oxford and New York Berg Publishers), 241-255.

—— (2004), 'The mobility of thought: reflections on Blanchot and Benjamin', *Interventions: The International Journal of Postcolonial Studies* 6(2), 216-228.

Foucault, M. (1983), 'Preface', in G. Deleuze and F. Guattari (eds).

Goodchild, P. (1996), *Deleuze and Guattari: An Introduction to the Politics of Desire* (London: Sage).

Hardt, M. (1993), *Gilles Deleuze: An Apprenticeship in Philosophy* (Minneapolis: University of Minnesota Press).

Harrison, P. (2007), 'How shall I say it...? Relating the non-relational', *Environment and Planning A* 39(3), 590-608.

Harvey, D. (2006), 'The geographies of critical geography', *Transactions of the Institute of British Geographers* 31, 409-412.

Hinchliffe, S. (2000), 'Entangled humans: specifying powers and their spatialities', in J.P. Sharp, C. Philo and R. Paddison (eds).

—— (2008), 'Reconstituting nature conservation: towards a careful political ecology', *Geoforum* 39, 88-97.

Honig, B. (1993), *Political Theory and the Displacement of Politics* (Ithaca: Cornell University Press).

Hutchens, B. (2004), *Levinas: A Guide for the Perplexed* (London: Continuum).

Jones, O. (2005), 'An ecology of emotion, memory, self and landscape', in Davidson, J., Bondi, L. and Smith, M. (eds).

—— (2008), 'Stepping from the wreckage: geography, pragmatism and anti-representational theory', *Geoforum* 39, 1600-1612.

Katz, C. (2003), *Levinas, Judaism, and the Feminine: The Silent Footsteps of Rebecca* (Bloomington: Indiana University Press).

Levinas, E. (1969), *Totality and Infinity* (Pittsburgh: Duquesne University Press).

—— (1978), *Existence and Existents* (Pittsburgh: Duquesne University Press).

—— (1981), *Otherwise than Being or Beyond Essence* (Pittsburgh: Duquesne University Press).

—— (1987), *Time and the Other and Additional Essays* (Pittsburgh: Duquesne University Press).

—— (1989), 'The servant and her master', in S. Hand (ed.), *The Levinas Reader* (Oxford: Blackwell).

—— (1996), 'Meaning and sense', in A. Peperzak, S. Critchley and R. Bernasconi (eds), *Basic Philosophical Writings* (Indianapolis: Indiana University Press).

Lorimer, H. (2005), 'Cultural geography: the busyness of being 'more-than-representational', *Progress in Human Geography* 29(1), 83-94.

Massumi, B. (1992), *A User's Guide to Capitalism and Schizophrenia* (Cambridge: MIT Press).

McCormack, D. (2003), 'An event in geographical ethics in spaces of affect', *Transactions for the Institute for British Geographers* 11, 488-507.

—— (2005), 'Diagramming practice and performance', *Environment and Planning D: Society and Space* 23, 119-147.

Nash, C. (2000), 'Performativity in practice: some recent work in cultural geography', *Progress in Human Geography* 24, 653-664.

Patterson, M. (2005), 'Affecting touch: towards a "felt" phenomenology of therapeutic touch', in J. Davidson, L. Bondi and M. Smith (eds).

Patton, P. (2000), *Deleuze and the Political* (London: Routledge).

—— (2005), 'Deleuze and Democracy', *Contemporary Political Theory* 4(4), 400-413.

—— (2007), 'Utopian political philosophy: Deleuze and Rawls', *Deleuze Studies* 1, 41-59.

Popke, J. (2009), 'Geography and ethics: non-representational encounters, collective responsibility and economic difference', *Progress in Human Geography* 33(1), 81-90.

Protevi, J. (2001), *Political Physics : Deleuze, Derrida and the Body Politic* (New Brunswick: Athlone Press).

Saville, S. (2008), 'Playing with fear: parkour and the mobility of emotion', *Social and Cultural Geography* 9(8), 891-914.

Simpson, P. (2008), 'Chronic everyday life: rhythm analysing street performance', *Social and Cultural Geography* 9(7), 807-829.

—— (2009), '"Falling on deaf ears": a post-phenomenology of sonorous presence', *Environment and Planning A* 41, 2556-2575.

Smith, N. (2005), 'What's left? Neo-critical geography, or the flat pluralist world of business class', *Antipode* 37, 887-99.

Spinney, J. (2006), 'A place of sense: a kinaesthetic ethnography of cyclists on Mont Ventoux', *Environment and Planning D: Society and Space* 24, 709-732.

Thrift, N. (1983), 'On the determination of social action in space and time', *Environment and Planning D: Society and Space* 1, 23-57.

—— (1997), 'The still point: resistance, expressive embodiment and dance', in S. Pile and M. Keith (eds).

—— (1999), 'Steps to an ecology of place', in D. Massey, J. Allen, and P. Sarre (eds), *Human Geography Today* (Cambridge: Polity Press), 295-321.

Thrift, N. (2000), 'Entanglements of power: Shadows?', in J. Sharp, P. Routledge, C. Philo, and R. Paddison (eds), *Entanglements of Power: Geographies of Domination/Resistance* (London: Routledge), 269-278.

—— (2004), 'Summoning Life', in P. Cloke, M. Goodwin and P. Crang (ed.).

—— (2008), *Non-Representational Theory: Space, Politics, Affect* (London: Routledge).

Thrift, N. and May, J. (eds) (2001), *Timespace: Geographies of Temporality* (London: Routledge).

Whatmore, S. (2002), *Hybrid Geographies: Natures Cultures Spaces* (London: Sage).

Wilford, J. (2008), 'Out of rubble: natural disaster and the materiality of the house', *Environment and Planning D: Society and Space* 26, 647-662.

Wylie, J. (2002), 'The heart of the visible: an essay on ascending Glastonbury Tor', *Geoforum* 33(4), 441-454.

—— (2005), 'A single day's walking: narrating self and landscape on the South West Coast Path', *Transactions of the Institute for British Geographers* 30(3), 234-247.

—— (2006), 'Depths and folds: on landscape and the gazing subject', *Environment and Planning D: Society and Space* 24(4), 519-535.

Index

Index of Names

Adey, P., 11, 16n11, 24n13, 208
Adorno, T., 161, 166, 168, 175
Agamben, G., 42, 167, 189, 201
Ahmed, S., 225n6, 228, 234, 237
Althusser, L., 184, 323, 333
Amin, A., 11, 24n13, 256, 322, 332, 343
Anderson, B., ix, 1-35, 8, 13, 14, 14n8, 16, 16n12, 22, 23, 37, 84, 85, 86, 93, 100, 104, 105, 183-199, 248, 252, 263, 265, 266, 285, 294, 295, 321, 326, 330, 335, 341, 342, 343
Aristotle, 167
Austin, J.L., 131, 142, 142-144
See also: Derrida, J., and Austin, J.L.

Badiou, A., 106, 127, 147, 148, 150, 157, 158, 195, 284, 289, 295, 295-300, 296n3, 298, 322, 330
Barnett, C., 11, 11n12, 108, 249, 292, 309,
Baudrillard, J., 117, 119, 122
Beckett, S., 150, 154, 155, 156, 156n2, 157, 158
Benjamin, W., 10, 175
Bennett, J., 6n3, 11, 13n7, 17, 23, 24, 85, 105, 208n1, 212, 265, 266, 308
Bergson, H./Bergsonism, 38, 39, 39-42, 40, 41, 41n1, 44, 45, 49, 149, 164n5, 209, 325
See also: ontology, and encounter (Bergson, H.)
Bingham, N., 14, 37, 46, 51, 56, 262, 277, 341
Bissell, D., ix, 8, 16, 17, 18, 25, 79-95, 88, 105
Blanchot, M., 104, 106, 109, 150, 155, 156, 157
Bourdieu, P., 10, 184, 184n2, 185, 224, 225, 225n6, 228, 231n11, 312
Butler, J., 9n4, 10, 142, 321, 322
Braun, B. 51, 56

Braidotti, R. 325

Caruth, C., 171, 171n10, 172n11, 176
Castree, N., 37, 56, 100, 103, 343
Cavell, S., 131, 132, 142-144 142, 143, 144
Cloke, P., 58, 242, 244
Colls, R., 14n8, 23
Connolly, W., 6n3, 18, 23, 24, 84, 202, 210, 216, 248, 251, 253, 255, 257, 352
Conradson, D., 24n13, 242, 244, 246, 255
Cosgrove, D., 5, 5n2
Crang, M., 5n2
Cresswell, T., 5n2, 9n5, 75, 100, 343

Darling, F., 56, 57, 58, 59, 62, 63, 65, 66, 69,
Darling, J., ix, 11, 18, 23, 241-260, 244, 256, 257
Davidson, J., 80, 225n5
Davies, G., 48, 100
De Certeau, M., 10, 142
DeLanda, M., 13n7, 352
Delbo, C., 162, 164, 170, 169-173
Deleuze, G./Deleuzian, 149, 193
    and Bergson, H./Bergsonism, 40, 45, 49, 305, 325
    Deleuzian, 24
    and Derrida, 106-108
    and Foucault, M./Foucault, 126, 322, 329, 353-354
    and Guattari/Deleuzo-Guattarian, 13, 59-62, 284-286, 285n1, 293-295, 345
    influence on NRT, 2-3, 100, 101, 104-105, 322, 344-345, 353
    and Kant I./Kant's Critical Philosophy, 323, 325, 328, 329-330